hidden valley

bergie pan

lion last stop
leopard tree
north tree

fox den

captain's jackal den

cheetah hill
north pan
communal hyena den

twin acacias

eagle island
old camp

bones and the blue pride

midpan

north bay hill

star's second den

dune woodlands

deception valley

east dune

CRY
of the
KALAHARI

CRY
of the
KALAHARI

MARK *and* DELIA OWENS

HOUGHTON MIFFLIN COMPANY   BOSTON   1984

The authors are grateful for permission to quote the following:

"Voices of jackals calling..." and "Quick, ere the gift escape us!..." from "Bridge Guard in the Karoo" by Rudyard Kipling from *Rudyard Kipling's Verse: Definitive Edition*. Reprinted by permission of the National Trust and Doubleday & Company, Inc.

"It is not easy to remember..." from *Sing with the Wind* by Winston O. Abbott, published by Inspiration House Publishers, South Windsor, Connecticut. Published by permission of Winston O. Abbott.

"I look down now. It is all changed..." from "Milkweed," copyright © 1962, reprinted from *The Collected Poems* by James Wright. By permission of Wesleyan University Press. This poem originally appeared in *The Minnesota Review*.

"When I look behind...," "and my tribe is scattered...," and "In a rising wind..." from "The Layers" by Stanley Kunitz, *Poems of Stanley Kunitz*, published by Atlantic–Little, Brown, copyright © 1979 by Stanley Kunitz. First appeared in *American Poetry Review*.

"Only when we pause to wonder..." and "You could feel the rain..." from *Fields of Wonder* by Rod McKuen, copyright © 1971 by Montcalm Productions, Incorporated.

"Never did we plan..." and "An ecologist must either harden his shell..." from *A Sand County Almanac* by Aldo Leopold, copyright © 1966 by Oxford University Press. Reprinted by permission.

"The things which will not awaken..." from *To Those Who See* by Gwen Frostic. By permission of Gwen Frostic.

"From his slim Palace in the Dust..." Reprinted by permission of the publishers and Trustees of Amherst College from *The Poems of Emily Dickinson*, edited by Thomas H. Johnson, Cambridge, Mass.: The Belknap Press of Harvard University Press, copyright 1951, © 1955, 1979, 1983 by the President and Fellows of Harvard College.

"Green fields are gone now..." from "Green Fields" by Terry Gilkyson, copyright © 1956 by Blanchard Music, Inc. International Copyright Secured. All rights reserved. Used by permission.

"The ecologist cannot remain a voice crying in the wilderness..." Reprinted by permission of The Plennum Press from *Breakdown and Restoration of Ecosystems* by M. W. Holdgate, copyright © 1978.

Endpapers drawn by Lorraine Sneed.
Maps prepared by Larry A. Peters.

Library of Congress Cataloging in Publication Data

Owens, Mark.
  Cry of the Kalahari.

  Includes bibliographical references and index.
  1. Zoology—Kalahari Desert. 2. Zoology—
Botswana. 3. Kalahari Desert. I. Owens, Delia.
II. Title.
QL337.K3O95 1984   591.9681'1   84-10771
ISBN 0-395-32214-6

Printed in the United States of America

P 10 9 8 7 6 5 4 3 2 1

We dedicate this book to
Dr. Richard Faust
and to
Ingrid Koberstein
of the Frankfurt Zoological Society
for all they have done for the animals
of this earth.

And to Christopher, who could not be with us.

Contents

	Prologue	I
1	The Jumblies	5
2	Water	21
3	Fire	36
4	The Cry of the Kalahari	48
5	Star	70
6	Camp	83
7	Maun: The African Frontier	99
8	Bones	111
9	The Carnivore Rivalry	131
10	Lions in the Rain	141
11	The van der Westhuizen Story	164
12	Return to Deception	171
13	Gone from the Valley	189
14	The Trophy Shed	195
15	Echo Whisky Golf	200
16	Kalahari Gypsies	208
17	Gypsy Cub	222
18	Lions with No Pride	235
19	The Dust of My Friend	245
20	A School for Scavengers	253
21	Pepper	265
22	Muffin	275
23	Uranium	279
24	Blue	284
25	Black Pearls in the Desert	292
26	Kalahari High	309
	Epilogue	313

Appendixes
A Conservation of Migratory
   Kalahari Ungulates   319
B Conservation of Kalahari
   Lions   322
C Conservation of Brown
   Hyenas   324
D Latin Names   325
Notes   327
References   329
Acknowledgments   331
Index   335

CRY
of the
KALAHARI

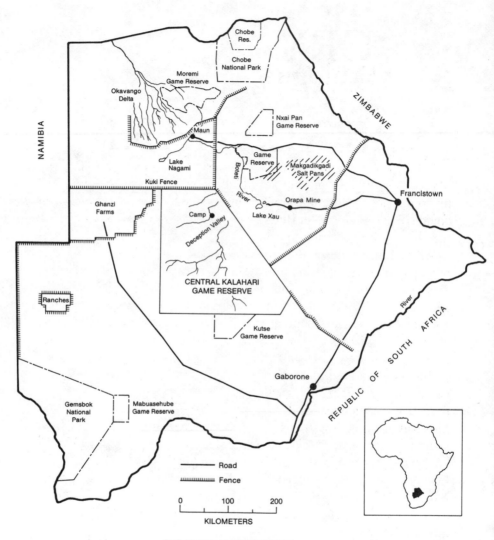

THE REPUBLIC OF BOTSWANA

Prologue

Mark

MY LEFT SHOULDER and hip ached from the hard ground. I rolled to my right side, squirming around on grass clumps and pebbles, but could not get comfortable. Huddled deep inside my sleeping bag against the chill of dawn, I tried to catch a few more minutes of sleep.

We had driven north along the valley the evening before, trying to home on the roars of a lion pride. But by three o'clock in the morning they had stopped calling and presumably had made a kill. Without their voices to guide us, we hadn't been able to find them and had gone to sleep on the ground next to a hedge of bush in a small grassy clearing. Now, like two large army worms, our nylon sleeping bags glistened with dew in the morning sun.

Aaoouu—a soft groan startled me. I slowly lifted my head and peered over my feet. My breath caught. It was a very big lioness—more than 300 pounds—but from ground level she looked even larger. She was moving toward us from about five yards away, her head swinging from side to side and the black tuft on her tail twitching deliberately. I clenched a tuft of grass, held on tight, and froze. The lioness came closer, her broad paws lifting and falling in perfect rhythm, jewels of moisture clinging to her coarse whiskers, her deep-amber eyes looking straight at me. I wanted to wake up Delia, but I was afraid to move.

When she reached the foot of our sleeping bags, the lioness turned slightly. "Delia! S-s-s-h-h-h—wake up! The lions are here!"

Delia's head came up slowly and her eyes grew wide. The long body of the cat, more than nine feet of her from nose to tuft, padded past our feet to a bush ten feet away. Then Delia gripped my arm and quietly pointed to our right. Turning my head just slightly, I saw another lioness four yards away, on the other side of the bush next to us . . . then another . . . and another. The entire Blue Pride, nine in all, surrounded us, nearly all of them asleep. We were quite literally in bed with a pride of wild Kalahari lions.

Like an overgrown house cat, Blue was on her back, her eyes closed, hind legs sticking out from her furry white belly, her forepaws folded over her downy chest. Beyond her lay Bones, the big male with the shaggy black mane and the puckered scar over his knee—the token of a hurried surgery on a dark night months before. Together with Chary, Sassy, Gypsy, and the others, he must have joined us sometime before dawn.

We would have many more close encounters with Kalahari lions, some not quite so amicable. But the Blue Pride's having accepted us so completely that they slept next to us was one of our most rewarding moments since beginning our research in Botswana's vast Central Kalahari Desert, in the heart of southern Africa. It had not come easily.

As young, idealistic students, we had gone to Africa entirely on our own to set up a wildlife research project. After months of searching for a pristine area, we finally found our way into the "Great Thirst," an immense tract of wilderness so remote that we were the only people, other than a few bands of Stone Age Bushmen, in an area larger than Ireland. Because of the heat and the lack of water and materials for shelter, much of the Central Kalahari has remained unexplored and unsettled. From our camp there was no village around the corner or down the road. There was no road. We had to haul our water a hundred miles through the bushveld, and without a cabin, electricity, a radio, a television, a hospital, a grocery store, or any sign of other humans and their artifacts for months at a time, we were totally cut off from the outside world.

Most of the animals we found there had never seen humans before. They had never been shot at, chased by trucks, trapped, or snared. Because of this, we had the rare opportunity to know many of them in a way few people have ever known wild animals. On a rainy-season morning we would often wake up with 3000 antelope grazing around our tent. Lions, leopards, and brown hyenas visited our camp at night, woke us up by tugging the tent guy ropes, occasionally surprised us in the bath boma, and drank our dishwater if we forgot to pour it out. Sometimes they sat in the moonlight with us, and they even smelled our faces.

There were risks—we took them daily—and there were near disasters that we were fortunate to survive. We were confronted by terrorists, stranded without water, battered by storms, and burned by droughts. We fought veld fires miles across that swept through our camp—and we met an old man of the desert who helped us survive.

We had no way of knowing, from our beginnings of a thirdhand Land Rover, a campfire, and a valley called "Deception," that we would learn new and exciting details about the natural history of Kalahari lions and brown hyenas: How they survive droughts with no drinking water and very little to eat, whether they migrate to avoid these hardships, and how members of these respective species cooperate to raise their young. We would document one of the largest antelope migrations on earth and discover that fences are choking the life from the Kalahari.

*   *   *

I don't really know when we decided to go to Africa. In a way, I guess each of us had always wanted to go. For as long as we can remember we have sought out wild places, drawn strength, peace, and solitude from them and wanted to protect them from destruction. For myself, I can still recall the sadness and bewilderment I felt as a young boy, when from the top of the windmill, I watched a line of bulldozers plough through the woods on our Ohio farm, destroying it for a superhighway—and changing my life.

Delia and I met in a protozoology class at the University of Georgia and it didn't take long to find out that we shared the same goal. By the end of the semester we knew that when we went to Africa, it would have to be together. During this time we heard a visiting scientist tell of Africa's disappearing wilderness: More than two-thirds of its wildlife had already been eliminated, pushed out of its habitats by large ranches and urban sprawl. In the southern regions, thousands of predators were being trapped, shot, snared, and poisoned to protect domestic stock. In some African nations, conservation policies and practices were virtually nonexistent.

These were frightening reports. We became determined to study an African carnivore in a large, pristine wilderness and to use the results of our research to help devise a program for the conservation of that ecosystem. Perhaps, also, we simply wanted to see for ourselves that such wild places still existed. But if we didn't go immediately, there might be little left to study.

Going to Africa as part of our graduate programs would mean years of delay, and since we had not finished our doctorates, we knew there was little chance of our getting a grant from a conservation organization. We decided to take a temporary, if prolonged, leave from university and to earn the money needed to finance the expedition. Once a study site had been chosen and our field research was under way, we thought surely someone would grant us the funds to continue.

After six months of teaching, we had saved nothing. I switched jobs and began operating the crusher at a stone quarry while Delia worked at odd jobs. At the end of another six months we had saved $4900, plus enough money for air fares to Johannesburg in South Africa. It was not nearly enough to begin a research project. But it was late 1973 and the Arabs had just pulled the plug on cheap oil; prices were skyrocketing. We had to go then, or not at all.

Trying desperately to scrape enough money together, we piled everything we owned—stereo, radio, television, fishing rod and reel, pots and pans—into our small station wagon and drove to the stone quarry one morning, just as the men were coming off the night shift. I stood on top of the car and auctioned it all away, including the car, for $1100.

On January 4, 1974, a year after we were married, we boarded a plane with two backpacks, two sleeping bags, one pup tent, a small cooking kit, a camera, one change of clothes each, and $6000. It was all we had to set up our research.

* * *

This book is *not* a detailed account of our scientific findings; that is being published elsewhere. Instead, it is the story of our lives with lions, brown hyenas, jackals, birds, shrews, lizards, and many other creatures we came to know, and how we survived and conducted research in one of the last and largest pristine areas on earth. The story was taken from our journals and is all true, including names and dialogue. Although each chapter is written in one voice, we developed every phase of the book together.

1

The Jumblies

Mark

They went to sea in a Sieve, they did,
In a Sieve they went to sea:
In spite of all their friends could say,
On a winter's morn, on a stormy day,
In a Sieve they went to sea!
. . . .
   Far and few, far and few,
     Are the lands where the Jumblies live;
   Their heads are green, and their hands are blue;
   And they went to sea in a Sieve.
                 —Edward Lear

SLEEPLESS, I rested my head against the thick double windowpane of the jet, staring into the blackness of the mid-Atlantic night. The world turned slowly below as the plane reached for the dawn of Africa.

With careful grace the cheetah strolls onto the plain. Head erect, its tail a gentle vane turning easily on the wind, it glides toward the stirring herd. Alert, the antelope prance back and forth, but do not run. The cat is hungry and begins loping forward.

The plane met and passed the dawn. Soon it was standing on asphalt, disgorging its passengers near a hazy city. Customs officials in short pants and spotless white shirts with bold black epaulets called orders and waved clipboards. We filled out long forms and questionnaires, waited in crowded halls, and gazed through chain-link fences. Plenty of time to daydream.

A perfect union of speed, coordination, balance, and form, the cheetah accelerates toward the dashing antelope and singles one out.

Others veer aside and the ageless footrace between predator and prey begins.

A smaller plane, a shorter ride—we had been traveling forever. On a train this time, again we stared numbly past our reflections in a window. Miles and miles of thornbush, all of it the same, rushed by in time with the clickety-clack of the rail sections as the train swayed along. "Clickety-clack, clickety-clack, you can't get off and you'll never go back; clickety-clack . . ."

The cheetah is a blur across the plain. Fifty, sixty, seventy miles per hour, the living missile streaks toward its target. At this moment, as it draws near the flashing rear quarters of its prey, the awesome beauty of their contest is inescapable. Each is a sculptor who, using eons of time as its maul and evolution as its chisel, has created, in the other, something of such form, such vitality, such truth that it can never be duplicated. This relationship is the best Nature has to offer; the ego of the natural world.

It is the moment of truth for the gazelle. The cheetah, still at full speed, reaches forward a clublike paw to destroy the balance of its prey. The antelope cuts sharply, and what was ultimate form is suddenly perverted. At seventy miles per hour the fence wire slices through the cheetah's nose, shatters its jaw, and snaps its head around. Before its momentum is spent on the mesh, its elegant neck is twisted and broken, the shank of a splintered white bone bursts through the skin of its foreleg. The fence recoils and spits the mutilated form, ruptured and bleeding, into the dirt.

With a hissing of air brakes the train lurched to a stop and interrupted the nightmare. We shouldered our backpacks and stepped down onto the sandy station yard in the black African night. From behind, the diesel rumbled and the car couplings clanged as the train pulled away. Standing alone by the ramshackle station house at two o'clock in the morning, it was as though we were in a long, dark tunnel. At one end a grimy sign beneath a dim yellow light read GABORONE BOTSWANA.

The quiet darkness seemed to swallow us. Alone in a strange country with too little money, all of it stuffed in the pocket of my backpack, we suddenly felt that the challenge was overwhelming: We had to find a four-wheel-drive truck and a study area and accomplish enough solid research to attract a grant before our money ran out. But we were exhausted from traveling, and before worrying about anything else, we needed sleep.

Across a dirt road from the train station, another weak light bulb dangled over the tattered screen door of the Gaborone Hotel—a sagging building with flakes of paint peeling from its walls, and tall grass fringing its foundation. The rooms were eight dollars a night, more than we could afford.

As we turned and began to walk away, the old night watchman beckoned to us from the hotel. A flickering candle cupped in his hands, he led us through the bare lobby into a small courtyard choked with weeds and thornscrub. Smiling broadly through teeth like rusted bolts, the old native patted my pack and then the ground. We bowed our thanks and within minutes we had pitched our small pack tent next to a thornbush and settled into our sleeping bags.

Morning came with the chatter of native Africans moving like columns of army ants through fields of tall grass and thornbush toward the town. Most of them wore unzipped and unbuttoned western shirts and dresses or pants of mix-matched, bright colors. Women swayed along with bundles balanced on their heads—a pint milk carton, a basket of fruit, or fifty pounds of firewood. One man had slabs of tire tread bound to his feet for sandals, a kaross of goatskin slung over his shoulders, and the spotted skin of a gennet cat, the tail hanging down, set at a rakish angle on his head. These people eked out their livelihoods by hawking carvings, walking sticks, and other artifacts to travelers through the windows of railroad coaches. They lived in shanties and lean-tos of corrugated tin or cardboard, old planks or mud bricks. One was made entirely of empty beer cans.

Looking over the scene, Delia muttered softly, "Where the devil are we?"

We made our way toward the haze of wood smoke that covered the town of Gaborone, which sprawled at the foot of some rocky hills. It is the capital of Botswana, known before its independence in 1967 as the British Bechuanaland Protectorate. Architecturally, it is a crossbred town: One avenue of small shops and a few three-story office buildings of Western design rise from a mishmash of mud-and-thatch huts called *rondavels*. Dusty paths were crowded with Africans in European clothes and Europeans in African prints.

It is an interesting blend of cultures, but nothing happens very fast in Gaborone, and for two months after our arrival in Botswana we were stuck there. Day after day we walked from one isolated government department to another, trying to arrange residence and research permits and meeting with people who might know something about a suitable study site. We were determined to find a place—one far from fences—where the behavior of the predators had not been affected by human settlements.

From all accounts, the best places for the type of study we had in mind were in the remote regions of northern Botswana, but none of the Wildlife Department personnel had ever been to the most inaccessible of those areas. Without anyone to guide us, the expedition seemed more difficult and risky than we had supposed. Even if we could find our own way into such an undeveloped part of Botswana, setting up and supplying a research camp would mean moving food, fuel, and other supplies over vast tracks of uncharted wilderness. Besides, practically the entire northern third of the country was under water from the heaviest rains in its recorded history. The only road to the north had been impassable for months.

One of our most immediate problems was how to find a vehicle among the population of battered four-wheel-drive trucks that rattled around the town. The best we could afford was an old thirdhand Land Rover with a concave roof, bush-scraped sides, and drab grey paint. We bought the "Old Grey Goose" for 1000 rands ($1500), overhauled the engine, installed a reserve gas tank, and built flat storage boxes in the back. Covered with a square of foam rubber, the boxes would also serve as our bed.

When we had finally finished outfitting the Grey Goose, it was already early March 1974; we hadn't been in the field yet, and we had only $3800 left, $1500 of which would be needed to get us home if we failed to get a grant. Every delay meant lost research time. If we were to have any chance of convincing some organization to fund us before our money ran out, we had to find a study site immediately and get to work. So, despite warnings that we would not get through to the north country, early one morning we headed out of Gaborone into the rolling thornbush savanna.

A few miles outside town, with a bone-jarring crash, we left the only pavement in Botswana behind us. As I swerved to dodge the ruts and chuckholes, the narrow dirt road led us deeper and deeper into the bushveld. I took a deep, satisfying breath of wild Africa; our project was finally under way. The sense of freedom and the exhilaration were almost intoxicating, and I reached across to pull Delia over next to me. She smiled up at me—a smile that washed away the tensions that had built over the long, frustrating weeks of preparation. Her eyes spoke her total confidence that we could handle any challenge that confronted us, and her confidence was itself a challenge.

Our destination, the village of Maun, lay where the waters of the Okavango River delta reach the sands of the Kalahari Desert, more than 450 miles to the north. There was only one narrow gravel road to follow through a territory that offered little shelter, except an occasional cluster of native huts. Because of the flooding, no one had driven the road for weeks. As we crawled north at ten to fifteen miles per hour, the savanna grew wetter and wetter until we were churning through deep black mud.

Near Francistown, the last large village on the east side of Botswana, we swung northwest toward Maun, still more than 300 miles away. Whole stretches of the road had completely washed away. In places I waded ahead through shallow lakes more than a mile across, searching under the water for firm ground with my bare feet as Delia followed in the Land Rover. Dodging ruts three or four feet deep, we passed the mud-caked hulks of trucks bellied-up in muck like dinosaurs in a tar pit. They had been abandoned for weeks. Time after time the Grey Goose sank to its undercarriage. Using a high-lift jack to raise it, we piled thornbush, stones, and logs under the wheels. Another few yards and we were down to the axles again.

At night, slapping at swarms of mosquitoes, we would squat next to a mudpuddle and wash the crust off our faces, arms, and legs. Then we would fall asleep on top of the boxes in the back of the Land Rover. We kept the truck parked in the middle of the roadbed because if I had driven off its crown, we would have become hopelessly mired. We had met only two or three other vehicles in several days, so it was unlikely that someone would need to pass by in the night.

In the morning we would be on our way once more. Dazed with fatigue, we would spin forward, sink, dig out, and spin forward again. Some days we made no more than a mile or two. But we had to keep going. Though we didn't talk about it, we both had the desperate feeling that if we couldn't even make it to Maun, we would surely fail in the field. Yet failure was an option we simply could not afford. We had invested all our savings—our dreams and our pride—in this venture. There was no reason to turn around; there was nothing to go back to.

Occasionally we saw goats, cattle, and donkeys drinking and wallowing in mudholes along the way. They were the only signs of animal life in the flat monotony of overgrazed thornscrub. It was depressing and disconcerting that we had come all this way to find in these remote areas no herds of wild antelope. Perhaps after all, we had chosen a country in which little wildlife was left. Even then we knew that much of Africa had been grazed to death by domestic stock.

Eleven days after we had set off from Gaborone, hollow-eyed and covered in mud, we stopped on the one-lane bridge over the Thama-lakane River. On its banks was Maun, a village of reed-and-straw huts, donkeys, and sand. Herero tribeswomen had spread their lavish skirts, made of yards of different materials, on the emerald riverbanks to dry, like great butterflies fanned out in a riot of reds, yellows, blues, greens, and purples.

Delia's eyes were red and her face and hair spattered with grey mud. Her hands were deeply scratched from piling rocks and thornbush under the mired truck. But she grinned and gave a rebel yell. We had made it!

On sand tracks that ran between *rondavels* we drove to Riley's, a large compound including garage, general store, hotel, and bar, where we bought gasoline and a few supplies: lard, flour, mealie-meal, and sugar. Perishables such as milk, bread, and cheese were not available in northern Botswana, and when we arrived even staples were in short supply because no transport trucks had been able to get through in weeks. The people of the village were hungry. We avoided the eyes of the begging children, embarrassed that we had nothing we could give them, yet knowing we were wealthy by comparison.

Officials in the Department of Wildlife in Gaborone had advised us to ask professional hunters about a good place to start our research. One of the names we had scribbled in our journal was "Lionel Palmer— Maun." Lionel was well known at Riley's, where we asked for direc-tions to his home. We made our way along deep sand tracks and through more mudholes until, about four miles north of the village, we found the Palmer homestead. Over the river hung tall fig trees with orange, red, and yellow bougainvillea spilling over their tops. Red-eyed bulbuls, grey hornbills, hoopoes, and a myriad other birds flitted about the canopy above the garden.

Lionel Palmer, deeply tanned, his dark hair brushed with grey, was dressed in baggy jeans, a cowboy shirt, and a bandana. He sauntered out to greet us, holding a glass of whisky in his hand. The oldest and most experienced professional hunter in the area, Lionel held considerable social positon in Maun. He was famous for his parties, where bedroom furniture sometimes ended up on the roof, and once a Land Rover had been hung in a fig tree—and for his capacity for Scotch. Once, after several days of intoxication, he woke up with a stabbing earache. The doctor at the clinic removed a two-inch-long sausage fly—a reddish-brown, tubelike, winged insect—which had taken up residence in Lionel's numbed ear while he slept off his drunkenness in a flowerbed. For a week Lionel carried the fly's carcass bedded down in a cotton-lined matchbox, proudly showing it to everyone he met, whether or not he knew them.

Sitting with us on the patio overlooking the river, Lionel suggested a few areas in northern Botswana where flooding was not too severe and where predators unaffected by man could be found. One, the Makgadikgadi Pans, is a great tract of remote bushveld wilderness more than 100 miles east and south of Maun. The pans are the remnants of an enormous inland lake that dried up some 16,000 years ago.

"Go ninety-nine miles east of Maun on the Nata road and find a palm tree broken off at the top. Look for an old spoor that runs south from the main track. There's no sign, but that's where the reserve begins. Nobody goes out there much—there's bugger-all there, except miles and miles of bloody Africa."

Most game reserves in Botswana are large tracts of totally undeveloped wilderness. There are no paved roads, fast-food stands, water fountains, campgrounds, restrooms, or any of the other "improvements" found in parks and reserves in more developed countries.

Two days later we found two faint tire tracks at a broken palm, turned off the main road, and left all traces of civilization behind us. Immediately we had a sense of being in Africa, the real Africa, the one we had always dreamed about. The vast untracked savanna, broken only by occasional isolated trees, made us feel frail, minuscule, vulnerable. It was beautiful, exciting—but also a little intimidating.

About thirty miles south of the main road, the track we had been following led us to the edge of a vast plain. Then it disappeared. Delia noted our compass heading, the mileage, and a lone thorntree we thought we might be able to recognize again. With no chart or guide, and with only fifteen gallons of water and the barest minimum of essential food, we set off across the Makgadikgadi.

The savanna was very rough, the grass tall and heavy with ripe seed, and it was hot. We made no more than three miles per hour for the rest of the day. Gradually the front of the Grey Goose was buried under a thick moving carpet of grass seed and insects that completely obscured the headlights and hood. Every quarter of a mile or so we had to brush off the front of the engine and cool the boiling radiator by pouring water over the top.

Around midmorning of the second day we came to an immense network of saucer-shaped salt pans interlaced with crescents of grass savanna, touches of woodland, and wisps of palm islands. Some pans were filled with brackish, unpotable water and flowery masses of orange, purple, green, and red algae; others were covered with a thin salt crust. We were at the edge of an alien world—no roads, no trails, no people. A shimmering mirage drew the tops of the palms into the sky.

"Whatever you do, don't drive across those pans or you'll go down like a bloody rock," Lionel had warned us. "The salt crust'll look firm, but it won't be, 'specially with all the rain we've had lately. Underneath there's nothing but mud for God knows how deep. Game Department lost a whole truck in one of them last year. No matter how much time you think you'll save by crossing, go around."

While I was skirting these enormous irregular depressions, Delia sketched a map of our route, noting compass headings and odometer readings at regular intervals, so that we would be able to find our way back to "Lone Tree."

Itching from grass seed and insects, I drove toward a large pan that looked as if it might contain enough fresh rainwater for bathing. We were coming over the rise above it—suddenly the truck dropped from under us. The chassis cracked like a rifleshot and we were thrown from our seats hard against the windshield. The engine stalled and a haze of dust rose in front of us. When it had cleared, the hood of the Land Rover stood at ground level, buried in a large antbear hole that had been hidden in tall grass. After checking to see that Delia was all right, I jacked the truck up and began shoveling a ton of sand under the wheels. When we were finally able to back out, I crawled under the Goose to check for damage. There were several new cracks in the chassis, one near a motor mount. Another bad hole could tear the engine loose. Still, we were lucky; if only one of the front wheels had gone in, it could have broken off.

I was sharply aware that if we lost the service of the Grey Goose in some way, our chances of ever leaving the Makgadikgadi alive were not good. I didn't trust my limited knowledge of mechanics, and we hadn't been able to afford all the backup spare parts we should have been carrying for an expedition like this. Furthermore, no one knew where we were or when to expect us back. Lionel knew only that we had left Maun headed for one of several areas he had mentioned.

We didn't discuss these risks, but they lingered in the backs of our minds. We washed in the brackish water of the pan, and after we had dried in the wind, our faces felt stretched tight, like overblown balloons.

For the rest of the day I walked ahead of the Land Rover, checking for holes in the long grass while Delia drove. Several times I stepped into rodent burrows, hoping that they weren't also the home of some poisonous snake. We carried no anti-venin, since it would have to be refrigerated.

That second night we camped next to a small tree not more than six feet high, the only one for miles around. We had been irresistibly drawn to it and had actually driven quite a way off course to get to it. Though we slept inside the truck, the tree gave us a vague sense of security. Our early primate ancestors would probably have been similarly pleased to find even this mere seedling on a nearly treeless plain, after they left the safety of the forests to venture onto the vast savannas millions of years ago.

We climbed a low rise late in the afternoon of the fourth day, and I was walking ahead. Suddenly I stopped. "My God! *Look* at that!" The sounds and smells of animals, tens of thousands of animals, carried to us on the light wind. For as far as we could see, the plains beyond were covered with zebra and wildebeest, grazing placidly near a large water hole. Fighting zebra stallions bit and kicked each other, puffs of dust rising from their hooves. Wildebeest tossed their heads and pranced and blew their alarm sounds. The great herds stirred, and my skin tingled at the spectacular display of life. If we never saw another sight like this, the months of working in a stone quarry and the hawking of all our belongings would have been worth this one glimpse of what much of Africa must once have been like.

We watched for hours, passing the binoculars back and forth between us, taking notes on everything we saw—how the herds mingled and moved, how many drank, how many fought—as though this signified in some way that our research had begun. We pitched camp near the top of the ridge, so we could watch for cheetahs or lions preying on the herds. When it was too dark to see, we sat heating a can of sausages over the kerosene lantern inside the Grey Goose and discussed establishing our research in the Makgadikgadi.

We went on watching the herds all the next day and into the evening. Then reality returned: Our water was getting low. Frustrated, anxious to get some solid field work done, and hating to leave the zebra and wildebeest, we began the long drive back across the plains. Following the reciprocal of compass headings and using the schematic diagram Delia had made of our course eastward, we would return to Lone Tree, get our bearings, then drive on to the Boteti River, thirteen miles farther west, for water.

For two days we retraced our course, but somewhere we went wrong. A great salt pan, unfamiliar to us, a dazzling white depression more than a mile across and miles long from north to south, blocked our way. Standing on the roof of the Land Rover and scanning with binoculars, we couldn't see any way around it.

After driving north, and then south for some distance along the bank, I decided to see how firm the surface was. I was more and more concerned about our dwindling gasoline and water supply. Perhaps with caution, we could drive across the pan instead of laboring over miles of rough terrain to skirt it. I dug a test hole with the spade. The clay beneath the salt crust seemed surprisingly dry and solid, and no matter how hard I jumped on it with my heels, I could barely make an impression. Next, I slowly drove the front wheels of the Grey Goose onto the pan; the crust held firm. Finally, I brought the full weight of the truck onto the surface, which was as hard as concrete pavement. So in spite of Lionel's warning, we decided to make the crossing.

Starting the run, I accelerated quickly. By driving fast in four-wheel drive, I hoped to skim over any soft spots we might encounter farther out.

I bent over the steering wheel, scanning the white salt crust ahead for dark patches, a sign that the pan had not dried out completely. But there was none. It was like driving over a billiard table, and I began to relax. Then, about 800 yards from the edge, we saw some timbers and poles sticking at odd angles from a depression in the grey, cracked surface. We got out to investigate. What could have made such a hole? And where had the timbers come from? There were no tracks or any other clues. Puzzled, I looked into the deep, ragged pit, to the place where the ends of the posts converged and then disappeared into an abyss of mud. My throat suddenly tightened—someone had tried, unsuccessfully, to save his truck. I glanced quickly at ours.

"My God! The truck's sinking! Get in—hurry—we've got to get out of here!"

Its wheels were slowly settling through the salt crust into a pocket in the softer clay beneath. The surface was giving way; in seconds our truck would break through.

I tried to drive forward, but the engine stalled. The wheels had sunk too deep. Working frantically, I restarted the motor and jammed the gearshift into low-range four-wheel drive. Spinning and throwing clay, the Land Rover churned forward until it heaved itself up onto the firm surface again. I quickly shifted to high range for better speed, spun around, and raced to the safety of the grass bank at the edge of the pan. We sat staring at each other and shaking our heads in dumb relief. I was furious with myself for having tried to make the crossing, but I had endangered us even more by stopping the truck in the middle of the pan. After consulting our sketch map, we headed north. It took an entire afternoon to drive around the rest of the pan.

On the morning of the fourth day of the return trip, we finally reached the west edge of the Makgadikgadi plain, and slipped beneath the cool, refreshing canopy of riverine forest. Spider webs were drawn like fishing nets from tree to tree, and their hairy black-and-yellow architects scrambled over the hood of the truck as we ploughed through heavy sand toward the river. Kudu watched from deep shadows.

At last we stood on the high banks of the Boteti River. Deep-blue water gently caressed its way around the lilies, hyacinths, and other water plants nodding in the sleepy current. At the top of a tall fig tree, a pair of fish eagles threw back their heads and called to the sky. We ran down the steep bank and plunged into the cool water.

Climbing the riverbank after our swim, we saw something red lying in the grass. It was a fifty-gallon drum—a great find! We had looked for one in Maun, but they were almost impossible to get in northern Botswana; everyone needed them. By lashing this one to the top of the Land Rover and filling it with water, we could greatly increase our range and endurance while searching remote areas for a study site. The drum looked sound enough. We never stopped to wonder why it had been abandoned.

In the late afternoon, heavy splashes began sounding from the river below. After living for weeks on mealie-meal, raw oatmeal, powdered milk, and an occasional tin of greasy sausages—so pale and limp we called them dead man's fingers—we both craved a thick, juicy piece of meat. Fresh fish would be scrumptious! I found a tangle of old fishing line the previous owner had left in the Land Rover, made a hook with a pair of pliers, and fashioned a spinning lure from the shiny top of a powdered-milk tin.

Delia had watched skeptically while I made my fishing rig, and now she began baking mealie bread in the three-legged iron pot. Heading down the riverbank, I snatched a corn cricket from the grass, put it on the hook, and threw it into the water. It was almost dusk, and the surface of the river was jumping with big fish. In a few seconds, shouting and laughing, I hauled in a beautiful bream, and then a big catfish.

Delia rolled the fillets in mealie-meal and flour before frying them, and soon we were sitting by the fire stuffing ourselves with big chunks of steaming mealie bread and tender white pieces of fish. Afterward, we sat high above the quiet waters of the river, talking over our Makgadikgadi adventure. Africa was seasoning us.

The next day we caught and ate more fish. Then we stocked up with river water, hauling jerricans up and down the steep bank. After filling the red drum, we set it on its side, and lashed it to the roof of the Land Rover. By noon we were on our way back into the Makgadikgadi to look for predators.

Four days later, back on "Zebra Hill," the thousands of antelope we had seen a week before were gone. We drove around for hours without finding a single one. And without prey, no lions, cheetahs, or other carnivores were likely to be around. It was depressing. We had seriously thought of settling in the Makgadikgadi to do our research, but considering the tremendous mobility—with no apparent focal point—of these great herds, and presumably of the predators, how would we locate and stay in contact with our study animals? We drove back to Maun for more supplies and more advice.

Over the next several weeks we made reconnaissance trips to Nxai Pan, the Savuti Marsh, and other areas on the fringes of the Okavango River delta. The marsh, pans, and forests all had alluring varieties of antelope and predators, but by and large these places were still flooded. The water would limit our operations severely. Often, as we crossed the *malopos*—reed-choked, swampy waterways—from one palm island to the next, water ran in over the floor of the truck, shorting out the motor. We spent hours digging the Land Rover out of black muck.

Discouraged, we turned back to Maun. With each unsuccessful reconnaissance and return for supplies, our operating funds were shrinking. Again it was the hunter Lionel Palmer who finally suggested, "Why don't you try the Kalahari? I've seen a place called Deception Valley from the air...has lots of game. 'Course, I've never hunted there myself; it's miles inside the game reserve."

On the one-to-a-million-scale map of Botswana, we could quickly see that the Central Kalahari Game Reserve was one of the largest wildlife protectorates in the world, more than 32,000 square miles of raw, untracked wilderness. And the wilderness didn't stop at the reserve boundaries; it extended for 100 miles farther in nearly every direction, interrupted only by an occasional cattle post or small village. According to Lionel, there was not one road, not a single building, no water, and no people, except for a few bands of Bushmen, in an area larger than Ireland. It was so remote that most of it had never been explored and the Botswana government had not opened it for visitors. Consequently, no wildlife research had ever been done there. It was just what we had been looking for—if we could get there and solve the problems of surviving in such a remote and difficult environment.

After puzzling over the featureless map for some time, we finally devised a route into the Kalahari and, having done so, decided that we would go there without telling the Department of Wildlife. They probably would have refused us permission to work in such an isolated area, and they would learn about us soon enough, anyway.

The Grey Goose loaded with gasoline and other supplies, our red water drum lashed to the top, we set out for the Kalahari in search of Deception Valley. It was by then late April of 1974. Nine miles east of the village, we found a spoor running south to the Samadupe Drift on the Boteti River. There the water shallowed up, and a corduroy ford of logs was laid across the bottom. Tumbling over the logs and stones, the river fell away to easy swells and swirls between reed banks and a great avenue of giant fig trees. Cormorants dove, and trotters padded from one lily pad to another. Spurwing geese and egrets passed low above the water, their wings singing in the air.

We stopped on the ford for a last swim, and I cut off Delia's shoulder-length hair; it would take too much water to keep it clean in the desert. Her long locks fell to the water, eddied, and drifted away on the current.

For a moment I could see the reflection of her laughing face in the water—the way she looked the first time I saw her. I paused and held my hand to her cheek and then began snipping again.

After we had crossed the river and climbed a steep ridge of heavy sand, the track narrowed to two wheel marks bordered by thick acacia thornbush. For the rest of the day we slogged through heat, dust, and deep sand, on either side of us the dense scrub raking the sides of the Land Rover with a screeching that set our teeth on edge. In the late afternoon the spoor dead-ended. We found ourselves in a small dusty clearing where tumbleweeds blew past a crumbling mud hut and a tin watering trough for cattle. We sat wondering where we had gone wrong.

A gnarled, knobby old man—all elbows, knees, and knuckles— with a gnarled, knobby walking stick, appeared from the bushes. His wife and four spindly boys, dressed in little more than leather sashes, led a line of bony cattle through the blowing dust toward the watering trough.

I waved. "Hello!"

"Hello!" one of the boys shouted back. They all laughed.

"Ah, they can speak English," I thought.

"Could you help us? We're lost." I got out of the truck and began unfolding our map.

"Hello," said the boy again. They all crowded around. "Hello-hello-hello—"

I put away the map and took another tack.

"Ma-kal-a-ma-bedi?" I asked, holding my palms out and up, hoping they could understand the name of the fence line that would point us into the Kalahari. The skinniest and most garrulous boy scrambled onto the hood of the Land Rover and pointed back up the spoor; the three others joined him. We drove up the track, all of us laughing. The boys bounced on the hood, their twitching fingers showing us the way.

A few minutes later they all began pounding on the truck simulta-neously. I stopped. They jumped off and pointed east through the bush. At first we did not understand. Then, standing next to them, we could see a faint line leading away through the savanna, an old survey cut-line running east. It was our only option. We were determined not to go back to Maun without having found Deception Valley.

We thanked the boys, gave them a paper bag of sugar, and set off.

"Hello-hello-hello!" They shouted and waved as we disappeared into the bush.

Early the next morning we came to the fence—weathered posts and five strands of wire running straight across our path as far north and south as we could see. We turned south, and hours later the barrier was still beside us, a great scar across the savanna. It was irritating to us then, but someday it would give us cause to hate the very sight of it.

We slept along the fence that night. The next morning, with the truck churning on through the sand, our backs became sweaty against the seat, and we were covered with a layer of dust and grass seed. Suddenly, the fence ended: There was nothing left but sand, thornscrub, grass, and heat. Two wheel tracks continued on through the grass, becoming fainter . . . and fainter . . . until, like a distant memory, they disappeared. Now we were driving over mostly flat grass savanna and occasional low sand ridges covered with lush green bushes and stands of trees. Was this the Kalahari Desert? Where were the great shifting sand dunes?

There was no way to be certain of our position. We consulted the map and calculated the number of miles we must have come south from Maun. Then we turned west. We would drive twenty more miles. If we hadn't found Deception Valley by then, we would have to find our way back to Maun.

Eighteen . . . nineteen . . . nineteen point six . . . and just as we were about to give up all hope, we crested the top of a large dune. Below us lay the gentle slopes and open plain of Deception Valley, an ancient fossilized river channel meandering through forested sand dunes. Herds of springbok, gemsbok, and hartebeest grazed peacefully on the old grass-covered riverbed, where water used to flow. The blue sky was stacked high with white puffs of cloud. Deception was incredibly serene and all we had hoped it would be. It was May 2, 1974, almost five months since we had left the United States, and we had found our place in Africa. Home, as it would turn out, for the next seven years.

The gentle dune face led us into the valley. We crossed the dry riverbed, and the springbok hardly bothered to lift their heads from the grass as we passed. On the western edge, we found a solitary island of acacia trees that would provide shelter and give a panoramic view. It would be a good campsite.

On the move for months, lugging our shelter with us wherever we went, we had begun to feel like a turtle with a steel shell. Already it felt good to be putting down roots.

It didn't take long to set up our first base camp: We tied our cloth sacks of mealie-meal and flour in an acacia tree to keep them safe from rodents, stacked our few tinned foods at the base of the tree, and arranged the pots and pans along a limb. Then we gathered firewood. With no shelter other than our tiny pack tent, we would sleep in the Grey Goose for the rest of the year.

Delia made a fire and brewed some tea while I unloaded our old red drum and rolled it beneath the acacia tree. It held the only water for thousands of square miles.

2

Water

Mark

... if one advances confidently in the direction of his dreams, and endeavors to live the life which he has imagined, he will meet with a success unexpected in common hours... If you have built castles in the air, your work need not be lost; that is where they should be. Now put the foundations under them.
—*Henry David Thoreau*

Squeezed by pressure from above, hour by hour, molecules of water sweated through the flakes of rust. Outside they coalesced: A drop grew. Swollen and heavy, the droplet ran along the battered rim. Finally losing its grip, it splattered quietly into a furrow of thirsty sand and disappeared. Above, on the rim, another drop had already taken its place.

Days passed. The drops continued their march from the rust, to the rim, to the sand. The drum's wound opened still farther... the drops came more quickly, spat-spat-spatting into the dark stain hidden from the sun.

A near total silence crept in on me when I opened my eyes and gazed at the Land Rover ceiling. A moment's confusion; where was I? I turned to the window. A gnarled acacia tree loomed outside, its limbs held up in silhouette against the greying sky. Beyond the tree, in soft, easy lines, the wooded sand dunes descended to the riverbed. Morning, our first in Deception Valley, grew in the sky far beyond the dunes.

Delia stirred. We listened to Africa waking around us: A dove cooed from the acacia, a jackal* wailed with a quavering voice, and from far away to the north the bellow of a lion rolled in, heavy and insistent. A lone kestrel hovered, its wings fluttering against a sky turning fiery orange.

Grunts and snorts sounded from outside—very close. Quietly, slowly, Delia and I sat up and peered through the window. Just outside camp was a herd of at least 3000 springbok, small gazelles with horns a foot long turned inward over their heads. Their faces were boldly painted with white and black bars running from their eyes along their muzzles. They looked theatrical, like marionettes, as they grazed the dew-sodden grasses, some of them only fifteen yards from us. A few young females stared at us with deep, liquid eyes while quietly munching stems of grass. But most of the herd grazed, their stomachs rumbling and tails flicking, without looking our way. Easing ourselves up against the back of the front seat, we sat in our bed and watched two yearling bucks lock horns in a sparring match.

Though they hardly appeared to be moving, within twenty minutes the antelope had drifted more than 100 yards away. I started to speak, to try to express what I was feeling, when Delia pointed to the east. A black-backed jackal, first cousin to the American coyote, but smaller and with a sly, foxy face and a saddle of black hair over his back, trotted into our tree island and began sniffing around last night's camp-fire. Considered vermin and shot on sight in most of Africa, jackals usually run at the first sign of man. This one walked up to a tin coffee mug left near the coals, clamped the rim in his teeth, and inverted the cup over his nose. Looking right and left before ambling calmly out of camp, he surveyed our few belongings, and then shot us a glance, as if to say, "I'll be back for more later."

It's difficult to describe the excitement and joy we felt. We had found our Eden. Yet we were very anxious not to disturb the intricate patterns of life that were going on around us. Here was a place where creatures did not know of man's crimes against nature. Perhaps, if we were sensitive enough to the freedom of these animals, we could slip unnoticed into this ancient river valley and carefully study its treasures without damaging it. We were determined to protect one of the last untouched corners of earth from ourselves.

---

*Appendix D is a list of the Latin names for the mammals, birds, and snakes mentioned in the text.

Hoofbeats, thousands of them—the air shuddered. The springbok herd was charging south along the riverbed. I grabbed the field glasses, and we kicked out of our sleeping bags and jumped from the Land Rover into the tall, wet grass. Eight wild dogs bounded down the valley after the antelope. When they were abreast of camp, two of the predators veered straight toward us.

Delia quickly reopened the back door of the truck, but by then the dogs, their gold-and-black patchwork coats wet with dew, stood no more than five yards away. Their bold, dark eyes looked us up and down. We stayed still; several seconds passed while they leaned toward us, their noses twitching, their ragged tails raised. Then, with dark muzzles held high, they began to come closer, putting one paw carefully in front of the other. Delia edged toward the door. I squeezed her hand—it was not a time to move. The dogs stood hardly more than an arm's length away, staring as though they had never seen anything quite like us before.

A growl rose deep from the chest of the one with a swatch of golden fur dangling from his neck; his body trembled and his black nostrils flared. Both of them spun around and, rearing up, placed their forepaws on each other's shoulders, as if dancing a jig. Then they loped away, following the rest of the pack.

We pulled on our clothes, started the truck, and followed. Working as a team, the pack split the herd into three smaller groups and began to run each group, alternately, around the riverbed. The lead dog spotted a yearling that apparently seemed vulnerable. After being chased nearly a mile, wild-eyed and breathing heavily, the springbok began to zigzag sharply. The dog seized his prey high on its hind leg and dragged down the ninety-pound buck. Eight minutes later it had been consumed, and the dogs trotted to the shade of a tree island, where they would rest for the day. It was not our last encounter with "Bandit" and his pack.

Back in camp, we rolled up our sleeping bags, dug some powdered milk and raw oatmeal from the food box beneath, and washed it down with swigs from the canteen. After breakfast we set out to explore the dry riverbed as a study site.

The springbok had quieted down after the hunt and were again grazing in a line stretched across our path. Moving slowly and stopping when any of the animals began to show alarm, I eased the truck gingerly through the herd. We would have driven around them, but they were everywhere, so we took great care not to drive more than about three miles per hour or to make any sudden movements or noises that might frighten them. They did not yet have a negative impression of man or his vehicles, and we took every precaution to avoid giving them one.

Deception Valley is the remains of an ancient river that last flowed through the Kalahari about 16,000 years ago, at a time when rainfall was much more generous than it is today. But the land and its weather have always been fickle, and as it had done at least three times before, the climate turned arid, leaving the body of the river mummified in sand. The ancient riverbed remains in remarkable detail, a narrow ribbon of grassland wandering through dunes. It is so well preserved that when we drove along it we could easily imagine water flowing where grasses now wave in the wind.

Because it often receives somewhat more than ten inches of rainfall, the Central Kalahari is not a true desert. It has none of the naked, shifting sand dunes that typify the Sahara and other great deserts of the world. In some years the rains may exceed twenty—once even forty—inches, awakening a magic green paradise.

On the other hand, as we would learn later, all moisture received is soon either lost by evaporation, absorbed by the sands, or transpired by the vegetation. More dramatically, sometimes it may hardly rain at all for several years. So there is seldom a surplus of moisture: no secret springs, no lakes of standing water, no streams. The Kalahari is unique in this respect, a land of great contrasts, a semidesert with no oasis. It has no seasons as we know them. Instead, there are three distinct phases: the rains, which may begin anytime from November to January and last through March, April, or May; a cold-dry spell from June through August; and the hot-dry season, from September through December, or until the rains come again. We had arrived in Deception Valley after the rains and before the cold-dry season.

The sandy slopes, covered with grasses and thornbushes, rose from the old riverbanks to the dune crests, more than a mile away on both sides. The dune tops were capped with woodlands of *Combretum*, *Terminalia*, and *Acacia*, which, together with the mixed bush and grass communities, reached deep into the sand and prevented it from shifting.

Different zones of vegetation lay like tiers of a cake between the riverbed grasslands and the dune crest woodlands, and each community had something special to offer the various birds and animals there: Various species of plains antelope, mostly springbok and gemsbok, grazed short, nutritious grasses on the riverbed. Steenbok, duiker, hartebeest, and eland fed on taller, more fibrous grasses and leaves along the duneslope. Farther upslope, near the dune top, giraffe and greater kudu browsed on the leaves and fruits of the woodlands. Surely the concentration of antelope in our small area would attract predators such as lions, leopards, cheetahs, jackals, and spotted hyenas.

On our first drive, we began to name some of the landmarks that would help us pinpoint important observations and determine the ranges of the animals we would be studying. Clumps of acacia and ziziphus trees looked like small, round islands in the river of grass, and soon we knew them by such names as Eagle Island, Tree Island, Bush Island, and Lion Island. A bush-covered sand tongue protruding into the riverbed became Acacia Point; a bush where a family of bat-eared foxes slept became Bat Bush. Eventually each prominent feature had a name, for orientation and for easy reference.

Deception Valley seemed the perfect place to launch our predator research. Unlike the Makgadikgadi, the fossil river habitat provided a specific focus for the antelope prey populations, enabling us to locate our predator subjects consistently for observation.

The difficulties and dangers of operating in such an isolated area were obvious. Unlike research projects in most other parts of Africa, there was no nearby place to get water and food, and there were no dwellings, no contact with other humans, and no one to rescue us in an emergency. In fact, if we died, it was unlikely that anyone would realize it for months. Although we did not consider this remoteness a disadvantage, we would have to deal with some serious logistical problems: We faced an expensive 140-mile round trip to the Boteti River when the water in our drum ran out. Despite our efforts to ration ourselves to a gallon each a day since leaving Maun, we had finished half of the water in the Land Rover's inboard tank and much of that in our jerricans. I was glad we had another full drum. Especially after our Makgadikgadi experiences, we were very aware that the two absolute essentials for survival in the Kalahari were water and the truck.

In spite of these difficulties, we were certain that our predator research, along with an ecological analysis of soils, plant communities, and seasonal rainfall and humidity patterns, would reveal the dynamic workings of the entire fossil river system. Such a broad approach was necessary because no one had ever done research in this area and there was no background information available. We felt privileged to be breaking this new ground, but it would involve a lot of hard work apart from watching predators.

Considering our critical shortage of operating money, our primary subject would have to be a species that was fairly easy to observe, so we wouldn't have to go chasing all over the countryside, burning up precious gasoline to find it. It should also be an animal about which little was known, which would help to make our research more attractive to potential sponsors.

For days, in the mornings and afternoons, we sat on top of the Land Rover watching the antelope herds at various points on the riverbed, waiting for signs of a predator. Though we saw lions, jackals, cheetahs, and wild dogs, for one reason or another none of these would do. Since they all had been studied in other parts of Africa, we worried that we would be unable to attract a grant by researching them. Both wild dogs and cheetahs were rare and highly mobile in the Kalahari; they would be difficult to find and observe regularly. Furthermore, the cheetahs we saw were very wary—apparently with good reason. We heard later that Bushmen often chase them off their kills and appropriate the carcasses for themselves.

To help us decide which predator we should study, we took notes on any that we could find. In the process, we came to realize something that set the direction of our research for years to come: Night belongs to the carnivores of the Kalahari.

*   *   *

Purple-black and darker than the night, the supple dunes slept beside the ancient river. The sky sparkled with points of starlight and meteors streaked through the atmosphere. Below, the grasses, dry and tan before the dry season, reflected the celestial light, as if the river moved again.

I switched off the motor and the spotlight stabbed into the darkness. Eyes, thousands of them, shone like globes of phosphor. Behind the eyes the springbok herd rested, the vague curves of their horns and the bold white stripes on their faces showing above the grass. Some began to stand, nervously raising and lowering their heads. I swung the light away to a tree. Another eye, a larger one, stared down like a shiny marble from the treetop. A giraffe was browsing on acacia leaves.

We soon learned to recognize the various animals at night by the color and movements of their eye reflections, and by their height above the grass. A jackal's reflected yellow and jogged along just above the grass-heads. A lion's eyes were also yellow, but larger and higher above the ground, and they swung from side to side slightly as the animal walked.

We were driving back from making observations one night, watching along the spotlight's beam and trying to make out the dim forms of our camp trees. Suddenly, eyes that we had never before seen were reflected in the light; they were emerald-green and wide-set. A dark, bearlike form covered with long, flowing hair moved through the outer reaches of the light. It stood quite tall at the shoulder and had a large, squarish head, but its hindquarters were dwarfed, as if stunted, and it had a long bushy tail. It was walking quickly away from us. I eased my foot down on the accelerator pedal, and we strained through the cracked and yellowed windshield to hold it in sight. Moving faster, it seemed to glide across the savanna like some dark, shaggy ghost. Then it was gone.

Back at camp, we thumbed through our guidebook to the larger animals of Africa. Aardwolf? Spotted hyena? Aardvark? It certainly hadn't been a cat. None of the descriptions or pictures seemed to fit. We hadn't had a good view of it, but whatever the animal was, it was not common. Sitting cross-legged on our bedrolls in the back of the Land Rover, we paged through the book and back again, a flickering kerosene lantern hanging between us. Smaller than a spotted hyena, larger than an aardwolf, the wrong range for a striped hyena—but from its proportions, it definitely was a hyena. We finally decided it could only have been *Hyaena brunnea*, a brown hyena, one of the rarest and least-known large carnivores on earth.

What a stroke of luck! Here was an endangered species that had never before been studied in the wild and about which practically nothing was known. Anything we learned would be a contribution to science and important in the conservation of rare and endangered species. It seemed the perfect animal on which to focus our research.

Though they were entirely nocturnal, and secretive, we continued to see the brown hyenas, if only for fleeting seconds, when we jolted along over the riverbed. Their habits would make them difficult to study, but we grew more and more curious about them. Every night, beginning at dusk, we combed the riverbed with the truck and spotlight, looking for them. Left—right, right—left, I swung the light for hours as we drove slowly along. It was a frustrating business. Jackals, bat-eared foxes, korhaans, plovers, and wild cats were everywhere in the thick grass. Occasionally we did pick up those wide-set emerald eyes, but they were always at the outer limits of the spotlight, moving quickly away into the darkness.

*   *   *

Early one morning near the end of May, after a long disappointing night searching for hyenas, we reached camp stiff and sore, aching for sleep. Next to the fireplace stood a jackal, his feet planted solidly apart, his muzzle buried deep in our black iron stewing pot. Daring yellow eyes peered over the rim at us; gravy dripped from his whiskers. He finished licking the pot clean, cocked his leg, peed on it, and trotted casually out of camp. As he disappeared into the night, we recognized, from his black, anchor-shaped tail patch, the jackal we had named "Captain," a big, broad-chested male whom we saw often. He had a saddle of jet-black hair spiked with silver and a full, bushy tail.

Several nights later we sat watching a gemsbok carcass that had been abandoned by lions after they had eaten their fill. We were hoping a brown hyena would come by to scavenge. By 3:30 A.M., no matter how hard I tried, I couldn't stay awake. Leaving Delia to watch for hyenas, I quietly unrolled my sleeping bag on the ground beside the Land Rover. After setting my shoes in the grass next to me, I slipped into the bag and bunched my shirt up to form a pillow.

I had just dropped into a sound sleep when my head suddenly bounced on the hard ground. I sat up and fumbled for my flashlight. Five yards away a jackal was in high-speed reverse, dragging away the shirt he had yanked from beneath my head. "Hey! Drop that!" I was half amused, half annoyed, and still half asleep. I struggled out of my sleeping bag, scolding myself for having raised my voice; I might have frightened off a hyena. Then I began looking in the grass for my shoes. They, too, were gone.

This was more serious, because I didn't have another pair. I hobbled after the jackal, the sharp stubble spiking my bare feet. In the beam of my flashlight I could see his beady eyes fixed on me as he hauled my shirt away through the grass. My feet nicked and stinging, I finally gave up the chase and huddled in the Land Rover until dawn, when I recovered the slobbered toe of one shoe and the tattered remains of my shirt. Captain, the pirate jackal, had struck again. Later that day

I spent several hours sewing a pair of moccasins from two pieces of faded canvas.

At breakfast that morning the same thought struck Delia and me at the same time. While driving around the riverbed for hours each night, waiting for these timid brown hyenas to get used to us, why not find out what we could about jackals? They had never been studied in an area like the Kalahari, so anything we learned about them would be new.

Every sunset, we began parking the truck on Cheetah Hill, a bush-covered bulge of sand that intruded onto the riverbed north of camp. With field glasses, notebooks, and a tin of corned beef, we would sit, each cradled in a spare tire atop the roofrack, watching the nightlife begin in Deception Valley.

Sometime before dusk Captain would usually stand from his favorite resting place near North Tree, point his muzzle to the sky, and call to his jackal neighbors. Then, his ears cocked, he would listen to the high-pitched wavering cries that answered from up and down the valley. He would scratch and shake his thick coat, the silver hairs along his black saddle shimmering in the fading twilight. After a deep stretch, he would be off, his keen muzzle spearing through the grasses as he jogged along hunting for mice. From Cheetah Hill we would note his direction, then move to intercept and follow him.

Steering and changing gears with my left hand, holding the light out the window with my right, I would try to keep pace and contact with Captain, staying from fifteen to twenty-five yards behind him. If we followed any closer, he would look back at us, obviously disturbed; any farther away and we would lose him in the grasses. Meanwhile, Delia, holding a flashlight over a notebook and compass on her lap, made notes on his behavior, the directions in which he traveled, the distances, and habitat types. The binoculars lay on the seat between us so that either of us would grab them and describe what was happening. With some practice, our technique worked very well. We could see the species of bird, and often the kind of rat or mouse, that Captain caught before it had been chomped down. And by driving to the spot in the grass where he had been nosing about, we usually discovered a line of frenzied termites or ants he had just lapped with his tongue.

One night in early June, soon after the theft of my shoes, we were following Captain when, without warning and with incredible speed, he took off after a steenbok fawn. I accelerated, and we managed to keep him in sight during a long chase that involved several complete circles, before he disappeared. But after that, no matter which direction we took, we could not spot the lantern we had hung high on a limb in camp to guide us back after dark. Captain had led us out of the stretch of riverbed that was familiar to us, and he had changed directions so frequently that he had scrambled our mileage and compass readings. We could not backtrack with any certainty. We were lost.

With less than a quart of water on board the Land Rover, we could not risk driving farther from camp. So we stopped for the night. From the roofrack the next morning, we could see North Tree on the riverbed about a mile away; our camp was about the same distance south from there. Heading back, I decided that I would tap the red water drum that day and fill a jerrican, and from then on always keep it with us in the truck. There had been no rain for weeks. Each day was cloudless, the savanna drier than the day before. If we got lost again without water, it might be more serious than having to spend one night away from the shelter of our tree island.

Back in camp, while Delia began to make our breakfast, I took a wrench from the toolbox and carried one of our empty jerricans and a length of siphon hose to the drum. I laid the wrench in the bung to unscrew it. A hollow, empty sound rang from deep inside.

It just couldn't be... I dropped the wrench and pushed the drum. It toppled over and rolled aside—empty. All that was left of our water was a patch of dampness in the sand.

"Delia! There's no water in this blasted drum!" I stooped to look at the rusted bottom, and then I kicked the damned thing. Delia was as shocked as I was. In a small voice, she asked, "Mark, what are we going to do? Can we make it to the river?"

The Boteti River was nearly a full day's drive away through the heat, sand, and thornscrub. We had less than a quart of water in the truck, and we needed much more than that just to keep the motor cool. If it ran dry and seized up, we would be in a lot worse trouble. How could I have been so stupid! Each day I had measured the level of gasoline in the large inboard tank in the Land Rover to make sure we had enough to get to Maun. I thought we had plenty of water—but I should have *looked*! We stared at the damp spot as if to will the water out of the ground. I fought off a growing uneasiness. This was just the type of situation we should have avoided.

"We'll have to go tonight when the air is cool, so we won't have to use so much water in the radiator," I said, putting my arms around Delia. There was nothing else to do.

That afternoon we climbed into the Land Rover to leave for the river. I fumbled with the ignition key, switching it over and over. "Come *on*, you son of a bitch!" My throat was tight with anger and rising fear. There was nothing but a deadpan click. I jumped out, ran to the front, and threw up the hood. "Try it again!" I called to Delia, then listened to the motor, trying to find out what was wrong.

From the time we had driven out of Gaborone on our way north to Maun, what I had feared most was being stranded somewhere far away from help with a broken-down Land Rover, which I could not fix because I lacked the talent, tools, and spare parts. Up to this point we had had continual minor problems with the worn-out vehicle: corroded battery cables, a discharging generator, punctured tires, and broken exhaust lines. I had always been able to make these relatively minor repairs. But now, as I peered under the hood and Delia worked the ignition switch, the tension settled heavy in my chest. The click that comes from a dead battery now had become a deeper, more ominous clunk, a sign that something much more serious had gone wrong.

I kept my worst fears to myself and went to work. It was dark by the time I discovered that the starter had come apart. Its ring gear had fallen into the flywheel housing and had jammed the motor, so hand-cranking was going to get us nowhere.

I found a piece of heavy wire and bent the end into a hook. We crawled under the truck. Delia held the flashlight for me as I slid the long wire into the housing, trying to feel my way past the flywheel. But through the wire everything felt the same; I could only guess where the engine was jammed.

About midnight we came out from under the truck. My knuckles and forehead were bleeding and smeared with grass chaff, oil, grease, and dirt. It seemed hopeless. I could not be certain that I had even touched the ring gear. I heaved on the crank. The motor turned slightly, and then stopped solidly.

We put some wood on the fire, and while we warmed ourselves and rested, I tried to think through our predicament. If we were going to survive this, the starter gear had to come out. We were both thirsty, but neither of us would drink. I looked at Delia, sitting beside the fire, her head on her arms. I felt utterly frustrated. I couldn't think of anything else to try, and time was running out. Only a quart of water was left for the two of us.

Back at the truck I moved the crank counterclockwise until it stopped, then clockwise just slightly. I was sure the wire was long enough and must be touching the gear. For the rest of the night I bent the end into hooks of different shapes and inserted it at different angles, poking and probing into the housing while Delia held the light. When I could feel nothing move, I repositioned the engine and probed some more.

It was sometime after sunrise when I heard a clank. I hurried to the crank and turned the motor—it was free! We would rest through the day and set off at dusk.

This predicament made us realize that staying in an area as remote as Deception Valley was out of the question. Assuming that we were able to get to the river and return, how long before another crisis? And how many such crises would there be before a real disaster? The hard truth was that we just did not have enough money to do research in such a place. What little we had left would soon be consumed in trips to Maun for supplies, especially since we had no way to haul and store adequate quantities of water and gasoline. We would have to find a less remote place to work, a place less wild...and less free. It was a bitter realization. Two years of planning and working to save enough money for the expedition, five months of reconnaissance and study in Africa, all seemed to have drained away with our water into the sand. Though we had been in the Kalahari for little more than a month, we had already developed a deep attachment to this old riverbed and its animals, especially the ones, like Captain, whom we recognized.

We picked through a depressing breakfast of beans, a meal we had had three times a day for the last sixteen days, and then began loading the truck with our few belongings. The droning of an engine abruptly roused us from our apathy. A stubby green-and-white Land Rover bounced down the duneslope east of the riverbed, a long tail of dust rising behind it. We stood and watched it approach, completely amazed that there was another truck in the area. Before it had rolled to a stop a freckled, ruddy-faced man in baggy shorts, knee socks, and a knit shirt stretched over his round belly launched himself from the driver's seat. His thin, greying hair was slicked straight back over his sunburned head, and his eyes crinkled with a smile. Kalahari sun, wind, and sand had etched his face deeply.

"Hello! Name's Berghoffer, Bergie Berghoffer. You can call me Bergie. Someone in Maun told me that you two were out here somewhere, and when I crossed some truck spoor miles east of here, I figured it was you." Rummaging in the back of his Land Rover, he called over his shoulder, "I reckon by now you might be needing some of this." He set out some brown paper packets of goat meat, a bucket of mealie-meal with eggs buried inside—to keep them cool and protect them from breaking—potatoes, and coffee. While we were thanking him for about the twelfth time, he finally put up his hands and said with a wink, "It's only a pleasure . . . I'm half Yank myself, you know."

We were to learn later that Bergie had roamed the Kalahari for twenty-three years, living in bush camps while drilling mineral test holes for the Botswana Department of Surveys and Lands. He lived a nomadic existence, moving his camp from one area to another, usually far outside the game reserve. "I'd much rather never find anything out here except the animals—not sure I'd tell anyone if I did." He said wryly. "I'm bloody glad someone is finally here to study the wildlife. No one ever has, you know. The Kalahari needs someone to be her champion."

Bergie had a special affection for "Yanks" because his father was an American who had traveled to the Republic of South Africa with the Bill Cody Wild West Show. There he had met and married a woman of British extraction and had settled in the republic. Bergie reckoned it was his father's yearning to travel that coursed through his own veins and kept him on the move and in the bush for most of his life.

"I'm sorry . . . wish we could offer you some tea or coffee," I apologized. "But we've got a problem." I showed him the empty drum.

"Well, a bit of bad luck, that." He frowned, rubbing his chin. "Not to worry about the coffee, but what are you chaps going to do for water?"

I explained that we would be heading for the river, then Maun, and that we wouldn't be able to come back to Deception. "Oh . . . I *am* sorry about that . . . that really is a bit of a bugger." Looking out over the riverbed, he sighed.

"Tell you what." He brightened. "You take this, just to be sure you get there." He hefted a jerrican of water from the back of his Land Rover. "Now I'll have that cup of coffee, if you don't mind, missus."

Despite our protests, Bergie would take none of the water with him to insure his safe return to his own camp. He had scarcely finished his coffee when he held his hand out to me.

"Okay, Mark, okay, Delia, I must be on my way—I'll be seeing you." Then he was gone, his Land Rover disappearing over the eastern dune.

Blind luck and Bergie's generosity had provided us with more than enough water to get to the river—or to Maun, for that matter. We decided to spend one more night in Deception Valley. We hated to leave it, and besides, we were exhausted from the trials of the previous night.

Not an hour before we were to leave Deception Valley for good the next morning, Bergie was back, this time with a large flatbed truck and his drilling crew of eight natives. They unloaded a folding wooden table, two chairs, a heavy iron fire grate, a gas burner complete with cylinder, a small cooking tent with a big fly sheet, four drums of water, and some gasoline. Bergie was like a genie. He waved his arms and shouted orders to his crew, and as if by magic, a small camp appeared.

Almost before we had realized what was happening, he was gone, a wisp of dust vanishing across the dunes again. We stood in the middle of our instant camp staring after this whirling dervish of the Kalahari. In one swift gesture of unbelievable kindness, Bergie had made it possible for us to stay in Deception Valley, at least for a time, while we fought to get more research data and a grant.

We resumed our research, but it was not easy to observe and follow animals without a starter on the truck. Every day near dusk, when we found a jackal sleeping in the grass on the riverbed, we would park nearby, switch off the motor, and wait for it to get up and begin foraging. As soon as it stood and began to stretch, I would sneak to the front of the truck and crank the engine while Delia tried to keep her eye on the jackal moving away through the tall grass. The hand-cranking created such a terrific racket that it drew the attention of every creature for half a mile around. And whenever we were sitting with lions at night, it was a little disconcerting for me to turn my back on them in the dark, knowing that they were watching as I heaved on the crank.

Two weeks after Bergie had given us our camp, he was back with more water. While Delia was making coffee, he quietly took me by the arm and led me to his truck. "Now listen lad, if you expect to keep Delia in the Kalahari, you must spoil her a little. Every woman needs a hot bath!" He turned and pulled a tin tub from the back of his truck. "And does she have a looking glass?" Reaching through the window, he brought out a mirror. From the expression on Delia's face when she saw these gifts, I knew Bergie had been right.

His camp was so far away that we only saw him rarely, but he had an uncanny sense of timing. Weeks would pass, and just when we were getting low on water, he would appear, always with more gifts of goat or wildebeest meat, eggs, potatoes, brawn—a meat gelatin, or headcheese—and other luxury items that he brought either from his camp or from Gaborone. Even if we had had a way of getting these things, we could never have afforded to buy them.

One day he took us much farther south along Deception Valley than we had explored. After an hour of riding in his Land Rover, with its stiff, super-reinforced leaf springs—he was very proud of them—our kidneys ached and our necks were stiff from the pounding. We finally stopped on a dune overlooking a large, perfectly round clay pan. Because of the slate-grey soil in its bottom, it looked as if it were covered with water, an illusion so complete that in later years we saw migratory water birds, once even a pelican, drawn to its surface during drought. Bergie told us the Bushmen had named the valley after this pan, with a word in their language meaning "deception," and also because when traveling along the riverbed, one is deceived into believing every bend is the last. Below us, the old river channel continued winding away into the Kalahari beyond this Real Deception Pan, as we often called it.

"I've been this far and no farther," Bergie said. "Beyond here no man knows." None of us spoke for a long time. We listened to the wind singing in the grasses and looked over the great expanse of wilderness stretching for hundreds of miles beyond. "You know," Bergie said, "there's only one thing that really frightens me out here, and that's fire."

3

Fire

Mark

Voices of jackals calling
And, loud in the hush between,
A morsel of dry earth falling
From the flanks of the scarred ravine.
—*Rudyard Kipling*

THE RAINS of 1974, which flooded many parts of the country, had been the heaviest ever recorded in Botswana. They had ended in May, but in their wake, the grasses of the savanna stood taller than a Bushman's head, like a field of golden wheat hundreds of miles across, bowing in the wind. By July, when we had been in Deception for three months, the dry-season sun had turned the wheat to straw, the straw to tinder. Some said the sun's rays passing through a dew drop could set it off.

"Grass number 27: base, 9.2 centimeters; dry canopy, 57.2; green canopy, 14.3..." We had been at it all morning, measuring the basal areas, canopies, and species composition of the grass and herb communities along our vegetation sample lines from riverbed to dune top.

I stood up to rest my cramped knees and noticed a curious grey cloud rising from the eastern horizon. Billowing thousands of feet into the higher atmosphere, its top was sheared off by winds into a vaporous tail that slowly drifted south. Far away—how far we could not tell—the Kalahari was burning.

As we stood watching the ominous cloud, a strong wind, gusting to thirty miles per hour, struck us full in the face, tugging at our clothes and bringing tears to our eyes. Only miles of dry grass stood between us and the fire.

We continued to follow Captain and the other jackals every night, always conscious of the eerie glow on the eastern horizon. There was still so much savanna between the fire and our camp that it would take several weeks to reach us. Before then we would have to develop some plan to protect ourselves, the Land Rover, and camp.

The July nights were bitterly cold. We had not expected temperatures that dropped from a daytime of seventy degrees to fourteen above zero just before dawn. We had no winter clothes—there had been no room for them in our packs when we left home. Following the jackals, we could hardly bear the frigid air, and after holding the spotlight out of the window for just a few minutes, my arm and shoulder would be numb. The truck had no heater, so I cut holes in the sides of a coffee can, turned it over a candle, and set it on the floor. We put socks on our hands, sleeping bags across our laps, and ate cans of stew heated on the exhaust manifold. Still we cramped with the cold and could stand no more than three or four hours in the Land Rover before heading back to the campfire.

At first most of the jackals looked alike, especially at night, so we decided to immobilize and collar some for easier identification. On our first supply trip to Maun, when I had fixed the starter on the truck, Norbert Drager, a German vet in the village, had given us some buffalo-collaring material and a carbon dioxide darting rifle full of leaks and covered with rust. I sealed the rifle with tire patches so that it would hold a charge and, using the buffalo material, made some lightweight collars that fastened with small bolts.

One very cold night in mid-July, we managed to immobilize a jackal near camp. Animals sometimes develop hypothermia when drugged in such weather, so when we had the collar in place, we gently carried our subject into camp, where he could recover near the warmth of the campfire. Then we kept watch from the Land Rover to protect him from larger predators until he fully regained his coordination.

Hours passed, the night grew colder, and our coffee-can heater could not keep us warm. By 1:00 A.M., Delia had had enough. Casting a dim yellow light ahead of her with our weak flashlight, she dragged her sleeping bag and our flimsy foam mattress into the small mess tent. On the riverbed there was a rodent boom—so many furry creatures that for several weeks we had had to eat supper with our feet on tin cans to keep them from crawling up our legs. In spite of the occasional rat or mouse that would scurry over her, Delia was determined to be warm and get some sleep.

I sat in the truck, now and then turning the spotlight on the jackal, who was beginning to flop around, trying to stand up. The light warmed my hands and I wanted to leave it on, but I didn't want to run down the battery. So I sat with my binoculars resting over the steering wheel, shivering in the dark and studying the distant glow of the grass fire, wondering how far away it was.

Two weeks had gone by since we first noticed it. Since then it had broadened and intensified into an orange-red corona stretching across the entire horizon from north to south. Now, in the quiet of night, with the air still and moist, the vivid colors had nearly faded from the sky. The fire seemed to have gone to sleep. But I knew that in the morning, the heavy winds would return and send an immense curtain of grey smoke into the atmosphere.

The little camp Bergie had given us wasn't worth much materially, but it was all we had in the world, and we could never have afforded to replace it. If the fire destroyed it, we would be wiped out financially and our research would be finished. Furthermore, the roots we had put down in that tree island were already something very vital. In the short time we had been there, Deception Valley had become our home.

We worried, too, about Captain and the other animals. Surely some of them would die in the fire. And the sample plots along our vegetation transect lines would be incinerated. After the burn there might be little left to study.

Now, sudden fountains of color surged into the night sky, and then, mysteriously, drained away to a small, dim smudge, only to flare up again minutes later: The fire was dune-walking. Each time it descended into the shelter of an interdunal valley where there was less tinder, its intensity diminished, but as it crept back to the wooded crests, added fuel and wind renewed it. With growing concern, I began to realize what a giant it really was. The Kalahari was burning along a front more than fifty miles long from north to south.

I hung the spotlight outside on the mirror bracket of the truck, and turned it on. It was about 3:30 A.M. and the jackal was recovering nicely. After switching off the light, I sat blowing through my hands to warm them. But then something urged me to have another look. I flipped the switch again and saw *seven lions standing over the jackal.*

Startled by the light, the two females and five subadults jumped back and turned away. But seconds later they were back, eyes fixed on their prey. I hit the starter and drove past the tent, where Delia was sleeping soundly. Ignoring the noise of the truck, the lionesses refused to leave the jackal. Their surprise and confusion gone, they stalked toward him, their heads low, their tails flicking from side to side.

I drove quickly by the jackal, getting between him and the lions. They turned to avoid the truck, and I nudged one of them gently in the rear with the bumper. The lioness grunted once and then wheeled and spat at the headlight. They tried to move around the truck, but I blocked their way, turning them around and heading them at a walking pace into West Prairie, the grassland off the riverbed beyond camp. By cutting the steering wheel left and right and holding the bumper close behind them, I kept them moving away. I didn't like manipulating the lionesses in this way, but my major concern at the moment had to be the jackal, because we had rendered him temporarily helpless.

I was about 400 yards west of camp when, through the rearview mirror, I noticed a weak glow winking on and off somewhere behind the truck. It took some seconds before I realized that it was coming from the mess tent.

Delia had been unaware of the lions. Awakened by the truck, she thought I was following the jackal from camp to make sure he had completely recovered. The Land Rover hadn't been gone long when she heard the padding of heavy feet on the ground outside the tent. The canvas walls shook once. Then a heavy rush of air sounded at her feet. She slowly raised her head. Framed in the doorway—there were no zippers left on the tent flaps—and just visible in the starlight, were the massive heads of two male lions looming over her toes.

She held her breath as the lions smelled the floor of the tent, puffs of air blowing from their nostrils, their whiskers skimming the nylon of her sleeping bag.

She moved her feet. The lions froze, looking straight into the tent. She could hear the Land Rover moving farther away. She slowly reached for the flashlight on the floor beside her. The lions were standing dead still; they seemed to have stopped breathing. She began to raise the flashlight to the screen window above her head. The lion on the left moved against the tent, and the walls trembled again. With the flashlight held to the window, Delia hesitated to thumb the switch, afraid of the noise it would make. Finally she nudged it once—in the silence, the click was like a shot.

The lions didn't budge. On-off, on-off, on-off, she signaled again and again. Moments later she let out a long, slow breath when she heard the Land Rover's engine racing and the peculiar squeak in its bumper as it rattled toward camp.

As I neared our tree island, I swung the spotlight back and forth. Everything looked normal. Still, the dim flashlight continued to blink on and off. I rounded the tent and then jammed on the brakes, gripping the steering wheel hard: Two black-maned lions stood shoulder to shoulder, their heads buried in the doorway. Delia was trapped like a mouse in a shoebox.

I had to do something to distract the lions without jeopardizing her by intimidating them in the wrong way. An inept move would increase the danger. A story had been circulating in Maun when we were last there about a woman being dragged from her sleeping bag by lions in Chobe, a park in northeast Botswana. For once I wished I had a firearm of some sort. At least I might have frightened the lions away from the tent by firing it in the air.

Maybe I could use the Land Rover to herd the males, as I had the females. Slowly, I drove toward the lions. Standing firmly in front of the tent's flap, they both looked at the truck, their eyes round, ears perked, tails twitching. At least their attention was focused on me now, not Delia. As I drew nearer they seemed to grow, until they stood as tall as the truck's hood, their shoulder muscles bunched and tense. They held their ground. I stopped.

After a few more seconds, the males began to blink their eyes. They settled back on their haunches and turned back to Delia. I let out the clutch and drove forward again, this time with my head out the window and slapping the side of the Rover to keep their attention on the truck. When I was very close, they finally turned away. Laying their ears back as if annoyed, they put their noses to the ground and walked off in the direction of the females. Just beyond camp they began to bellow, their crescendos rolling up and down the valley, and the lionesses answered from the bush savanna farther west. The jackal had escaped during all the commotion.

I quickly slipped into the tent to lie next to Delia. Half frightened and half excited, she chattered for a while and then buried her head in my chest. Soon we were sound asleep. I woke once, when a rat dropped onto my forehead. I slapped it against the wall of the tent and shivered, and after a while fell back to sleep.

Some nights later, we were following one of our collared jackals when the eastern sky flushed angry red. "Mark, the fire's almost here!

We've got to get back to camp and get ready!" I felt sure it was still quite far away and that it would be a waste of time to tear down camp so soon. But Delia insisted, and I finally succumbed to her pleas and turned the truck around.

Before I had even brought the Rover to a full stop, Delia leaped from her seat. She began collecting pots, pans, and bags of flour and mealie-meal, hauling everything she could drag or carry toward the truck. I tried to reason with her. "Look, Boo, it's not going to come tearing over the dunes within the next few minutes, or even before morning."

"How do you know!" She snapped, as she struggled with a heavy bag of onions. "You've never seen a range fire in the Kalahari—or anywhere else."

"We'll be able to hear the fire way off, you'll see the flames in the grass—there'll be sparks. Where are we going to eat, sleep, and work if we pack up before the fire's even here?"

But it was no use trying to stop her. I half expected her to smother me with a wet blanket at any moment. Now she was staggering toward the truck with another armload of boxes, some clothing, and a jerrican. The pile in the back of the Land Rover was growing by the minute. I began to sneak things out the side door as she piled them in through the back. "Now look, dammit! When the fire gets here we'll know it. Get hold of yourself!"

"I'm not taking any chances!" She shouted back.

I had just managed to hang the onions back in a tree on the dark side of the truck when one side of the mess tent collapsed. Delia was hauling up tent stakes like a gopher in a turnip patch.

"What are you doing?" I begged.

"I'm putting the tent in the Land Rover."

I stomped over to the truck and shoved all the stuff out the back onto the ground. "Stop!" I yelled, standing between her and the pile. "If it'll make you feel better, we'll do something constructive—like make a firebreak around camp."

I tied a fallen tree to the back of the Land Rover with a heavy piece of old cotton rope. A few circles around camp and I had flattened a swath through the tall grass.

When I had finished I began making up our bed in the back of the truck. By now it was long past midnight.

"What are you doing?" Delia was standing behind me.

"I'm going to sleep. I know we're perfectly safe. Anyway, you're so damned stubborn . . . no matter what, you'll watch that fire all night."

Much later, cold, stiff, and contrite, she crawled into her sleeping bag and huddled against me. I put my arms around her and drifted back to sleep.

About midmorning, Bergie's flatbed came churning over East Dune and into camp. Laughing, he slid from the driver's seat. "What's all this?" he asked, as he eyed our disheveled camp.

We asked him about the approaching fire.

"Well, I guess you'll likely survive for a while yet." He chuckled. "That fire's still thirty miles east of here—passed my camp day before yesterday."

Delia glanced at me and smiled faintly.

Then Bergie frowned. "Make no mistake, though, its bloody-minded. Even with a tractor to make a break and my crew to fight it, we had a time. Take care when it gets this side—it's no small fry."

"How did it get started?" I asked.

"Mon, the bloody Bushmen set these fires every year, you know. They can hunt—track—better with the thick grass burned away. And it's easier for them to collect bauhinia nuts, one of their staple foods. I suppose you can't blame 'em too much, but the fire sure raises Cain with the trees in the woodlands. Dries out the lower leaves the animals need for browse in the dry times. And the Bushmen aren't the only ones to blame. The safari hunters set the veld alight, too, though you'll never hear 'em admit it."

He turned to the back of his truck. "Had a few things extra around camp I thought you could use." He set a gunny sack full of goat meat, eggs, and mealie-meal in the grass next to the mess tent. Our near-empty water drums boomed as we filled them from the drums on his truck. Delia brewed some coffee.

With a last sip from his cup he was on his feet and saying goodbye. "I've got three weeks' leave due me. I reckon I'll go to Johannesburg for a spell to visit my daughter and her family. But you can bet I won't be able to stand the city life for long—I'll be back in ten days or so. Probably beat the fire here; it dies down a lot each night so it's still a couple of weeks away from Deception."

We begged him to spend a few days with us on his return, so that we could show him some of the things we had learned. "All right, all right—I'll pop over directly when I get back. Okay Mark, okay Delia—I'll be seeing you."

Two weeks went by, but still Bergie had not come back. Day after day we listened against the wind, imagining dozens of times that we could hear his truck approaching. Maybe it was a tent line humming in the wind, or our ears buzzing from the silence, that fooled us again and again. It's like that in the bush, when you've waited a long time for a friend.

Was he ill? Had his truck turned over somewhere along the spoor east of Deception? Worried, we finally drove along the track he took to our camp, but we found nothing. We decided he must have stayed longer than he had expected in Johannesburg.

A few days later, on a frosty morning in early August, I opened the back door of the Land Rover and crawled from bed. The sickly sun cast a frail, sallow half-light over the old riverbed. The birds were silent. Hordes of insects—ones that usually came out only at night—swarmed in the air or crawled through the trees and over the ground in the eerie quiet. The ashen skeleton of a grass leaf, incinerated by intense heat, settled on the back of my hand. I looked up, and the air was full of them, floating in, softly covering everything, like black snow. From north to south, the veil of smoke in the eastern sky boiled skyward for thousands of feet. The fire was almost on us. I felt small and threatened. It looked larger, more powerful than I had imagined. I knew we probably should have packed up camp days before and gone to Maun.

I hurriedly piled into the Land Rover pots, pans, bags of flour, mealie-meal, and everything else that would fit, and Delia flattened the mess tent and its fly sheet to the ground. But if the flames hit the valley in midafternoon, with the humidity at rock bottom and the winds blowing easterly at thirty to forty miles an hour, it would be almost impossible to keep the camp from burning. Aside from our personal safety and that of the animals, we were most concerned about our data books—the record of our research—and about the Land Rover. Tying the dead tree to the truck again, I dragged it around camp to broaden our firebreak. With a spade and axe, we cleared away as much grass and dead wood as we could. Delia set pans of water near the mess tent, and I cut branches for beating at the flames. There was little more we could do.

The morning wore on, the winds blew harder, and the roar from the fire grew louder. More and more ash rained into camp and swirled across the ground in the churning air. By midafternoon, driven by the heavy desert winds, the first flame reached the top of East Dune. It paused for a moment, licking at the tall grasses and lower branches of a tree, then leaped quickly to the top, turning the tree into a thirty-foot torch. Another flame crested the dune, then another. A line of fire invaded the woodlands, and whole trees exploded like flares.

The intense heat created its own wind, a wind that fed oxygen to the flames and spurred them down the duneslopes toward the riverbed at an incredible speed, sweeping them through grass and bush as far north and south as we could see. Nothing could have prepared us for that sight.

"Our break will never stop it!" I yelled above the roar. I dropped the branch in my hand and ran to the Land Rover, tied the fallen tree to the hitch again, and dragged it round and round camp to widen the firebreak.

When the flames reached the riverbed, 1000 yards from camp, they dropped and spread out in the grasses. An immense cloud of seething white smoke erupted from the savanna, and the fire, with flames eight to ten feet high, swept down the valley. I had hoped that our truck spoor across the riverbed, 400 yards from camp, would slow the advance, but the fire only paused for a moment and then surged toward us again. I could see immediately that the break around camp was still far too narrow.

Once again I dragged the tree, this time in large figure eights. When the flames were about 200 yards away, I ran to the edge of our break and knelt to set a backfire. My hands fumbled with the matches, trying to get one lit. It was impossible in that wind; I turned my body to form a shield and felt heat on the back of my neck. I fought off an urge to stand up and run. Finally I touched off the whole box and stuffed it into the grass.

But it was too late—the backfire could not burn fast enough against the strong wind. I sprinted to the truck and drove along just ahead of the fire line, pulling the dead tree. If I could break the fire's momentum, we could beat it out around camp when it reached our backfire and break.

I made several passes just ahead of the flames. But they were still moving much too quickly toward Delia and camp. At a spot where the fire had slowed down a little in the flattened grass, I drove directly into the flames, straddling the fire line with the truck and driving as fast as I could while dragging the tree. After about fifty yards, I turned out of the fire and looked back. It was working. There were gaps in the fire line, and it had been slowed. Before it could rebuild momentum, I swung around and made another pass, and then another.

On the third sweep, the smoke from smoldering grass was so thick that I could hardly see. Suddenly Delia appeared in front of the truck, beating at the flames, her branch above her head. I slammed on the brakes and missed her by a foot. She jumped back and I sped away.

When I was turning for another pass at the fire, she came running toward the Land Rover, screaming and waving, her face white.

"Mark! My God, you're on fire! The truck's on fire! Jump! Get out before it explodes!" I looked back. The tree, the cotton rope, and the undercarriage of the Rover were in flames.

A fifty-gallon tank of slopping gasoline was riding against the back of my seat; its overflow pipe ran through the floor of the truck and came out ahead of the right rear wheel. Stamping on the brake and turning off the motor, I lunged from the door as flames leaped up around both sides of the Land Rover. Then I ran the thirty yards to where Delia was standing. Together we waited for the explosion.

"All our data books, our cameras, *everything* is inside!" she cried.

Then I remembered the old fire extinguisher clipped to the ceiling over the front seat. I got back inside the burning truck, but the trigger of the extinguisher was frozen with rust. I threw it out the window, started the motor, and jammed the truck into gear. Holding the accelerator flat to the floor, with the engine racing, I slid my foot off the clutch pedal. The Land Rover lurched forward with a shudder that shook every part of it. The flaming rope and tree broke free and, miraculously, most of the burning grass dropped from beneath. I stopped the truck over a small patch of bare calcrete rock and threw sand into the undercarriage to put out the rest of the flames.

We poured pots of water over the mess tent and flailed with branches and tire innertubes while the fire continued to burn around the edge of camp. Flames crawled over the ground we had cleared, working their way across the break along single stems of grass. One of the guy ropes to the tent fly sheet was on fire; I cut it free. We dragged a plastic jerrican of gasoline and a box of Land Rover parts farther into camp. Sparks showered over us. Beating at the flames, we struggled to breathe, choking and gasping in the acrid smoke and hot, deoxygenated air. Time and the fire seemed to be standing still. We barely had the strength to raise our branches for another feeble swat at the flames.

In minutes, or seconds—I don't really know how long—the main fire had passed. We had slowed it just enough to divert it around camp. After mopping up some of the remaining small pockets of flame, we were finally safe.

We sank to our knees, coughing and heaving in exhaustion, our lungs burning. When we were able to look up, we watched in a stupor as, one by one, the other tree islands along the valley became torches of orange flame. North Tree and Eagle Island were burning wildly.

Our lips, foreheads, and hands were blistered, our eyebrows and lashes singed. We would be coughing up ash and soot for days after, and the charcoal had invaded the pores of our skin so deeply that it was impossible to wash it away. For weeks, wherever we drove or walked, a grey cloud would envelop us. On windy nights, the Land Rover would be filled with a gritty haze that dimmed the light of our kerosene lantern to a murky yellow glow. We slept with bandanas tied across our faces.

After the fire passed us it marched on across the dune tops into the Kalahari, lighting the night sky like a spectacular sunset. Behind it, the cool pink glow of burned-out trees and logs remained, until the fire's crimson was lost in the blush of dawn.

At sunrise the next morning we sat staring over the blackened Kalahari. Tendrils of white smoke crept from burned-out stumps. Fragile tufts of grey ash—all that remained of grasses along the dunes and riverbed—would soon crumble to a powder before the winds. Whole trees, big ones, had been completely consumed, leaving an embroidery of white ash against the blackened sands. We felt as though we were the only inhabitants on a volcanic island that had formed in the night. Lava and ash had not yet cooled, and flares from within the earth still flickered through the molten surface. Our research had been incinerated.

Around noon Bergie's big white truck came growling over the blackened face of East Dune and lumbered toward camp. Delia quickly began stoking the campfire for coffee.

The four-ton Bedford rolled to a stop, and the Africans who worked with Bergie climbed down and stood in a ragged line.

"*Dumella!*" I greeted them.

"*Ee*," came their hushed reply.

"Where is Mr. Bergie? How is he?" Delia asked. They all looked at the ground, coughing and scuffing their shoes in the dust.

"Khaopheli," I asked the foreman, "what's the matter? Where is Mr. Berghoffer?" They hung their heads in embarrassed silence.

"Mr. Bergie no coming back," Khaopheli said softly, still looking at his feet.

"Why not? You mean he's still in Johannesburg?"

"Mr. Bergie dead." I could scarcely hear him.

"Dead! What do you mean—that's impossible!"

He lifted his face, patted his chest, and muttered, *"Pilo*—heart."

I sat down on the bumper of the truck, my head in my hands. Though we had known him only a short time, he had been like a father to us. I kept shaking my head, still trying not to believe it.

"We take camp...Mr. Bergie's things," Khaopheli mumbled. I nodded and turned to stare over the Kalahari that Bergie had loved so much. The crew immediately began loading their truck with our only table, the two chairs, the tent, and other pieces of equipment.

"But Mr. Bergie would have wanted us to have these things," I protested. Khaopheli insisted that the government would have to decide what to do with them. When they began rolling the water drums toward their truck I flatly refused to give them up, explaining that I would contact the Department of Surveys and Lands for them to be formally loaned to us. They finally relented and drove away. The few trees we had saved from the fire, the Land Rover, the drums, a sack of mealie-meal, and other foodstuffs were all that was left.

We had never felt such utter despair. We couldn't even tell Bergie's family how much he had meant to us; we didn't know his daughter's name. A book we had planned to give him lay on the ground next to where the tent had stood, its pages fluttering, gathering ash in the wind.

After a while we drove across the charred riverbed. Dense clouds of soot and ash swirled about us, filling our eyes, noses, and throats. Everything was black. At the top of West Dune we stood on the roof of the Land Rover: Complete destruction ranged as far as we could see in every direction.

A month or so earlier, we had gambled our return airfares to the United States on more supplies for the project, trying to keep our research alive while waiting for word of a grant. None had come, and now we had less than $200 left. Our research was finished. Somehow we would have to earn the money to get home.

We stood staring despondently over miles of blackened dunes. Delia, with tears rising in her eyes, put her head on my shoulder and said, "Whatever brought us to this place?"

4

The Cry of the Kalahari

Mark

The earth never tires,
The earth is rude, silent, incomprehensible at first,
Nature is rude and incomprehensible at first,
   Be not discouraged, keep on, there are divine things well
      envelop'd,
I swear to you there are divine things more beautiful than
   words can tell.

—*Walt Whitman*

THE SANDS were laid bare by the harsh, persistent winds. An ashen debris of cremated leaves tumbled over the dunes before the gale—a black wind howling over the Kalahari.

The fire had burned away the privacy of steenbok, korhaans, jackals, and others who lived in the grasses. Bat-eared foxes skulked nervously about, or tried to hide behind stubble an inch or two high, their big ears drooping sadly. There was no refuge.

But there was warmth and darkness not far below the blackened surface, and hidden in the soil, left by the heavy rains of months before, there was moisture, the universal ingredient of all life. Long chains of water molecules, linked through capillarity, were being drawn from deep beneath the dunes by the hot, thirsty winds above.

A tiny grass seed lay waiting in its moist, subterranean bed. Bursting with life, it finally split. A pale sprout emerged from within, and then, turning and shouldering its way between sand grains, it grew toward the surface. When, straining with turgor, it shoved aside a cinder and reached toward the sky, the sprout was not alone. It stood with millions of others, a hint of green on the sands.

Within three weeks after the fire, luxurious shocks of short grass blades had flushed from the blackened bases. Herds of springbok and gemsbok ambled over the faces of dunes near the riverbed, cropping away at the succulent greenery.

* * *

We stood together for long minutes on the dune, after Bergie's crew had left, and then drove silently back to camp. We had been living on nervous energy, not thinking about the reality of our situation. There were certain facts we had to face: Our money was gone and there was little chance for a job in Maun, because the safari hunting companies hired mostly native laborers, for only two or three dollars a day. Prospects in South Africa weren't much better. It would take us months to make enough to get home, find jobs with higher wages, and earn enough to come back again.

We still had several weeks of food and fuel stocks left in camp, and if we stayed longer in Deception Valley perhaps we could find out if a brown hyena study was really possible. By rationing our food and water, we could keep making research observations until we had just enough of everything to get us back to Maun. Foolish as it may seem, we decided to do it.

The day after the fire had passed our camp, curious about the animals' reaction to the fire, we drove west to the fire line to watch them. Few antelope or birds showed alarm: A small herd of gemsbok galloped onto a stony area, which had only a little grass, and let the flames pass by. A group of springbok milled about, pronking—leaping stiff-legged—high into the air 100 yards from the fire. Plovers and pheasantlike korhaans flushed ahead of the burning grasses, squawking.

Most of the animals stayed surprisingly calm. A family of five bat-eared foxes slept in the grass until the flames were several hundred yards away, and then they roused themselves, evidently not because of the approaching danger but because of the thousands of insects flying and crawling for safety. They stood, yawned, and stretched as usual, then began foraging through the grass, eating one large grass-hopper after another. There were always a few bare spots or areas of short, sparse grass on pans and fossil riverbeds. When the fire burned too close, lions, springbok, gemsbok, and hartebeest strolled through these avenues of escape to places that had already burned. Many creatures, including squirrels, foxes, meerkats, mongooses, snakes, and even leopards, took refuge in underground burrows and waited for the fire to pass. Since the fire line moved so fast, there was little danger of asphyxiation inside the dens. The only victims of the fire were some rodents and insects and a few reptiles.

Captain, the black-backed jackal, wasted no time in taking advantage of the situation. At his usual fast trot, he bounded over the incinerated dunes, crunching up dead grasshoppers, beetles, mice, and snakes. He also fed on unfortunate insects and rodents who had lost their cover and were scurrying about fully exposed on the barren sands.

Our fears that our research would be ruined by the fire proved groundless. In fact, the burn opened interesting new lines of investigation and made it easier to observe and follow animals. We tried to measure how quickly the grasses were growing back and how the diets and movements of the jackals, bat-eared foxes, and antelope were changed. Like the other animals, we tried to take advantage of the situation. There was a great deal to learn about jackals, and a great deal to learn *from* them.

Our decision to stay in Deception until our supplies ran out did not come easily. For many weeks, even before the fire, we had been living mostly on mealie-meal, oatmeal, and pablum mixed with powdered milk. I had lost nearly thirty-five pounds and Delia had lost fifteen. We were persistently weak and lethargic, and I was sure Delia was anemic.

In late July, some days before the fire hit camp, I had been awakened by the sound of the truck's back door opening. I found Delia outside on the ground, doubled over with severe stomach pains. Though this had been happening for several weeks, she had managed to keep it from me. I was sure that her sickness was due not only to our lack of a proper diet but also to the stress of not having the funds either to continue our research or to go home. I lay awake that night trying to think of some way I could get something more substantial for her to eat.

The next night we were following a jackal through the sandveld, when a steenbok suddenly appeared in the spotlight. Quite naturally, with no second thoughts or feelings of guilt, I drew my large hunting knife from its scabbard, slipped off my shoes, and slid quietly from the Land Rover, ignoring Delia's whispered protests. Taking care not to get between the steenbok and the spotlight, I stalked toward the twenty-five-pound antelope. Its large emerald eyes glowed in the light, its nose twitching, trying to take my scent, and the veins stood out on the inside of its big ears, perked to catch the slightest sound. I felt a heightened sense of awareness, the sand cool on my bare feet as I stepped quietly around the grass stubble, my eyes locked on the animal. At the same time, I stood apart from myself, watching with interest this unfamiliar part of my nature that had so long lain dormant.

Finally, trembling and sweating heavily, I crouched no more than five feet from the steenbok, the knife raised in my right hand. Gathering myself, I sprang forward, driving the blade toward a spot just behind its shoulder. But it had sensed me, and at the last second it dodged me and ran away. I sprawled in the sand, pricking my arms, legs, and belly on devil's keys—sharp, three-cornered thorns. Stinging all over and feeling foolish, I walked back to the truck empty-handed.

There were other similar attempts to get protein for us, none of them successful, and we continued to subsist mostly on cereal. Delia's condition did not improve.

* * *

Studying the jackals through July and early August, we realized that we could identify them by the distinctive black tail patch each carried midway down its tail. There was no need to risk injuring or alienating them with the immobilization rifle. Since both they and the brown hyenas scavenged, we guessed that the jackals must compete with browns for carrion. If we followed them often enough, surely they would lead us to the reclusive hyenas.

Each evening all along the valley, the jackals called to one another in a type of reveille before beginning their night's hunt. Unlike black-back jackals in the Serengeti (also called silverbacks there), where mated pairs stay together year round, Kalahari jackals often foraged alone in the dry season. We took compass bearings on their cries to help us locate individuals to follow. Captain's hoarse voice, which sounded as though he had chronic laryngitis, made him easy to identify.

In the three months since we first set up camp in Deception Valley, there had been no rainfall, and there was no drinking water in the entire Central Kalahari. Captain and the other jackals survived on the moisture in the rodents and birds they killed, on maretwa (*Grewia* spp.) berries they picked with their teeth, and on wild melons they found scattered over the duneslopes.

Like an American coyote, Captain was a superior hunter and an opportunist par excellence. In the cool of the evenings, just after sunset, he often trotted along the riverbed below Cheetah Hill, pausing now and then to lap up termites from a column hauling grass stems toward its colony. A dive into a green clump of grass usually paid off with a large, horny grasshopper, a spider, or a beetle, which he quickly chopped up and swallowed. Then he would rush forward with his nose to the ground, curling his lips up tightly, and showing his front teeth as he deftly nipped at a scorpion. Snorting, shaking his head, and grinning widely—lest it sting him on the nose— Captain would toss the arachnid into the air. On the third try he'd bite it in two and swallow the brittle prey. As he trotted along, he often dodged this way and that, hopping off his back feet to snatch large sausage flies and flying termites from the air, crunching them up like appetizers.

By about eight-thirty or nine o'clock, when colder air rolled off the backs of the dunes into the valley, the insects stopped moving. Now Captain turned to the more rewarding task of catching mice. Trotting from one clump to the next, he would wade into the grass, holding his head high and cocking his ears forward. After smelling out a mouse's exact location, he would rear to full height on his hind legs, his paws drawn together against his chest. Then, launching his feet like javelins into the grass, he'd pin the rodent to the ground and seize it from under his paw. If he missed, his prey would sometimes jump straight up, only to be snatched in midair—three or four quick chomps and it was down. He might kill as many as thirty or forty mice in a three-hour hunting period, catching one in every four attempts. Even when his sides were bulging, like garbage bags, he still went on hunting, catching mice and rats in small holes that he dug in the sand with his forepaws and covered over with a quick push of his snout.

Captain was having his usual round of successes one night, trotting across North Pan, killing and burying one rodent after another. Now he was prancing around another patch of belly-high grass, sticking his nose in one side, then the other. He was just about to rear for the strike when he looked back along his line of caches. What he saw made him bristle: Another jackal was trotting from one cache to the next, uncovering and gobbling up his rats. She stood in full view, pilfering the stores of the male who was dominant in this part of the valley.

Captain rushed at the little thief, but she held her ground, her sleek head lifted high, her blonde neck and rufous shoulders standing proud. He was almost upon her when his charge fizzled. Somewhere inside him a circuit had shorted, he was powerless to attack her. It was as though she stood behind an invisible shield.

Instead of routing her as he normally would have, he began trying to impress her: His neck arched, his chest swelled, his ears perked forward, and his nose twitched. He strutted toward the slender female until they stood face to face. Slowly, gently, he touched his muzzle to hers. She stood stiff and tense. Captain's nose roamed from her nose to her cheek, up the side of her face to her ear, along her neck, then light as a whisper over her shoulder. Suddenly he flung his body around and bumped her hindquarters. She stumbled sideways, regained her balance, and froze under his roving nose. Then she abruptly sidestepped him and trotted off, coyly depositing her scent on an herb upwind from him. He smelled her mark for long moments as he watched her disappear into the bush of Cheetah Hill. Then he followed after her.

An indelible impression had been made, and Captain met the female again the next night. After a ritual greeting of nose-sniffing and gentle hip-nudging, the two stood, their necks crossed, sealing the bond. They hunted together, with Mate—as we called her—in the lead. She often paused to mark territory by cocking her leg against a shrub, or to advertise her femininity by squatting to scent-mark. Captain followed in her steps, and, watching her attentively, covered her every mark with his own, telling the other jackals that this female belonged to him.

While the night was still warm, together they plundered the insect population on the flat riverbed. Then, as the evening cooled, they took to mousing at a rodent colony in the sands of Cheetah Hill, poking their noses into one warren after another, snorting and snuffling, trying to find the most likely place to begin excavating. Captain suddenly began to dig excitedly, his front feet churning the sand like a waterwheel, spraying it through his spraddled hind legs, his tail waving like a flag as he tore at the burrow. Mate watched him for a while, and then trotted to another colony nearby, where she began her own dig.

Burrowing frenetically and biting big chunks of sod and sand away, Captain was getting close to his quarry. But the excavation had become too deep for him to keep a proper check on the warren's other exits.

He began digging in short bursts, then quickly backing out of the hole and looking from one escape route to another, intending to catch the rat either in the burrow or when it ran into the grass above ground.

After eons of predation, rats apparently have figured out such strategies, and this one wouldn't leave the warren until the last moment. Now the hole was so deep that precious seconds were lost to Captain each time he backed out to check the other exits. It was at this point that he displayed a touch of jackal genius and showed us a behavior pattern that, to our knowledge, has never been seen before in mammals.

Faced with the problem of digging underground and keeping watch at the same time, Captain stood on his hind legs, his head outside the hole, and began drumming with his forepaws on the ground near the entrance, snatching quick glances at one exit after another. After a short pause, it was back underground for another burst of four or five strokes of actual excavation. Then he stood and began "sham digging" again. From the vibration, the rat must have thought the jackal was getting very close, for it scurried from one of the exits. Captain leapt from the hole, snatched the rodent in his jaws, and ate it, his eyes closed and ears twigging sweetly.

By 10:30 P.M. it had become quite chilly and the rodent population had holed up for the night; bird hunting would be more profitable for the jackal pair. Abruptly they quit hopscotching from one grass clump to another in search of rats and mice. Now they headed back to the riverbed, where they began jogging much faster, circling and zigzagging, their noses glued to scent trails on the ground.

Near Acacia Point, Mate stopped, her front paw held to her chest, her tail rising. She stalked forward, step by step, her nose pointing straight ahead, ears perked, toward a kori bustard not fifteen yards ahead. At twenty-five pounds, with a twelve-foot wingspread, a male kori bustard is one of the heaviest flying birds in the world. In the Kalahari, lone jackals usually didn't tackle such large prey, but with Captain to back her up, the odds were better for Mate.

The big turkeylike bird flared its wings and its neck and tail feathers and made a short, threatening dash toward her. The cock outweighed her by about ten pounds, but without hesitation Mate charged him. The bird bluffed her once and then took to the air, his enormous wings whipping up swirls of dust. He was straining to gain height when Mate leaped more than six feet up and caught him by the thigh. Both were suspended in the air for an instant, the jackal clamped beneath the kori, the broad wings beating against their combined weights. Then, in a tumble of feathers, they crashed to the ground. While Mate fought to hold on, Captain sped in and crushed the kori's head in his jaws.

Both jackals began feeding furiously. Their tails wagged aggressively, and they glared at each other with blazing eyes, their faces smeared with blood and feathers. They had been feeding for two or three minutes when a brown hyena began circling at a distance, obviously wanting to challenge the jackals for their kill, but wary, as always, with us sitting nearby.

The hyena moved closer. We sat perfectly still, not making a sound. We could just make out a blaze of white on her forehead and hear her feet stir the grass. Then she charged. Captain and Mate scattered. The hyena seized the bird, hoisted it off the ground, and began moving quickly into the bush of West Prairie. We tried to follow, but she loped away and disappeared.

Captain and Mate had lost their meal, but they would easily find enough to eat among the insects, mice, birds, and snakes in the grass of the riverbed, and there was always the chance of scavenging something from a larger predator. Unlike the Serengeti jackals, which sometimes kill prey as large as gazelles, Captain and Mate, even hunting as a team, would not be likely to tackle anything larger than the kori they had just killed. When we would next see them hunting together, however, their intended prey would be much more dangerous than the kori had been.

To conserve our dwindling supply of gasoline, we often followed Captain and Mate on foot, just after dawn, before they lay up to rest for the day. One morning early in September, Delia was taking notes while I described the pair's lazy journey back to Cheetah Hill. We were picking our way carefully through thornbushes along the riverbed, when the droning of a small aircraft engine sounded from over the dunes. It was the first plane we had heard since arriving in Deception. The area was so remote that the Botswana Department of Civil Aviation restricted pilots from flying over it. We were sure the aircraft would be coming to our camp because we were the only people for thousands of square miles. Excited by the prospect of seeing other people, we raced onto the riverbed, waving wildly at the little blue-and-white Cessna zooming by just above us. I stripped off my shirt and held it up so the pilot could see the wind direction.

The plane circled, lost height, bounced across the riverbed once, twice, a third time—then took off again. As it flashed by I could see

Norbert Drager, the German vet from Maun, his tense face bent over the control wheel. He circled low over the dunes, made another approach, bounced a few more times, and took to the air again. It was his first cross-country flight as a student pilot, and he was trying to land in a gusty crosswind. As he passed us the third time we could see his wife, Kate, beside him and their daughter, Loni, in the back. On his fourth try he came in much too fast, and the plane hit the ground hard, just missing a fox's den. Then it swerved toward a stand of brush. With its wheels sliding, and leaning down hard on its nose wheel, it managed to stop just a few yards shy of it.

"You get more out of a single landing than any pilot I've ever seen," I kidded him. Norbert was a slight, blond Bavarian with a broad grin who was in Africa with German Technical Aid.

"I've just about had it with flying," he grumbled, switching off the plane's systems. "It's ninety-nine percent boring and one percent sheer terror."

Kate climbed out with a large wicker picnic basket filled with homemade bread, small meat pies, fresh fish, cheese (all the way from Rhodesia), salad, and cake. Red napkins and a checkered tablecloth were neatly folded over the food. We must have looked like two vultures eyeing this feast. We thanked them profusely for this kindness, one of many yet to come from Maun people.

We ate the banquet under the old acacia tree. Delia and I, not having seen anyone since Bergie's crew had come with the news of his death, rattled on like ticker-tape machines about the fire and everything we had learned about jackals. When we had finally run down, Kate asked, "By the way, did you know your country has a new president?"

"No—why, what happened?" I asked.

"Nixon resigned because of Watergate about a month ago, and someone named Ford has taken his place." We had not read a newspaper or listened to a radio for more than six months.

Norbert, who was worried about the trip back to Maun, herded his family to the plane after only an hour. They all waved from the windows as the aircraft roared down the riverbed and lifted into the sky, leaving a tunnel of dust behind. When they were out of sight we walked quietly back to camp. Their brief visit had reminded us of our isolation, and now we felt a loneliness that had not been there before. The bundle of letters they had brought was not likely to offer any consolation. This was it: We had to hear soon about a grant or pack up and go home.

The mail lay on the butt of a fallen tree, daring us to open it. Delia picked it up, removed the string and began shuffling through the letters. "Here's something from National Geographic." Her voice was tense.

"Well, go ahead and open it—we may as well get it over with," I said glumly. We had been disappointed so many times, and this was our last hope. Delia tore off the end of the envelope and took out the letter.

"Mark! It's a grant! They've given us a grant!" She jumped around, waving the papers and cheering. At last someone believed in us—at least in the amount of $3800. We were a bona fide research team.

After a trip to Maun for supplies, we set about our studies with renewed confidence and determination: Sooner or later the brown hyenas would get used to us and, in the meantime, we would follow jackals until they did. Delia's stomach ailment immediately disappeared.

* * *

In September the hot-dry season came to the Kalahari. We were as unprepared for it as we had been for winter in July. Almost overnight, midday temperatures climbed above 110 degrees—then up to 116, in the shade of the fallen tree where we had posted our thermometer. The ground outside of camp was too hot for our thermometer, but it must have been over 140 degrees.

We withered, like the new tender grass, in the strong easterly winds that swept hot and dry across the valley. In the late afternoons, when the winds had quit for the day, our ears rang from the silence. Barren trees, a brown monotony of dead grass and bush, with a bleary simmering sky above: It was a different Kalahari from the one we had known. Moisture escaped our bodies so quickly that our skin stayed dry of perspiration. Our eyes felt scratchy; they seemed to shrink into our skulls, away from the heat.

We rationed ourselves to seven gallons of water each per week, for bathing, cooking, and drinking. The water from the drums tasted like hot metallic tea, and to cool it for drinking, we filled tin dinner plates and set them in the shade of the acacia. But if we didn't watch it closely, the water would quickly evaporate or collect bees, twigs, and soil. After washing the dishes, we took sponge baths in the dishwater, then strained the coffee-colored liquid through a cloth into the truck's radiator. A jerrican stored any excess for later use. We always shared a few cupfuls of fresh water a day with the birds, who flocked to our camp for shade, bread crumbs, and mealie-meal.

Our skin chapped and flaked, our fingers and toes split and bled. Day after day it was the same: the same T-shirt, ragged cut-offs, and holey tennis shoes, the same grey calcareous dust over everything, the same heat that sapped our strength away. We tried to sleep by lying in the back of the Land Rover with wet towels spread over us, but within fifteen minutes we were covered by a carpet of honeybees attracted to the moisture.

The grasses of the riverbed, duneslopes, and savannas had become dry and lifeless; the waterholes were dusty. There was no water left to drink anywhere in the Kalahari. Without sufficient moisture in their forage, the antelope herds had broken into smaller groups of fifteen or so and nearly all had left the fossil river, dispersing over thousands of square miles of range. By spreading out into the sandveld to browse from trees and shrubs, and to dig up fleshy roots and succulent tubers with their hooves, most of the antelope had survived the dry months without drinking. Soon after their departure from the riverbed, the lions, leopards, and other large predators had followed.

By October it had been more than six months since we had seen rain, or even a cloud. Then one afternoon we noticed furry cat paws of vapor tracking into the eastern sky. The hot wind died and a curious quiet fell over the valley. Worn out from heat and night work, we dragged ourselves from camp and stood in the open, watching the billowing vapor above. A lone springbok faced the tentative clouds, his head raised in the sweltering heat waves, as though praying for relief. But the white ghosts vanished before the sun.

Each afternoon they came back, but they only dissolved in the heavy heat that ran like molten glass over the bleary image of the dunes. We were constantly dizzy and could not concentrate enough to read, repair the truck, or perform even the simplest of tasks. We were irritable and our arms and legs seemed too heavy to drag around. At night we followed jackals anyway, always hoping to see a brown hyena. Each hot, sleepless day began at dawn, before the heat, when we worked on our soil sampling, grass transects, and fecal analysis. Three weeks of this were about all we could take, then we would collapse for a cool night of deep sleep.

Most of the jackals along the valley had paired, each pair having established a territory of about one square mile, including a portion of prime riverbed habitat and an adjoining section of duneslope bush savanna. Captain and Mate held Cheetah Hill, Bonnie and Clyde defended an area near "Last Stop," Gimpy and Whinnie roamed east of North Tree, Sundance and Skinny Tail owned North Bay Hill, and so on. They called every sunset and periodically throughout the night, and we could recognize each of the seven pairs by their voices or by their location in relation to camp.

Their thick, black saddles of long hair insulated them against the sun to a great degree. All Captain and Mate seemed to need for shelter was the patchy shade of a small leafless bush on Cheetah Hill, where they slept through the scorching days. Even the early mornings and late afternoons were hot now, and they hunted only at night, when it was cool. They had had no water to drink for months, and we often saw them fighting other pairs of jackals for the moisture in a single wild melon.

November clouds tantalized us: Their filmy curtains of rain smelled incredibly sweet and fresh—but always fell somewhere far away in the desert. None of the clouds was dark and heavy enough to challenge the great convective barrier of heat that rose from the baking riverbed.

One day the morning winds did not come; the air was utterly still, as though waiting. At midmorning clouds began building beyond West Prairie. Hour by hour they grew until they stood shoulder to shoulder, towering columns of water vapor too big for the sky. By midafternoon the purple-black sky was boiling with vapor. Daggers of lightning slashed across the clouds and thunder boomed through the valley.

After weeks of disappointment we were sure the storm would probably pass us by. But then an avalanche of black cloud tumbled over the shoulder of West Dune, sucking up a yellow blizzard of sand as it rolled toward camp. The stagnant air began stirring around us. We ran to the truck and backed it out from under the trees.

Thirty yards from camp I turned the rear of the truck to the coming storm. Seconds later, the sand and wind slammed into us. We pressed our shirts to our faces, trying to breathe in the grey air while the Land Rover rocked and creaked, the keys jingling in the ignition. Hail drummed on the truck's metal roof, and through the windshield we could see boxes, sacks, pots, pans, and other bits of camp rising into the air. The acacia tree was reeling like a crazed animal clawing at itself.

At last it was raining. Water was streaming through gaps in the window frames and trickling into our laps. "Smell it! Smell it! God, how wonderful! How beautiful!" we shouted over and over.

The storm came in surges, craggy fingers of lightning skittering over the low, black clouds, causing a ghostly blue glow to reflect off the rain and sand in the air. Much later that night, we finally drifted off to sleep between gusts of wind that shook the truck.

The valley was bright with sunshine when we opened our eyes the next morning. But it was not the same malevolent sun that had scorched the Kalahari for months. Soft, mellow rays caressed the backs of several hundred springbok, nibbling grass bases succulent with glittering droplets. The storm was only a smudge on the distant horizon. From camp we could see Captain and Mate and a pair of bat-eared foxes drinking from puddles on the spongy desert floor.

Our clothes, pans, papers, and other belongings were scattered over the riverbed. Delia recovered a pot fifty yards from camp and cooked a porridge of oats mixed with sorghum and *samp*—cracked corn. After breakfast we began picking up the pieces. The gasoline drum had been rolled halfway across the valley.

The storm painted the desert green again, and within a week the valley was full of antelope herds dropping their scrawny, floppy-eared fawns into the new sprouts of velvet. Flying termites swarmed after their queens. Bat-eared foxes scurried here and there with their fluffy kits, fattening themselves on the hordes of insects hopping, flying, and crawling everywhere. Everyone was gearing up to have their young and get them reared during this short and fickle period of abundance. Everywhere there was a sense of life renewed, of rebirth after long trials of heat and fire. Other storms soon followed, and with the beginning of the rainy season, daytime temperatures dipped to the mid-seventies and eighties and blue skies were filled with balmy breezes and brilliant white clouds.

Perhaps best of all, the same pride of lions that had trapped Delia in the mess tent months before returned to the valley. Their roars in the night and early morning, together with the calls of the jackals, brought the fossil riverbed to life again. We talked of coming back to the Kalahari one day to make a complete study of lions. But first there were the jackals and brown hyenas to reckon with.

At sunset several days after that first storm of the rainy season, we were having a hurried meal at the campfire before going out to find Captain and Mate. One of the other jackal pairs, Gimpy and Whinnie, began calling east of the Twin Acacias, their strident, quavering, and strangely melodic cries ringing through the valley. We fell silent, moved, as always, by the mournful sound. It seemed to come from the very heart of the desert—the cry of the Kalahari. The others began to join the chorus: Bonnie and Clyde, Sundance and Skinny Tail, and finally, Captain's deep, hoarse voice, together with Mate's clear song, from Cheetah Hill.

"Wait a minute...what's that?" Delia asked. High-pitched breathless squeals tried earnestly to mimic Captain and Mate.

"Pups!" We jumped into the Land Rover and drove toward the calls. After parking some distance away from the jackals, we peered this way and that, trying to see through the cover of bush. Then Mate appeared at the den opening and lowered her head. When she stepped aside, two powder puffs of fur with wriggling tails, short, fuzzy faces, and stubby black noses waddled into view.

Mate began licking the faces, backs, and bellies of "Hansel" and "Gretel," rolling each one over and over in the sand while the other stumbled beneath her on stumpy, uncertain legs. Captain lay nearby, his head on his paws. Then Bonnie and Clyde began to call from the north again. Before the calls had died away, they began to answer, Hansel and Gretel standing beside Captain and Mate, their tiny muzzles straining toward the sky.

Both parents participated in raising their pups, but they had no "helpers," as have black-backed jackals in other areas of Africa. Dr. Patricia Moehlman[1] found that some jackal subadults on the Serengeti Plains remain with their parents to assist in providing for the next litter. They help by regurgitating food to their mother and to their younger brothers and sisters, and by guarding the den. Although we did not see this among the jackals we knew, some other Kalahari pairs may have had helpers. This behavior is often difficult to observe and may not be detected until a species has been studied for perhaps several years.

In the early weeks, either Captain or Mate was always at the den to guard the pups from predators. After sunset each evening, Captain would walk over to Mate, Hansel and Gretel romping around his feet, biting his ears, legs, and the tip of his tail, and he would touch her nose with his. Then, lifting his feet high and stepping over the tumbling pups, he would trot away to hunt, leaving Mate to tend the litter. Once their father was out of the way, Hansel and Gretel would immediately begin pestering their mother, chewing her ears, rolling over her face, tumbling across her back, and pouncing on her tail. Mate was tolerant, but seldom took an active part in the play.

From the very beginning, adult behavior patterns were apparent in the pups' activities. They repeatedly practiced the holds, stalks, pounces, and killing bites that would make them successful hunters as adults. If their mother wouldn't cooperate, they attacked each other, or else the grass clumps and sticks within a few feet of the den.

The pups were about three weeks old when Captain began bringing them raw meat to eat. Wagging their tails, the youngsters would burst from the entrance and rush to their father, licking his lips hungrily, begging for food. He would open his jaws wide and regurgitate a slimy mass of partially digested mice and birds on the ground before them. As Hansel and Gretel gobbled up the steaming hash, Captain would settle under a nearby bush to rest and babysit while Mate trotted off to hunt.

As soon as the pups had been partially weaned on fresh meat, their parents began taking them for short excursions away from the den. The adults strolled along while the youngsters romped and played, smelling bushes, grasses, antelope droppings—everything they could get their noses on. They were learning more and more about their fossil river environment. One of the most valuable lessons for the pups on these early morning forays was how to kill and eat insects. This was an important predatory skill, for it allowed them to supplement their diet of milk and regurgitated meat while Mate was weaning them.

Now that Hansel and Gretel were better able to look after themselves, Captain and Mate began to hunt together again, leaving the pups to forage for insects near the den. One night the parents hunted an area that included the riverbed east of the Cheetah Hill sand tongue, the hill itself, and a slip of bush and woodland behind it. They moved along, each pausing frequently to cock a hind leg and scent-mark a low shrub or woody herb along the boundary of their territory.

They had just entered the duneslope woodlands when Mate began dancing around something on the ground ahead, her tail waving in the air. Captain rushed to her and found a nine-foot black mamba, one of Africa's most poisonous snakes, its body reared three feet off the ground and ready to strike. The mamba's tongue flicked in and out, its sinister coffin-shaped head drawn back like a crossbow ready to fire.

Captain feinted this way and that, trying to get past the snake's defenses. But the beady eyes tracked him like a missile. Wherever he moved, the mamba adjusted itself, waiting.

Mate moved around until she was opposite Captain, with the snake between them. She darted toward it, and for an instant it was distracted. With a motion too quick to follow, Captain lunged for the mamba, but it had recovered its attention and it struck. Several feet of its long ropy body sprang off the ground as Captain dodged away in a shower of sand, the lethal head barely missing his shoulder.

Instantly he was back on the attack, pouncing again and again, and each time the mamba struck, he jumped away. He would not let up; after each strike, the snake was taking a little longer to rear and prepare for another attack.

It was when the mamba was trying to recover after a miss that Captain managed to nip it hard on the back. Tired and injured, it tried to crawl away. But Mate blocked the retreat, and it raised itself and made another thrust, just missing Captain when he charged forward. Before it could escape, he bit it hard about three feet back from the head, and then again. It was writhing now. Finally, after several more bites, he held it for a split second and shook it violently, its coils squirming about his legs. Then he dropped it, and grabbing the dangerous head, crushed it in his jaws.

At that point the perilous hunt became a comedy. As soon as Captain seized the mamba's head, Mate grabbed its tail. In contrast to their supreme cooperation of just seconds before, they now began yanking at either end of the snake in a tug of war, each trying to run off with the prize. They glared, eyes blazing and ears laid back, along several feet of reptile. Their hackles bristled and their tails slashed as they seesawed back and forth, until finally the snake was yanked into two equal lengths of stringy white meat. Each began feeding feverishly; it took them nearly ten minutes to finish. Then they rolled in the grass, sniffed noses, rubbed faces, and trotted off together on a border patrol of their territory, their bellies round and bouncing.

*   *   *

Before we thought to tape a tiny calendar inside the cover of one of our field journals, we had lost track of the date. Judging from the time of our last supply trip to Maun, we guessed that Christmas 1974 must be near. Without the money or time to go to the village for the holiday, we picked a day and began preparations for celebrating it at camp.

One morning, after great deliberation, we selected and cut a half-dead broad-leafed *Lonchocarpus nelsii* tree from the dune woodlands and hauled it back to camp on top of the Land Rover. We decorated the tree with the thermometer, some red collaring material, a few syringes, and the dissection scalpels, scissors, and forceps, adding the hand scales, a lantern, a springbok jawbone, the defective fire extinguisher, and various paraphernalia from around camp. Once we had tied these onto the branches of the tree, we began to plan our Christmas dinner.

A flock of thirteen guinea fowl had found camp at the beginning of the rainy season. At least once, often twice, a day they took a stroll along our kitchen counter, which consisted of boards laid across the tops of water drums. They raked their horny feet through our tin dinner plates, scattered knives, forks, and spoons onto the ground, flipped the lids off the cooking pots, and devoured any leftovers lying about. And when they found a loaf of freshly baked bread, pieces flew off it as if it were being shot to bits by a Gatling gun. At first we thought they were cute, but the noise they made early in the morning, after a long night of following jackals, was difficult to tolerate, and they hogged all the mealie-meal we scattered on the ground for the other birds.

I finally decided to discourage the guinea fowl flock from coming into camp. That I reached this decision near Christmas, when we had gone for nearly four months without fresh meat, was perhaps no coincidence.

Early one morning, I propped a box up with a stick, weighted it with a stone, and sprinkled some mealie-meal under it. I tied a nylon fishing line to the prop and strung it along the ground to the opposite side of the Land Rover, where I hid behind the wheel. Not long after sunrise the guinea flock arrived with its piping gabble, raising the dust as it scratched and pecked its way into camp. Almost immediately one of the cocks spied the trail of mealie-meal leading beneath the box and, without hesitation, began his rapid-fire pecking along it, leading his entire flock toward the trap. I could already taste freshly roasted guinea.

Four plump hens and the cock crowded beneath the box, gobbling up mealie-meal as fast as they could. I flicked the fishing line. The box plopped to the ground in a cloud of dust and flapping wings. The guineas railed loudly. I jumped from behind the truck and hurried forward while the birds eyed me suspiciously.

The trap lay perfectly still, not a peep from inside. I looked around: *Thirteen* pairs of guinea eyes glared at me. I was dumbfounded. What had gone wrong? They couldn't be all that quick and crafty; after all, they were little more than a bunch of barnyard chickens. I'd get one the next time. I reset the boxtrap and ambled discreetly toward the Land Rover. By this time Delia was sitting up in bed grinning.

The guineas pecked their way back to the trap. This time only two ventured beneath it. I yanked the line and the box hit the ground. I quickly counted heads again. "One, two, three, four, five—Damn!" Thirteen squawking guineas and one snickering wife. On the third try the birds only pecked to the edges of the box—not one would go under it.

On our self-appointed Christmas morning, the guineas were back as usual, scattering our pots, pans, and dishes about with a loud clatter. We ignored each other as I sifted the weevils from some flour and baked a loaf of caraway bread in the bucket oven. Delia made a meat pie with the last of the rock-hard biltong Bergie had left us. Our Christmas dessert was another pie made with maretwa berries we had picked from the bushes of West Prairie.

Christmas was a hot day and, despite efforts to cheer ourselves, without family or gifts we were short on holiday spirit. We sang a few carols, and then, feeling somewhat lonely and let-down, we drove to the den and spent the afternoon with the jackals.

Hansel and Gretel, now about seven weeks old and three-quarters as tall as their parents, scampered out to us as we were parking the truck. Their saddles were beginning to gain definition, changing from a soft grey to a bold black. They had gained much skill at catching insects and even took a mouse occasionally, displaying fairly sophisticated hunting behavior. Captain and Mate roamed farther from the den now and brought back much less food.

That Christmas night, before Captain and Mate left to hunt, a strange type of jackal call rang out from the North Tree area. The jackals immediately jumped to their feet. With Hansel and Gretel trailing their parents at a distance, the whole family hurried toward the unusual nasal *weeuugh!...weeuugh!...weeuugh!* that echoed urgently over and over again.

By the time Captain and Mate arrived at the scene, six jackals had already surrounded a patch of tall grass, all of them voicing that strange call as they sprang up and down on their stiffened legs. Again and again they bounced into the grass and back out again, a split second later. Captain and Mate joined the ritual while Hansel and Gretel sat on their haunches, watching.

After about fifteen minutes of this, a leopard came slinking out of the thicket. His face and chest covered in blood, and still surrounded by springing jackals, he laid back his ears and walked away. The jackals followed for forty yards, calling, darting, and bouncing all around him. Then they trotted back to the grass patch to scavenge the springbok remains he had left behind.

Jackals are a favorite prey of leopards. The strange call and the associated jumping display probably allow them to keep an eye on the predator in tall, thick cover and to communicate the danger to other jackals. It serves the same purpose as the mobbing of a predatory snake by birds.

After a time, Hansel and Gretel entered the long grass area where their parents and other jackals were squabbling over the springbok carcass. Captain and Mate had been unable to hold off the invasion of Bonnie and Clyde, Gimpy and Whinnie, and two other pairs from outlying territories. Intense bickering around this large food resource enforced the strict social hierarchy that existed among members of the population throughout the year.

When Hansel and Gretel tried to join in the feeding, both of their parents turned on them, snarling aggressively, their mouths pursed in threat and their tails lashing. Apparently surprised and intimidated, the youngsters retreated a short distance. These were not the tolerant parents they had known. Both Captain and Mate had been testy of late and had rebuffed their attempts at play, but these serious threats were something new. They were being treated as competitors, and Captain directed most of his aggression toward Hansel. Gretel tucked her tail and sat on it, opened her mouth wide, and raised her front paw in submission. She would have to wait her turn to feed.

Hansel's saddle was black and becoming more and more distinct, and silver hairs were beginning to show through. He was close to adulthood in size and markings. He persistently moved in on the carcass, only to be rebuffed by Captain. Finally he had had enough. The two males faced each other snarling, hackles stiff as wire. Captain charged and hit Hansel with his shoulder. The youngster took the blow and delivered a hip slam in return. For a second they were a mass of boiling fur. When it was over, Hansel advanced boldly and fed next to his father. He had developed the competitive spirit essential for winning a place in the adult social hierarchy, where a jackal with higher status gains longer feeding time at carcasses, as well as superior mates and better breeding territories.

The encounter between Captain, Mate, Hansel, and Gretel was typical of the "parent-offspring conflict"[2] that occurs in many animal species, including humans, and is perhaps most visible at weaning. Anyone who has ever heard the screams of a young baboon when first turned out of its mother's nest, or seen the facial expression of a kitten spat at by its parent after weeks of being cuddled, fed, and groomed knows that such conflicts can be severe. The classical explanation for this behavior is that the parent is still caring for the juvenile by forcing it to become independent, which is necessary for its survival. A more recent theory argues that there is a time, after the young are weaned and have become larger and more demanding, when it becomes too costly for the mother to provide food, energy for defense, and other resources for her subadult young. Instinct advises her to devote the same efforts toward the production of new offspring. She is also encouraging her young to begin breeding themselves. This is a genetic benefit to her because her own genes will be passed on by the young she has just alienated, as well as by herself.

In each Kalahari dry season, probably because of low prey densities, jackal pairs break up and their breeding territories completely disintegrate. In the next mating season, near the beginning of the rains, different pairs establish new territories along the riverbed. As stated earlier, we saw no evidence that families stay together, with adolescents helping to rear the new litter, though this may occur when several good consecutive rainy seasons occur. But paired or not, the jackals in Deception Valley maintained a strict social hierarchy from year to year.

On this Christmas evening, all of the jackals eating at the springbok carcass suddenly paused, looking into the darkness to the east. Then they began feeding faster, almost frantically, seizing meat on the neck and along the spine and lunging backward to tear it free. I lifted the spotlight and swung it eastward. The large, wide-set emerald eyes of a brown hyena were watching from 125 yards away. Apparently she had heard the mobbing call of the jackals and knew a leopard was in the area and that there was a possibility of a kill. We sat unmoving, hoping, as we had so often before, that she would come to feed despite our presence.

The hyena circled the truck several times and stood watching for a long while. Finally, with her hackles standing up along her shoulders and back, she walked toward the carcass. From her udders we could see that it was a female. The jackals gobbled the meat even faster until, at the last moment, they leaped over the dead antelope and dashed out of the hyena's reach. The brown turned and looked at the Land Rover for several seconds before beginning to feed. Then, puffing and straining, she started shattering bones and ripping the flesh from the skeleton. The circle of displaced jackals pulled in around her, but whenever they tried to snatch a bite, she went thundering after them with gaping jaws.

After a time, most of the jackals moved off a few yards and lay down, except for Captain, who circled around and came slowly, almost nonchalantly in behind the feeding brown. The hyena freed a length of springbok leg, laid it at her feet, and continued to feed on the softer parts near the ribs. Lowering himself on wiry legs, Captain crept closer and closer to the unsuspecting brown hyena, until he was crouched with his nose to her rear. Still she continued to feed, unaware. Slowly he raised his muzzle to the base of the brown's flicking tail; he held it there for several seconds. Then, as the tail moved aside, he bit the hyena on the backside. She whirled to her left and Captain dashed to the right, seizing the springbok leg and a large swatch of dangling skin. It was almost more than he could carry, but by holding his nose high in the air he could run—and run he did.

Hair streaming, her jaws open wide near the tip of Captain's tail, the hyena chased him in great circles across the riverbed. Whenever it seemed he was about to be swallowed up, Captain would make a turn too sudden for the lumbering hyena to follow. On he ran, his muzzle sagging lower and lower with his heavy loot, until finally he dropped it. Panting heavily, he watched the hyena carry it back to the carcass. Once again the brown laid the leg at her feet and began to feed.

Little more than two minutes later, Captain was back, sneaking up on the hyena again. It looked like an instant replay: Captain chomped the brown in the rear, stole the springbok leg and fled, his tail flying, with the hyena in hot pursuit. But this time he escaped into the bush at the edge of the riverbed. The hyena sloped back to the carcass, licking her chops, ears laid back in evident disgust. Eventually she lifted all that remained of the carcass and walked with it into the thick bush of North Dune.

It was after midnight when we turned into camp. The headlights swung among the trees and fell on another brown hyena—standing near the water drums, not fifteen yards away! Unconcerned with us, she continued smelling her way through camp. Eyeing our bag of onions hanging from a tree, she stood on her hind legs, grabbed the net sack by the corner, and pulled. When a cascade of bulbs thumped over her nose and onto the ground in a shower of papery skins, she jumped back. After smelling one carefully and biting into it, she shook her head and sneezed. At the fire grate (the coals had died hours before) she took the water kettle by its handle and strutted from camp. Putting it down a few yards away, she jarred the lid off with her nose and lapped up the water from inside. Then she raised her tail and began to walk away, but before she disappeared she stopped and looked directly at us for several seconds. There was a small white star on her forehead.

5

Star

Delia

How I wonder what you are...
—*Ann Taylor*

WE HAD had a Kalahari Christmas after all: The brown hyenas had accepted us at last. We woke up early the next morning full of kick and ready to go, in spite of the late night's work. Sipping from enamel cups of steaming tea and talking over the experiences of the previous night, we strolled from camp north toward Acacia Point, as we often did on cool mornings.

"I don't believe it, look over there." Mark pointed to the edge of the heavy bushes on North Bay Hill about 300 yards away. A brown hyena was walking directly toward our truck spoor on a course that would intercept ours. She had apparently not seen us and was moving quite fast through the belly-high grass, apparently in a hurry to reach her bed before the sun rose any higher.

We stood perfectly still, not knowing quite what to do. If we started back to camp, our movements might frighten her. It was one thing for the hyenas—or any of the other animals—to accept us in the truck, but they were generally much more intimidated by the sight of us on foot. Very slowly we each sat down in a tire rut, expecting her to run away at any moment. When she reached the track, fifty yards away, she turned south and came directly toward us, the small white blaze on her broad forehead bobbing up and down. It was the same hyena who had taken the water kettle.

Without hesitating once, she steadily closed the distance between us, finally stopping just five yards away. We were exactly at her eye level. Her dark eyes were moist, perhaps because of the unfriendly sun. The sides of her face were battle-scarred, and a cape of fine blonde hair lay over her shoulders. Her long slender forelegs were boldly striped black and grey, ending with large, round feet. Her square jaws, capable of crushing or carrying away a fifty-pound gemsbok leg, were slightly parted.

Slowly putting one padded foot before the other, she reached her nose toward me, taking gentle wisps of my scent, her long whiskers twitching. Finally her face was no more than eighteen inches from mine. We stared into each other's eyes.

Courses in animal behavior teach that carnivores communicate fear and aggression through the postures of their ears, eyes, and mouth. Star had no expression on her face, and that in itself conveyed the strongest message of all. We had seen peaceful interactions between different species in the desert on many occasions: a ground squirrel smelling the nose of a mongoose; cape foxes actually denning in the same complex of burrows with a colony of meerkats, four tiny bat-eared foxes playfully chasing a small herd of hartebeest. And now, Star was communicating, through her curiosity and lack of fear, an acceptance of us into her natural world.

She stepped still closer, lifting her nose slightly and sniffing the edge of my hair. Then she sidestepped rather clumsily over her front feet and smelled Mark's beard. After that, she turned and walked on at the same even pace toward West Dune.

Star was enterprising and spunky, always ready to rally. Now and then, padding along the riverbed, she would dance a curious jig, leaping off her hind legs, tossing her head, and turning a half circle in the air. It was Star, mostly, who taught us the secrets of brown hyena society, and eventually some secrets about ourselves.

She and several of the other hyenas allowed us to follow them in the truck, as we had done with the jackals. But four or five hours, at the most, was as long as we could keep her in sight. As soon as she left the riverbed, tall grass and thick bush would close over her. Since we could never keep up with the hyenas all night, we still had no idea where they slept in the daytime. In the evenings we often searched the dark riverbed for hours before finding one to follow. Our entire study of brown hyenas was restricted to chance meetings with them on a narrow ribbon of riverbed grassland no more than 1000 yards across.

One night in January, a large pair of eyes was reflected in the spotlight, and trailing behind them was a long line of smaller eyes, all bouncing up and down. At first glance it looked like a female carnivore leading her cubs through the grass. But it was Star and, trotting single file behind her, five young jackals, including Hansel and Gretel, all apparently playing follow-the-leader. When Star stopped, they stopped; when she zigzagged, they zigzagged. From time to time she would wheel around, as if annoyed by her shadows and hoping to shake them off. When she reached Eagle Island, on the edge of the open riverbed, and lay down for a rest in the grass, the young jackals arranged themselves in a circle around her. Because brown hyenas and jackals are keen competitors, they always check on each other when meeting and often discover that the other has food. These inexperienced youngsters were apparently hoping Star would lead them to an easy meal.

A few minutes later, Hansel walked up to Star and put his small black nose up to her large muzzle, in what looked like a warm greeting between friends. In fact, he may have been assessing whether or not she had recently fed. Apparently Hansel found nothing interesting on Star's chops because he trotted away, as did the other jackals, each in a separate direction.

After resting for twenty minutes at Eagle Island, Star began to follow a zag-stitch course along the moonlit riverbed, walking at about three miles per hour. Now and then she would stop to lap up a few termites or leap into the air for a flying grasshopper. Suddenly she pivoted to the west and raised her nose high, analyzing the odors on the night air. Then she sprang forward and loped through the taller grasses of West Prairie, dodging around bushes and termite mounds and pausing only to retest the breeze. The scent led her for over two miles to the edge of the dune woodlands, where she stopped abruptly, peering into a dense thicket.

Two feeding lionesses and their cubs—low, dark forms in the grass—lay around the remains of a gemsbok, its belly torn open and flanks smeared red. The night air was heavy with the pungent odor of the gemsbok's rumen, which was probably how Star had smelled it two miles away. She circled widely around the area, then stood downwind. The kill was very fresh, not more than half an hour old, and the lions would not leave it tonight. She walked away north into the trees. For a scavenger, patience is the key to the pantry.

Early the next evening, we found Star headed straight for the lions' gemsbok kill, which was by now reduced to a rubble of white bones, strips of tattered red meat, and folds of skin. The lions were still there, sleeping on their backs, their legs sticking up from bloated bellies. Star flopped down beneath a bush to sleep, and to wait.

These were the same lions Mark had pushed away from the drugged jackal with the Land Rover; we had seen them quite often and believed they were permanent residents in our area. Tonight it was around eleven when they finally roused themselves and walked single file through the woodland of West Dune.

Star must have heard them leave. She stood up and circled the kill site three times, smelling and staring from different positions. This is one of the most dangerous situations brown hyenas ever face. They depend heavily on the leftovers from lions, and on finding such carrion before it is gobbled up by jackals or other brown hyenas, or by the vultures that arrive at dawn. But without being able to actually see the gemsbok carcass in the tall grass, Star had to rely primarily on her sense of smell to tell her whether all the lions had gone. It must have been difficult, in the confusion of odors from gemsbok remains and those of lion dung and urine. She took several tentative steps forward, then stood still, her nose raised, her ears perked to pick up any clue that might help her avoid walking into a lion. Fifteen minutes later, she had worked her way to within twenty-five yards of the gemsbok. After another long wait, she finally went to the carcass and began to feed.

After nibbling at morsels of stringy meat, tendons, and sinew, she opened her jaws wide and began to crush leg bones as thick as baseball bats and to swallow splinters at least three inches long. (We measured these later, by fecal analysis.) A brown hyena's teeth are veritable hammers specialized for processing bone: The premolars are flattened and enlarged, unlike the sharp, scissorlike cutting blades of other predators. Tilting her head to one side, Star wedged her teeth between the ball and socket of a hind leg until it tore free. Carrying it by the knee, she walked into the thick bush of the duneslope, where she tucked the leg under an acacia bush about 100 yards from the riverbed.

Star was uncanny in her ability to locate lion kills and to know when the remains—the most important part of a brown hyena's wet-season diet—would be abandoned. However, she also spent many long, lonely nights walking for miles and finding nothing to eat but a mouse or an old bone.

*   *   *

The few scanty reports on brown hyenas described them as solitary scavengers, real loners that ate only carrion or occasionally hunted small mammals. At first we thought this description was probably accurate: Star followed that feeding pattern and was always alone. But soon we began seeing some extraordinary behavior that made us question whether browns were indeed solitary creatures.

Any information about how many of them live in a group, whether or not they defend a communal territory, and why they associate together is important for the conservation of hyenas. But there is another reason to investigate their social life: Man is also a social carnivore, and by understanding the evolution and nature of societies of other predators, we can better understand our own sense of territoriality, our need for identity as part of a group, and our aggressive tendencies as competitors.

Later that night, following Star when she left the carcass, we noticed that she did not wander aimlessly over the range, but traveled on the distinct pathways she had used on previous nights. Some of these joined or crossed well-used game trails, such as Leopard Trail, a major route for gemsbok, kudu, giraffe, jackals, and leopards moving north-south along a string of temporary water holes at the foot of West Dune. Usually, however, the hyenas' paths were visible only as faintly divided grass or lightly compacted sand.

Star paused at a grass clump, smelling a small, dark blob at nose level on one of the stems. Then, in a most bizarre display, she stepped over the grass, raised her tail, and everted a special rectal pouch. By swiveling her hindquarters to feel for the stalk, she directed the two-lobed pouch to the stem and "pasted" a drop of white substance that looked remarkably like Elmer's Glue. After she had lowered her tail, she retracted the pouch and walked on. We took a sniff of the paste; it had a pungent, musty odor. Just above the white drop, a smaller rust-colored secretion was also smeared on the grass.

During the following weeks we saw other hyenas traveling the same trails that Star used, always alone, and often stopping to smell the paste that Star and others had left on the grass-stalks. Before moving on, they would add their own chemical signature to the stem, so that, in spots where paths crossed, a grass clump could have as many as thirteen scent marks, very much like a sign post at a highway intersection.

Late one night we were following a very timid female, about Star's size, that we had named Shadow. She was walking south along the riverbed on one of the hyena paths and pausing every hundred yards or so to smell a scent mark and then paste over it. She crossed South Pan through Tree Island and entered the thick bush, where we lost her. It was one o'clock in the morning, so we stopped for coffee on the edge of the riverbed before looking for another hyena. We were sitting on top of the Land Rover in the moonlight sipping from our Thermos cups, when Star came along. She crossed the first hyena's path and stood smelling Shadow's fresh paste mark for nearly a minute, her long hairs bristling. Then she changed course and followed quickly after her.

We managed to keep Star in sight until we could see Shadow walking back toward Star in the moonlight, the two dark forms moving silently through the tall, silvery grass. We stopped, Mark flicked on the spotlight, and the most unusual behavior we had ever seen between two animals began to unfold.

Star approached and Shadow crouched down until her belly was flat to the ground. She drew her lips up tightly and opened her mouth wide, showing her teeth in an exaggerated grin. Her long ears stuck out from her head like a floppy hat, and her tail curled tightly over her back. Squealing like a rusty gate hinge, she began crawling around Star, who also turned, but in the opposite direction. Each time Shadow passed beneath Star's nose, she paused to let her smell the scent glands beneath her tail. The hyenas pirouetted around and around, like ballerinas on a dimly lit stage.

The strange greeting continued for several minutes, even after Star tried to walk on down the trail. Each time she began to move away, Shadow hurried to lie in front of her, inviting Star to take another sniff under her tail. Like an aristocratic lady dismissing her attendant, Star finally stood with her nose held high, refusing to further indulge Shadow. Eventually she walked away, and Shadow departed in a different direction.

Several nights later we found Star again, but she was not alone. Tagging along behind her were two smaller hyenas who were only three-quarters her size and had finer, darker hair. We named them Pogo and Hawkins. They were romping behind Star along the riverbed near Cheetah Hill, playfully biting each other's ears, face, and neck. Whenever Star found a bit of carrion, they rushed to her, "grinning," squealing, and crawling back and forth under her nose, begging for food. In response to this performance, Star shared her find with them, and we naturally assumed that they were her cubs. But the next night we found Pogo and Hawkins with Patches, an adult female with tattered ears. If they were Star's cubs, why would they follow Patches?

By April we could recognize seven different brown hyenas in the immediate area. A large male we called Ivey had immigrated into the area quite a few months before. There were four adult females— Patches, Lucky, Star, and Shadow—and the youngsters, Pogo and Hawkins. But it was often difficult to identify the dark, shaggy creatures at night, and since brown hyenas are notoriously difficult to sex under the best of conditions, there were many times when we were unsure of both the sex and the identity of the hyena we were following.

Reluctantly, we came to the conclusion that we would have to immobilize and eartag as many of the browns as possible. This was a dismaying prospect, for it had taken many months to habituate the seven hyenas to our presence. If the darting alienated any of them, it could jeopardize our entire research program. Yet, unless we marked them, we could make mistakes in our observations of their social behavior.

Mark used a silencer for the darting rifle made by modifying a Volkswagen muffler, and we waited for an opportunity to dart Star, Pogo, and Hawkins. It came one night when the youngsters were following the older female to the remains of a gemsbok carcass. The cubs soon lost interest in the wrinkled skin, and in the bones that were probably too large for them to break, and wandered off, leaving Star to feed alone. By inching the truck slowly forward and stopping each time Star looked up, we moved to within twenty yards of her. Working slowly, with quiet movements, Mark estimated the dosage, prepared the dart, and slid it into the rifle. Star was very nervous, perhaps because of the odor of lions in the area. At the click of the rifle bolt she looked up into the spotlight and then dashed off a few yards. But after staring at the truck for a minute, she licked her chops and returned slowly to feed, her tail flicking—a sign that she had relaxed again.

Mark put his cheek to the rifle stock and took aim at her dark form. I gripped my notebook hard and looked away, afraid of what was about to happen and certain that she would dash off and we would never see her again. Months of hard work—everything, it seemed—were riding on this one shot. Mark brought his arm slowly up under the gun, but at the mere rustling of his nylon jacket, Star ran off again. This time she stared at the Land Rover for several minutes, and then began walking away.

We didn't move—not a muscle—for the next hour. Star was still in sight, lying down with her head on her paws, watching. After a while my back began to ache and my hips and legs to numb. I couldn't imagine how Mark must be feeling, one elbow on the steering wheel, the other on the door frame, and his cheek on the stock as he hunched forward over the gunsights.

Star knew something wasn't right with that carcass. When she stood again we could almost see her trying to decide whether to leave or come back to feed. Finally she lowered her head and plodded slowly toward the carcass.

Mark gently squeezed the trigger. There was a muffled pop, and we could actually see the dart fly from the barrel of the rifle. It slapped Star in the shoulder. She jumped back, whirling, twisting, and biting at the missile. Then she ran. Swearing under his breath, Mark quietly swung the spotlight after her. Otherwise, we did not move.

Star loped to the very limit of the light, where we could barely see her. There she stopped and looked around in the darkness, staring and listening, as if trying to figure out what had stung her in the shoulder. I was sure we had blown it and that she would never trust us again. But then her tail began flicking, and, unbelievably, she walked directly back to the carcass and started feeding again, without even giving us a glance.

Minutes later she slumped to the ground, and we let out deep sighs of relief. Pogo and Hawkins, who were much less wary than adult browns, came back to the carcass, and having briefly smelled Star, began to gnaw on the bones. Mark darted them, and within fifteen minutes the three hyenas were lying peacefully in the grass. Carrying our equipment box, we eased quietly from the truck, stretched our numb limbs, and began ear-tagging and measuring them.

"This *is* a female, isn't it?" Mark whispered, kneeling beside Star.

"I'm not sure. Are those things real?" I poked at two fleshy lobes that looked like testicles.

There we were—two students of zoology, with thirteen years of university between us—poking and prodding the confusing array of sexual and pseudosexual organs carried by this odd beast. Although female brown hyenas lack the enlarged pseudopenis (actually the clitoris) of spotted hyena females, they are equipped with fatty lobes or nodules located where the testicles would be if they were males. After a considerable period of investigation and consultation, we were still not sure if Star was male or female. Fortunately, Hawkins was well equipped with genuine testicles, which resolved our confusion: Star and Pogo were definitely female.

Star wasn't the least bit wary of us, several nights later, when she passed within fifteen yards of the truck, wearing her blue plastic ear tag. We followed her to a hartebeest carcass left by lions near the base of Acacia Point, where she began to feed. Fifteen minutes later Pogo and Hawkins joined her.

They had fed only briefly when all three raised their heads and stared into the darkness. Patches, holding her head high, was walking directly toward them in the spotlight. Pogo and Hawkins fed, but the two adult females glared at each other. Star lowered her head and ears, and every hair on her body stood straight out. Suddenly Patches rushed in and seized her by the neck, biting and shaking her violently. Star shrieked when Patches' teeth cut through her skin and the blood spread through her blonde neck hair. The two hyenas turned and stumbled through the dry grass, Star throwing her muzzle up and back, trying to break the grip. A cloud of dust shrouded the fighting pair.

Patches held on for almost ten minutes, flinging her adversary back and forth with such fury that Star's front feet were lifted from the ground. Blood dripped onto the sand. Between rasping breaths and screams came the sound of teeth grating through thick skin. Patches released Star briefly, found another grip near her ear, and held on. Inches away, Star's unprotected jugular surged with life. Again and again Patches changed her hold on Star's neck, raking her through the sand as if she were a rag doll. It was like watching someone being murdered in the street.

After twenty grueling minutes Patches suddenly released her. I was sickened by the sight of Star's neck, minced and shredded, with open penny-sized holes in her skin. For a moment I thought she would stumble to her forelegs, never to rise again. But then, as if she had been through little more than a mild scrap, she shook her long hair, flicked her tail, and side by side with Patches, walked over to the carcass. The two cubs had paid no attention to the ruckus; now all four fed close together, their muzzles almost touching. Even though there were no further signs of aggression, after five minutes Star walked away from the carcass and went to sleep nearby. She did not return to feed until Patches had left. Such neck-biting behavior had never before been reported, and it was one of the most intense and stirring encounters between animals that we would witness in the Kalahari.

For weeks we had seen hyenas, who were supposedly solitary, traveling and scent-marking the same trails and greeting each other with a bizarre crawling ceremony. We had seen Pogo and Hawkins foraging with two different adult females. And now, although the two females had just engaged in a pitched battle, we saw them feeding together at the same carcass. What a hodgepodge of signals they were giving us! This was most definitely *not* the behavior of a solitary species, in which males and females tolerate each other only long enough for courtship and mating. We became more and more sure that brown hyenas had some sort of unusual social system.

A few nights later, while looking for a hyena to follow, we found the resident pride of lions feeding on a gemsbok they had killed near Eagle Island, a group of trees north of the Mid Pan water hole. By 11:00 P.M. they had abandoned the carcass and were walking south along the riverbed. Within an hour Ivey, Patches, Star, Shadow, Pogo, and Hawkins were all either at or near the carcass.

During the rest of that night we saw more instances of their peculiar greeting and neck biting, sometimes preceded by a bout of muzzle-wrestling, when two hyenas would stand side by side, throwing their muzzles up high and hard against each other, each trying to get a grip on the other's neck.

But on the whole, the gathering at the carcass was very peaceful, as well as being organized. Usually one hyena—and never more than three—fed while the others slept, groomed, or socialized nearby. They alternated at the carcass; while one was carrying a leg off into the bush, another began to feed. Tonight it was six hours before the last hyena walked off into the bush, leaving only jawbone and scattered rumen content behind. This was leisurely feeding compared to the "scramble competition" between spotted hyenas in the Serengeti of East Africa. The spotteds crowd around a kill in a seething mass of bodies, competing with their fellows by eating as fast as possible. Dr. Hans Kruuk,[1] who studied them in the Serengeti, once saw a clan of twenty-one spotted hyenas consume in thirteen minutes a yearling wildebeest weighing about 220 pounds.

Unfortunately, we found the browns feeding together on a large carcass only once or twice a week during the rains. But after several months, a sketchy picture of their social organization began to emerge. We were sure that the seven hyenas in the area were not solitary animals but members of a clan.[2] Through muzzle-wrestling and neck-biting contests, each had gained a particular rank in the social hierarchy, which was displayed and reinforced in greeting. Ivey, the only adult male in the group, was dominant. The social order among the females ran from Patches at the top, down through Star, Lucky, and Shadow, to Pogo. Hawkins, the young male, was on the same social level as Pogo.

Usually when two brown hyenas met on a path, they would confirm their status through greeting, and then separate. Neck-biting followed only when the status was not well established, or when a hyena tried to rise through the ranks. Star seemed especially keen on social climbing; some nights she would arrive at a kill site and never feed. Her long hair on end, she would spend the entire time picking on lower-ranking females or challenging Patches.

One of the benefits of high status was obvious when several hyenas were feeding on a large carcass: If Ivey approached Shadow, they would usually eat together for only two or three minutes, after which Shadow would walk off and rest nearby until Ivey had left. Quite often there would be no obvious aggression on the part of the dominant hyena; it was as if the subordinate simply did not feel comfortable feeding with its superior and preferred to wait until later. As a result, dominants usually held priority at food sources.

At large carcasses, such as a gemsbok, even the low-ranking Shadow had a chance to feed. But at smaller ones, a steenbok or springbok, the competition was much more intense: The hyenas would feed faster, and only the first one or two arrivals would get anything to eat. When the carcass had been reduced to less than forty or fifty pounds, a dominant would often tuck all the ragged bits and pieces into a bundle and, tripping over dangling flaps of skin, carry it off, a trail of jackals tagging behind.

Lions, wolves, and other social carnivores usually sleep, hunt, and feed with at least some members of their group. But though the browns lived in a clan, they usually foraged and slept alone, only meeting other group members occasionally, while traveling along common pathways or at a kill. They have a limited repertoire of vocal signals, and none with which to communicate over large distances, as do the spotted hyenas. This may be because the dry Kalahari air does not carry sound very far, or perhaps because their territories are too large for clan members to transmit and receive even loud calls effectively. For whatever reason, brown hyenas do not attempt to communicate vocally over distances of more than a few feet, and they do not have the loud *whoo-oop*, or even the "laugh," of their spotted cousins.

This lack of a loud voice might seem to present a problem for animals who roam separately in a jointly owned territory as large as 400 square miles but must also maintain contact with other group members. However, the hyenas' well-developed system of chemical communication through scent-marking—*pasting*, as it is termed—probably takes the place of loud vocalizations. Spotted hyenas also paste, but not quite as extensively as browns, who use it as the most important means of transmitting information among individuals. Clan members appear to recognize the sex, social status, and identity of one another by the paste. For a social animal who must spend long hours alone searching for scattered bits of food, it is an ideal way for group members to stay in touch by "phoning long distance." The hyenas of the Deception Pan Clan also pasted extensively to demarcate their territorial boundaries.

So, the brown hyenas were a curious blend of social and solitary: They foraged and slept alone; they fed together on large carcasses, but carried away the remains for themselves at the first opportunity; they did not use loud vocalizations to communicate with each other, but did leave chemical messages. And, at least for a while, the females allowed the youngsters to follow them when they searched for food.

But when Pogo and Hawkins reached adult stature, at about thirty months of age, Patches, Star, Lucky, and Shadow no longer tolerated their tagging along on foraging expeditions. The subadults were forced to find food on their own. In order to secure a position in the pecking order, Pogo had to compete with the other females, and Star never let the young female forget for a minute who was boss.

On one unforgettable occasion we watched Star neck-biting Pogo for over two hours, chewing and shaking her by the neck for more than a quarter of an hour at a time. The youngster made loud humanlike screams, and it was difficult for us not to interfere. After Star had emphatically made her point, Pogo accepted her status and began displaying the deferential crawl of the subordinate in their greetings.

Hawkins had a different fate. Early one morning, while feeding on the remains of a lion kill, he looked up to see Ivey coming from the north. Hawkins walked slowly toward the dominant male, and he had begun to submit, when Ivey lunged forward, seized him by the neck, and shook him vigorously. Hawkins shrieked and struggled to free himself. Ivey chewed the youngster's ear, the side of his face, and his neck until blood ran through his blond neck hair. When Ivey tried to change his grip, Hawkins broke free. But he did not actually run away; instead, he loped in a large circle around the carcass until the dominant male easily overtook him. They faced each other, clopping their powerful muzzles together, each wrestling for a grip on the other's neck.

Ivey succeeded in again seizing the young hyena by the neck. This time he shook him violently and threw him to the ground.

Hawkins managed to get away, but again he made little real effort to escape, as though inviting another chance to challenge the other male. The stakes were high: The opportunity to remain in his natal clan and its familiar territory hung in the balance. Ivey soon caught him again. The contest continued for over two hours, with Hawkins taking all the abuse.

Finally, Ivey released the younger male. He walked to the water hole to have a drink and, panting heavily, lay down. Having followed him, Hawkins began to saunter back and forth in front of the older male, as if daring him to attack again. When Ivey ignored the challenge, Hawkins carried a stick to within eight yards of the old champion and made an obvious, if not terribly impressive, display of mutilating it. When that brought no response from Ivey, he began pacing back and forth in front of him again, drawing nearer and nearer, until he was within five yards. After resting, Ivey charged once more, and again began mauling Hawkins, who took several more minutes of this punishment before he finally broke free and started walking slowly toward the East Dune woodland. Ivey did not follow him.

Over the following weeks, Hawkins found it more and more difficult to forage and feed in the clan's home range without being harassed by Ivey. He took to roaming in the outskirts of the territory, and finally he disappeared. If he survived on his own as a nomad, someday he would challenge the dominant of another clan and perhaps become its only breeding male. Should he be unsuccessful in his bid for females and a territory, he would remain a solitary outcast, eking a living from marginal habitats away from the prime river valleys. His only chance to breed would be to mate with the occasional nomadic female or to sneak a copulation with a clan female.

Despite the fact that they always foraged alone, brown hyenas, we now knew, were social—and quite social, at that. But animals associate for some adaptive purpose, not because they enjoy being together. Lions, wild dogs, wolves, primitive men, and spotted hyenas hunting in a group are able to kill larger prey than can a single individual. Brown hyenas were scavengers, for the most part, and they rarely hunted. But since they did not hunt together, why did they live in a clan and share large kills left by lions? Why did they need each other? Why did they bother to socialize at all? There was a single answer for all these questions, as we were to discover.

6

Camp

Delia

I love not Man the less, but Nature more.
—*Lord Byron*

SHORTLY AFTER the fire, we found our original campsite too exposed
to the persistent winds, so we had literally molded a new one into the
interior of a tree island, cutting two or three dead branches to make
room for the faded tent given to us by a friend in Maun.

Our island was thick with ziziphus and acacia trees standing in long
grass and tangled undergrowth. The ziziphus had multiple trunks that
splayed, at about fifteen feet above ground, into hundreds of smaller,
thorny branches, which fell back toward the ground in a snarled tumble.
The flat-topped acacias and the drooping ziziphus interlaced, forming
a roof over our heads so dense and green that in the rainy season we
could barely see the sky. Camp was surrounded by the open plains of
the ancient riverbed, which stretched to the northern and southern
horizons. On the east and west, duneslopes crept gently to their wooded
crests.

Because we hated to disturb the small mammals and birds in the
island, we had left the dead wood, "sticky grass," and undershrubs in
place. Only narrow footpaths led from the kitchen, in an open alcove
at one end of the grove, to the tent. For the first year the island was
so overgrown after the heavy rains, and our camp so well hidden,
that now and then a giraffe's head would suddenly appear in the
canopy above us. After stripping some leaves from a thorny branch,
he would discover us and our few belongings tucked away below.
Curling his tail over his rump, the giraffe would clomp over the river-

bed for a way, and then look back as if he had imagined the whole thing.

And sometimes, during the rainy season of 1975, as many as 3000 springbok grazed so close that we could hear their stomachs rumbling.

Because this was a desert, every living plant was important to some creature. We became obsessed with saving the leafy branches and wilting grass, even to the point of always asking our rare human visitors to stay on the footpaths. I once became very annoyed when some visiting scientists cleared a large area in the center of camp for their sleeping bags. For months after they had gone, until the rains came again, their bivouac was the dusty "vacuum which nature abhors." Our behavior was perhaps a desperate attempt to fit in, to slip back into the natural world without being offensive or noticed. We felt as if we were guests who had been away for a long, long time.

*   *   *

I stood on the woodpile and watched until the heat waves swallowed the bleary image of the truck. It was early in the rainy season of 1975, and Mark was driving into Maun for supplies. Because he would be gone for three or four days, he had not liked leaving me alone in camp, but I had insisted that I remain at Deception to catch up on paper work. The drone of the engine faded from the dunes, making me one of the most remote people on earth. I stayed behind not only to transcribe notes, but also to experience the sensation of total isolation. Looking over the riverbed for a while, I let the feeling settle over me. It was comfortable.

But total solitude takes some getting used to. Although I was the only person for thousands of square miles, it took a while to shake the feeling that I was being watched. While making tea I talked to myself openly, but with the urge to look over my shoulder to make sure no one was listening. It never was being alone that bothered me, but rather the feeling that I might *not* be alone when I was supposed to be.

I walked to the kitchen, poked some life into the grey coals and moved the old enamel kettle over the fire for tea. It was coated in flaky layers of black—the history of a thousand campfires—and its worried old handle wore the tooth marks of the brown hyenas who had often stolen it from its perch. The kettle was our only source of hot water, and whether we wanted a sponge bath or a cup of coffee, it was always ready.

I made a simple bean stew for myself, and soon the pot was bubbling on the heavy iron fire grate that Bergie had given us. Then I kneaded bread dough, put it in the black three-legged mealie pot, and set it in the sun to rise. Later I turned a five-gallon pail on its side and set two dough-filled bread pans inside. Using the spade, I sprinkled glowing red coals beneath and on top of the bucket oven. In the midday temperature, with the wind blowing steadily, the bread would bake in seventeen minutes. It took as long as twenty-five minutes to cook if there wasn't any wind, and during the calm, cool, and more humid night, an hour.

Our food supply was limited by what was available in Maun, what we could afford to buy, and what would survive the long haul to camp in the heat. Sometimes even staples like flour, mealie-meal, sugar, lard, and salt could not be found in the general stores.

We had no refrigerator, so we could not store perishables of any kind for very long. Onions kept for several months if left hanging in the dry air, and the carrots, beets, and turnips that we bought occasionally from the gardens in Maun lasted for two weeks if buried under the sand and sprinkled with waste water, and if moved from time to time to fool the termites. Oranges or grapefruits stayed edible for up to two and a half months in the dry season, the rind gradually turning to a tough shell that protected the succulent pulp from dehydration. Nothing rotted in the dry months.

In 1975 we were given permission by the Department of Wildlife to shoot an occasional antelope, in order to analyse its rumen contents. We hated to do this, but it was important for the conservation of the animals to know which grasses or leaves they ate from season to season. Mark always took many hours to hunt, carefully stalking an individual separate from the others, to keep from disturbing and alienating the herd. These precautions paid off, for during our entire stay in Deception the springbok and gemsbok showed no more fear of us than when we had first arrived.

We learned how to make *biltong*—jerky—out of the antelope meat by soaking raw slices overnight in the washtub, in a concoction of salt, pepper, and vinegar and then hanging them up to dry on wire hooks. They dried in three days and could be kept for several months. Often our only source of protein were these biltong sticks, dipped in hot mustard, which were quite tasty. But we soon tired of this stringy meal, so I tried to prepare the meat in more creative ways. One recipe we devised was Biltong Fritters:

INGREDIENTS:

Two slabs of very dry biltong	desert-dried onion
desert-dried green pepper	short pastry

Pulverize the biltong in the bathtub with a five-pound hammer and a trailer hitch, then soak in water with onions and peppers for some time. Drain well and fry briefly in hot oil. Cut pastry into triangles, place one tablespoon of biltong hash on each and roll up. Fry fritters until crisp and golden brown.

Biltong also tasted good with Camp Cornbread (eggless), one of our standbys:

1/3 cup tinned margarine	4 tablespoons powdered milk
1/3 cup brown sugar	1 cup water
1 cup flour	1 cup mealie-meal (cornmeal)
3 teaspoons baking powder	salt to taste

Cream the margarine and sugar together. Mix in the powdered milk and water. Add the flour, mealie-meal, baking powder, and salt and mix well. Place in a greased pan and cook in bucket oven with moderate coals for 25 minutes (15 if the wind is blowing steadily).

When we didn't have meat, we ate various stews made of dried beans, corn, sorghum, and mealie-meal. Their insipid taste could be improved a bit with onions, curries, chile, and Mexican-style pastries, but it was often a matter of swallowing the food as quickly as possible and following it with a can of sweet fruit cocktail, if one was left in the tea-crate larder.

In September and October, before the rains begin, a female ostrich lays up to twenty large ivory-colored eggs, each about seven inches long and fifteen inches around, roughly equivalent to two dozen chicken eggs. Although we never robbed an undisturbed nest, we sometimes found an egg that had been abandoned after a predator had chased the parent birds away. Had we guessed then how highly prized such eggs would have been for a brown hyena, who had no water stored in drums or cans of fruit cocktail stashed away, we wouldn't have taken them.

With a hand drill, Mark would bore a quarter-inch hole in one end of an egg. Then, holding it between his knees, he would insert an L-shaped wire—first sterilized over the fire—through the hole and roll it back and forth between his palms, scrambling the white and

Moffet has just finished eating a porcupine, a staple food for desert lions in the dry season. Many of the lions had never seen humans before, and after they got used to us, we could often sit close to them with little danger.

Mark enjoys a rare bath in a gallon of water. Except during the short rainy season, we had to haul all of our water in drums from a cattle post fifty miles from camp.

Mark bores a hole in the shell of an ostrich egg, then shakes out some for breakfast. By taping over the hole and burying the egg to keep it cool and fresh, we could enjoy the egg for as long as ten days or two weeks. One ostrich egg is the equivalent of two dozen chicken eggs.

Pepper, a brown hyena cub, visits Delia in camp on one of the cub's first trips away from her clan's communal den.

Chief, a hornbill, searches for a snack in one of our pots.

Mark stalks a half-drugged lion with a syringe.

Mark treats Chary of the Blue Pride after immobilizing her. We usually darted lions and hyenas at night to avoid subjecting them to the severe heat and light of the desert day.

We use the truck and a canvas to drag a drugged lioness to shade for recovery.

Hansel and Gretel, two jackal pups, play-fight on the grass of the fossil riverbed in their first rainy season.

Gretel begs from Mate, her mother, who then regurgitates food for her. Carrying food in her stomach is a safe way for Mate to get food to her young without its being stolen by a brown hyena or another jackal.

Captain, Mate's mate, fights for a share of a gemsbok carcass abandoned by the Blue Pride lions.

Hansel gets soaked by a heavy rain while he sleeps on the riverbed, and then shakes off the cold water.

Still soaked, Hansel gives us a forlorn look before beginning the night's hunt.

Camp after a storm.

While searching the Kalahari's fossil riverbeds for lions, we often landed far from camp and spent the night under the wing of the plane.

Moffet made some unexpected visits to our toilet in the early morning.

Bimbo nibbles at Blue's chin, while Sandy rests on her back.

Bimbo and Sandy play-fight next to Blue.

Bones charges Blue as she tries to get a share of a kill she made herself.

Sassy bites at Spooky's belly in play.

Spicy and Spooky rest after playing.

Dusty and Sooty, brother and sister brown hyenas, greet each other at the clan's communal den.

As Star nurses her cubs, Pippin, her offspring from her previous litter, smells Cocoa, Pepper, and Toffee, his new half-brothers and half-sister.

Shadow, a brown hyena, competes with black-backed jackals for the remains of a lion's kill.

Brown hyenas dismember, carry off, and cache legs and other parts of carcasses for later meals.

Pepper stares up at Delia, who has just arrived for a night's observation at the hyenas' communal den.

yolk. I would shake enough for one meal into the frying pan, and then seal the hole with a Band-Aid before burying the egg under a shady tree. As long as the contents did not get contaminated, we could have scrambled eggs or an omelette every morning for the next twelve days or so.

The only danger in eating ostrich eggs was that we could not be certain how fresh one was until the drill bit actually broke through the thick shell. Unless Mark was very careful, the stinking juices from a very bad egg would erupt and squirt him in the face. Whenever he opened one, I left the kitchen, but his language left no doubt about whether or not it was rotten.

* * *

The morning Mark left for Maun slipped away before I could even start on the mountain of paper work stacked on the table beneath the ziziphus tree. We always had tapes to be transcribed onto data sheets and letters to write. Before leaving the kitchen I automatically moved the kettle to the edge of the fire grate to keep it from boiling dry. Conserving water had become second nature; if it hadn't, we would have spent much of our time and money trying to keep it in stock.

Whenever low grey-black clouds rolled in over the dunes, Mark and I would rush around placing pots and pans around the tent to catch the rain. Then we would zip the tent, topple gasoline drums onto their sides, stuff cloth bags of flour and all our data books into the front of the truck, cover the food shelf with canvas, set the equipment boxes up on blocks, and cover the fire with the half drum. Finally, after checking the tie-downs on the tents, camp would be secure.

As soon as the downpour eased, we would grab the pots and pans and pour the fresh rainwater into drums. Then we would scoop up as many as eighty gallons of the coffee-colored water that stood ankle deep in camp.

Later, when the mud had settled a bit, we would drive to the water hole on Mid Pan and, using pots and funnels made from plastic bottles, spend hours collecting more water. It was impossible to avoid including some of the springbok and gemsbok droppings that bobbed on the surface, but they would settle out in the drum and they did no real harm because we boiled all our drinking water. We hadn't always taken this precaution.

One month in the dry season of 1975 we both came down with

severe intestinal cramps, diarrhea, and lethargy that persisted for days. We kept getting weaker and had no idea what was causing the trouble. Our water supply was nearly finished, and I worried that if our illness worsened, we wouldn't be strong enough to make the trip for more. Without a radio, there would be no way for us to get help.

Mark dragged himself from the tent, and pausing frequently to rest, he consolidated into a bucket the last gallon or two from all the drums. As he tipped up the last one, feathers swirled into the pail, followed by the ripe ooze of a putrified bird, which had apparently drowned in the drum weeks before. From then on we boiled every drop that we drank, no matter how clean it looked, and kept the bungs plugged with rags.

Lionel Palmer, the hunter who had suggested that we work in the Kalahari, had recently loaned us a small trailer and had assured us that we could use it to haul water. The morning after we had found the decomposed bird, we pulled the trailer to the cattle post, where we now got our water, and loaded one drum inside the Grey Goose, one on the roofrack, and three in the bed of the trailer. A single drum filled with fifty gallons of water weighs nearly 500 pounds, so Mark chopped some log wedges and drove them beneath the spring blades of the Land Rover to help support the load.

On our return to camp, we were no more than a mile from the borehole when a tremendous screeching came from behind the truck. The Grey Goose lurched forward and slid to a stop. One of the drums had fallen through the floor of the trailer, wedging itself against the wheel. We heaved it up, placed bits of spare planking over the holes in the trailer bed, and roped the barrel in position before setting off again.

Two miles farther down the track, the second drum broke through. We blocked and secured it as best we could with fragments of the frayed rope and slogged on, now at a snail's pace.

We had gone perhaps another four or five miles when the truck staggered to one side, drums clamoring from behind. We ran to the trailer and found its tow-bar bent into an S shape. It had dropped to the ground and plowed a furrow through the sand when all three drums had tumbled forward.

Covered with dust, grass seed, and sweat, we had taken four hours to cover eight miles in the 120-degree heat. Exhausted and weakened from our illness, we sank to the sand in the shade of the boiling truck, our throbbing heads resting on our knees. I didn't see how we could go on. His jaw rigid, Mark stared silently across the savanna. The Kalahari would not give an inch—she never let anything come easily.

After a few moments, Mark pulled himself up and gave me a hand. Placing the highlift jack under the arms of the tow-bar, and with the trailer still hooked to the Land Rover, he jacked them until they were nearly straight. Then he fashioned a splint for the broken hitch with the Land Rover crank handle and small logs cut from a nearby stand of trees. We reorganized the drums and started off again.

At the end of nearly every mile, we had to stop and cool the Goose. While Mark poured water over the top of the radiator, I used a hairbrush to clean away the thick carpet of grass seed that blocked the air flow. By removing a spark plug, putting a hose to the opened cylinder, and racing the engine, we blew out the clogged grill. Before starting up again, Mark crawled under the Land Rover and, using a long screwdriver, cleaned away the charred straw that was smoldering on the undercarriage. Trucks in the bush often burn up when grass on the exhaust ignites.

Creeping along again, we suddenly smelled smoke. Mark slammed on the brakes and we jumped out. It was impossible to keep the overheated exhaust pipe clean for long, and now it had caught fire. Thick white smoke was pouring from underneath the truck. Mark grabbed the hose and a wrench, vaulted into the back of the Land Rover, and quickly opened a drum. Flames began to sneak through, showing orange against the billowing white. He sucked on the end of the hose, got the siphon going, and sprayed water onto the undercarriage. The smoke turned black, and a roil of steam and ash hissed back at him as the fire died.

Five hours and three flat tires later, we staggered into camp and collapsed onto our foam rubber pads under the stars.

The next day a freak storm appeared from nowhere and rained buckets of water all around us. We didn't even have an extra canteen to fill, and the Kalahari drank it all.

*   *   *

Though I was completely isolated from humans those four days Mark was in Maun, I was by no means alone. Late on the first afternoon, I put away our field notes, cut myself a slice of freshly baked caraway bread, and sat in our tea room, the alcove under the weeping branches of the ziziphus tree. In an instant, flocks of chippering birds crowded around. "Chief," a yellow-billed hornbill with a mischievous eye, watched from the acacia tree across the path. Then he fell from his perch, spread his wings, and swooped to a landing on my head, his wings fluttering around my ears. Two others sat on my shoulders, and the four in my lap pecked at my hands and nibbled my fingers. Another hovered in midair until he managed to bite off a chunk of crust. I divided the rest of the bread among them.

One of the first things we had done in our tree island camp was put out bread scraps and a small dish of water. Soon scores of birds — violet-eared waxbills, scaly-feathered finches, crimson-breasted shrikes, tit-babblers, Marico flycatchers — were twittering and preening in the trees. In the early morning, striped mice, shrews, and ground squirrels scurried around our feet to compete with the birds for food. But the hornbills were always our favorites.

The yellow-billed hornbill is an odd assemblage of parts: a hooked yellow bill that seems too large for his scrawny black-and-white body, a long black tail that looks like an afterthought, and seductive eyelashes that flutter over foxy eyes — a most beguiling companion. We could recognize forty of them by their natural markings or by dabs of black paint I applied to their bills while they were taking bread from my fingers.

Whenever I cooked, the "billies" crowded around the kitchen, perching on my head and shoulders, and even on the frying pan, picking up first one foot, then the other, fixing me with rude glares, as though they somehow knew it was my fault the pan was getting too hot. They found our leftover oatmeal and rice, stored in pots, by prying off the lids with their crescent bills. And when we sat down to eat under the ziziphus tree, we guarded our plates carefully lest our food disappear in a cloud of feathers. They also spoiled many a cup of tea by dropping their whitewash with uncanny aim from overhead branches.

One day while we sat writing under the trees, a small pearl-spotted owl dove from his perch and captured a scaly-feathered finch, a tiny bird with a distinctive black goatee. All the birds in camp, not just the finches, immediately rushed to the scene, bobbing up and down on the outer branches, giving their alarm calls. The captive finch screeched and flapped its wings wildly as it struggled in the clutches of the owl. Then one of the hornbills, hopping onto a branch just under the owl, reached up and snatched away the finch, who escaped to safety. It is impossible to say whether the hornbill was attempting to rescue the finch or to get an easy meal for himself; I prefer the former but believe the latter.

Another permanent companion in camp was Laramie the lizard, who nested every night in an empty Di-Gel box on the orange crate – bedside table. He was especially welcome because of his voracious appetite for the flies that invaded our tent. With never-ending patience and acute skill he tracked them down one by one, smacking loudly as he chewed them up. But termites were Laramie's favorite food, and I often fed them to him with a pair of forceps while he perched on the old tin clothes trunk next to our bed.

Since zippers on tents are notoriously short-lived, we rarely had a tent in which the doors and windows closed securely. Thus, mice were common visitors to our bedroom at night, and they often found their way into bed with us, especially in the cold-dry season. Feeling a slight pressure roaming between the blankets, we would bolt out of bed and trip around in the dark, waving dim flashlights and shaking out the blankets. When the mouse finally shot out from between the bed covers, we would throw shoes, flashlights, and books in every direction until he escaped.

We were quite accustomed to these intrusions, but one dawn I was half awakened by a very heavy pressure moving around over my legs. I imagined the world's largest rat crawling on our bed, and I began kicking with a frenzy. I sat up just in time to see a slender mongoose leap for the tent door. He paused for a few seconds, looking back, and we stared into each other's startled eyes. That was our introduction to "Moose."

Moose became the camp clown. He always remained aloof, maybe because I'd kicked him out of my bed. He would never accept a handout, but he was not above stealing everything in sight. One morning when we sat drinking tea under the ziziphus tree, Moose came sidling down the footpath, our pot of leftover oatmeal clattering behind him. Without so much as a glance in our direction, his head high and the handle in his mouth, he scraped his pot of porridge past us and straight out of camp, where he ate his breakfast in the morning sun.

Because the ever-present mice were always chewing their way into our food containers, we set traps in the kitchen area nightly. We did this reluctantly because there was always the possibility of killing some creature other than a mouse. Sure enough, when Mark and I neared the kitchen one dawn, I heard a loud snap and looked up to see a Marico flycatcher flopping about with the trap closed on his skull. Mark immediately freed the small bird, who stumbled about in ever-widening circles. I suggested that we should put him out of his misery, but Mark insisted that we wait to see what happened.

The flycatcher eventually stopped walking in circles and flew to a rather clumsy landing on a low acacia branch. From that moment on, Marique continued the normal life of a Marico flycatcher, except for three things: He was blind in the left eye, he lost all fear of human beings, and he took up the habit of "begging" from us by flapping his wings like a fledgling. Tamer than most parakeets, Marique would land on our heads, our plates, our books. He would stand on the path in front of us and shake his wings vigorously to demand food; we could almost imagine his hands on his hips and his little foot stomping. Probably because of our guilt over the accident, we always fed him, even if it meant dropping whatever we were doing to make a special trip to the kitchen.

When Marique took a mate, she also became very tame, though she did not beg. But when they reared their second brood (the first was lost in a storm), the babes soon picked up their father's habit of begging from us. And so the behavior was passed on, and for the rest of our years in the Kalahari, Marico flycatchers in camp would land at our feet and shake their wings for food. We could never refuse them.

Having wild creatures around us was one of our greatest pleasures, yet at times it was a mixed blessing. Early one morning, still groggy from sleep, I threw back the cover from the dilapidated tea crate. When I reached inside, looking for a tin of oatmeal, my breath caught. The long grey body of a banded cobra was coiled on the cans, inches from my hand. I'm not usually intimidated by snakes, but this time I snatched my hand back and let out a respectable howl. Fortunately, the cobra must have been as intimidated as I was because he did not strike, but instead slithered down among the tins. Mark appeared an instant later with the .410 shotgun. So far we had killed only a few of the most poisonous snakes that had visited us and those only because they insisted on living in camp. This one would be a real danger if allowed to stay. Mark aimed the gun into the box, and I imagined losing a month's supply of food along with the snake. But when we turned the crate on its side to remove the dead cobra, we found only one irreparably damaged tin—unfortunately, it was fruit cocktail.

Boomslangs, puff adders, black mambas, and other poisonous snakes frequently appeared in camp. That we had not been bitten was due mostly to our own private warning system. Whenever the birds spotted a snake they all landed on the branches above it, chirping, clucking, and twittering in loud alarm. Since there were sometimes as many as 200 birds in camp, the racket they made always tipped us off that a snake was on the prowl. The only problem was that they also mobbed owls, mongooses, and hawks, and once even a homing pigeon, complete with leg band, who had miraculously found his way to our camp. Sometimes they would keep it up for hours, or for several days, as in the case of the pigeon, and we would begin to prefer the quiet snakes to the noisy birds.

Small animals were not the only ones who made themselves at home in camp. Walking down the path to the kitchen at dawn, we often surprised two or three jackals who had slipped under the flap into the little mess tent. On hearing our footsteps they would ricochet around inside, searching for a way out, the tent walls billowing, until suddenly they would squirt out from under the tent door, ears back and tails bouncing.

Lions, leopards, brown hyenas, or jackals would wander into camp almost every night of the rainy season. After we bought a small mess tent, we tried to protect it, and the kitchen, with a barricade made of drums, thorn branches, spare tires, and the fire grate. Even so, we often got up several times a night to usher animals out of camp. Walking slowly toward them while talking to them quietly always worked with the hyenas and jackals, but sometimes the lions and leopards were less willing to leave.

One night we drove into camp and a leopard stepped out of the shadows in front of our headlights. Mark jammed on the brakes just as the cat gracefully sidestepped the truck. Totally unperturbed, he sauntered to the middle of camp and, with a single silent motion, jumped onto the water drums. He walked from one to the other, smelling the water inside until, apparently convinced that he could not get to it, he jumped down. Next he climbed swiftly up the acacia tree that slanted against the flimsy reed structure we had made for dry-season shade. There, as he slowly stepped onto the roof, his front paw caved through the reeds with a loud splintering noise. Lifting his feet high, as if walking in tar, and with his tail lashing about for balance, he continued across, stabbing through the ceiling with every step. Finally, gripping the tree with his hind feet, he managed to extricate his legs from the now sagging and tattered roof. He then jumped from the tree and walked to our tent, and after a good look around inside, climbed to a limb that hung over the door and settled comfortably in its crook. He closed his eyes and began casually licking his forepaw with a long pink tongue; obviously he intended to stay awhile. All of this was very entertaining, but it was now 2:45 A.M. and we needed to get to bed. Mark drove the Land Rover a bit closer, thinking that the leopard would leave, but he just peered down at us benignly, his tail and legs dangling down from his perch over the door.

We didn't want to frighten him away, and we couldn't quite bring ourselves to walk directly under him to get into the tent, so, leaning sleepily against the truck, we watched for about fifty minutes while he napped. Eventually he yawned, stretched, climbed down, and padded out of camp, his long tail trailing easily behind him. Stiff and tired, we began brushing our teeth next to the tent.

"Look who's back," Mark whispered several minutes later. I whirled around to see the leopard standing at the back of the Land Rover, his muzzle raised and his amber eyes staring. He apparently meant us no harm, so we finished brushing our teeth while he sat fifteen feet away, his head cocked to one side. We went into the tent, closed the flap as well as we could, and crawled into our bed on the floor. A few minutes later we could hear the soft top-top-top of leopard pads on the plastic ground sheet, and then a sigh as he settled down for a cat nap, just outside the door.

* * *

I was always very conscious of the fact that besides being a coworker, I was also a wife. In spite of all the dirt and grime and my ragged cut-off jeans, I tried to stay as feminine as possible. I usually put on a little make-up every day, and on our nights off, when we relaxed by the fire, I wore a blouse and a skirt of African printed cotton. On one occasion Mark had gone by himself to collect firewood, and since we did not plan to follow hyenas that night, I decided to fix myself up. I dug bright yellow curlers from the bottom of my trunk, washed my hair, and put it up in rollers.

As I walked through camp toward the kitchen, the hornbills swooped down into the branches just above my head, clucking loudly. I recognized it as their alarm call, and stopping short, I began to look around me. But I couldn't find a snake or anything else that could have alarmed the billies. I walked warily back to the tent to get the shotgun, and as soon as I was inside, the mobbing stopped. When I stepped out again, the racket began once more, as if on cue. While I searched for the snake, gun in hand, the hornbills repeatedly dive-bombed my head. With a pang of humiliation I suddenly realized the problem. I never did understand why, but from then on, whenever I curled my hair, I would either have to stay inside the tent or put up with the noisy objections of the hornbills.

*　　*　　*

Just before sunset on my first day alone in Deception, I filled a bowl with bean stew and sat down on the flat riverbed outside camp to eat my supper. The hornbills sailed over my head on their way to roost for the night in the dune woodlands. Soon afterward two nightjars flitted across the fading day and landed a few feet away. They waddled about, making low purring sounds as they searched for insects. The sky deepened. I lay back in the straw-colored grass, and pressing my fingers into the rough surface of the riverbed, as I had so many times before, I wondered how long the Kalahari would belong to the wild.

I sat up. Thirty springbok had wandered to within fifty yards of me while I was hidden in the grass. The male whistled alarm, and they all looked at me with tails twitching and necks stiffly arched. I stood, and they relaxed when the odd, hunched shape in the grass, which could have been a predator, turned into my familiar form. They resumed their grazing, but drifted away from me at an almost imperceptible pace, just the same. They could not know that I was here for them, and they disappeared over the dune.

I walked down the track from camp speaking softly to myself as the last traces of that day vanished altogether. There may not be a fine line between dusk and darkness to the eye, but there is to the mind. When I was half a mile from camp, I felt the night settling on my shoulders and along my spine. I began snatching quick glances behind me, and, like any good primate, I returned to my trees.

*　　*　　*

Over the next three days, while Mark was in Maun, I finished our backlog of paper work and fed nearly half of the bread to the hornbills.

I was still enjoying being alone in the wilderness, but more and more often I stopped my work and ran out of the tent, thinking that I had heard the truck. I stood listening for the distant drone of the engine over the eastern dunes, but there was only the wind. Mark would probably be back soon, so I baked him a lop-sided, eggless spice cake in the bucket oven.

The notes all neatly transcribed, I spent the fourth day cleaning camp. But I was losing interest in really accomplishing anything, and I sat with the hornbills for a long time, talked to myself a lot, and walked from camp again and again to listen for the truck. Maybe Mark would bring me something special—some chocolate from Riley's perhaps, or the mail, with a package from my mother. When he wasn't back by 5 P.M., I felt very let down.

In the early evening I was stirring my supper over the fire when seven lions came padding directly toward camp. My heart began doing flip-flops, and I quickly put the pot of stew on top of the hyena table and hurried deeper into the tree island. Peering through the branches, I could see the long, low forms gliding silently toward me, just 100 yards away. It was the same lionesses and their adolescent young we had often seen. But on the other occasions when they had visited camp, the truck had always been nearby; now I felt as vulnerable as a turtle without a shell. I tried to reason with myself: What difference would the truck or Mark have made, the lions weren't likely to do anything, anyway. Still, I felt trapped. Crouching low, I crept inside the tent and peeped through the window.

When they reached the edge of camp the lions began to play like giant kittens, romping and chasing one another over the woodpile and through the kitchen. Even when they were behind the bushes, I could tell by their sounds what they were doing. Then a pot hit the ground, and everything went quiet. They had probably found my stew.

It soon grew dark and for a while I could neither see nor hear them. *Where were they? What were they up to?* Suddenly their heavy thuds sounded on the ground just outside the tent. I sat back onto our bed, my thoughts racing. When we had last been in Maun one of the hunters had told of Kalahari lions flattening three of his tents one night while he and his clients huddled in the truck. I had thought the story an exaggeration; now I was sure that it was true.

I had to make a plan. My eye caught the tin clothes trunk. Moving very quietly, I opened it and piled its contents on the bed. If the lions began playing with the tent, I would get inside and close the lid. I sat in total darkness on the edge of the bed, one hand on the open trunk, and listened to the slaps, grunts, and pounding feet outside. Suddenly there was utter silence again. For minutes not a sound came from outside. They *had* to be there. I would have heard them move away. Huddled on the bed beside the trunk, I visualized all seven of them lying in a semicircle around the tent door.

Ages passed and still no sounds. Could they smell me? Should I get into the trunk or sit still? A twig snapped. The side of the tent began to balloon slightly. Then a rope sang out with a twang. Through the window I could see one of the lionesses pulling a guy line with her teeth. Soft footsteps in the leaves and loud sniffing noises: They were smelling along the base of the tent only inches from where I knelt.

Then there was a deep droning far away. The truck? God, let it be the truck! On quiet, damp nights I could sometimes hear it for three-quarters of an hour before it reached camp; then there would be long silences as it descended between dunes.

Once more all was quiet. Maybe I had imagined the sound. Soft footfalls moved along the side of the tent toward the door. I wondered what would happen if I stood up and screamed "Shoo—get out!" But I didn't budge. I was so much more brave when Mark was around.

Again the motor sounded from the distance—it had to be Mark. After an eternity the pitch changed completely, and the truck turned onto the riverbed and headed directly toward camp.

A noise just outside made me jump—a long brushing motion against the canvas side.

As Mark rounded Acacia Point he was surprised to see that no fire was blazing, no lanterns were lit; camp was completely dark. He switched on the spotlight and immediately saw the seven lions prowling around the tent. He drove quickly into camp, switched off the truck, and called out the window, "Delia...Delia, you okay!?"

"Yeah—yeah—I'm—I'm all right," I stammered. "Thank God you're back."

Mark's arrival having spoiled their fun, the lions left camp and filed slowly southward down the riverbed. I jumped up to give Mark a grand welcome, but then I remembered the pile of clothes on the bed, so I stopped to stuff them into the trunk. After all, there was no reason to tell him of my plan, which now seemed rather ridiculous.

"Are you sure you're all right?" Mark met me on the tent stoop and hugged me.

"Yes—now that you're here. And what about you? You must be starving."

We unloaded the truck and cooked a feast with the supplies he had brought from Maun. We ate goat meat, fried potatoes, and onions, and I chattered nonstop about the previous four days. Mark patiently let me talk myself out, and then he told me all the news from Maun as we snuggled by the firelight. A long while later, when we went to bed, I found a bar of chocolate under my pillow.

7

Maun:
The African Frontier

Mark

Quick, ere the gift escape us!
Out of the darkness reach
For a handful of week-old papers
And a mouthful of human speech.
—*Rudyard Kipling*

THE SUN was high over the river when the truck crossed the last sand ridge. Grey with dust and fatigue, we drove into the Boteti, opened the doors, and fell sizzling into the cool water. It was as if a fever had broken. Though we had been cautioned about big crocodiles and bilharzia, a debilitating parasitic disease picked up from infested rivers or lakes, nothing could have kept us from the water after the Kalahari heat. We lay with only our heads bobbing above the surface, letting the current rush over us, but we kept looking from one bank to the other for the telltale ripples of a croc.

It was March 1975, three months since my trip alone to Maun, and low on supplies again, we had set off for the village at dawn the day before. Besides restocking, we wanted to find an assistant, someone to take care of the dozens of camp chores that were eroding our research time. While trying to keep up with our expanding research project, it had become more and more difficult to cope with vegetation transects, scat collection and analysis, cartography, truck maintenance, hauling water and firewood, boiling drinking water, cooking, mending tents, and the myriad of other tasks associated with living in such a remote area. There just wasn't enough time or energy left for following hyenas each night, or enough hours left in the day to sleep.

But finding a native African who would live isolated in the Kalahari with very little water, no comforts, and lions roaming the valley would not be easy, especially considering the embarrassing wage we could afford to offer him. Only a very special person would do.

Now Maun was only thirty minutes away. We lay in the cool river, chatting about seeing friends in the village. Skeins of pygmy geese, ducks, and snow-white egrets passed low over our heads. After scrubbing our clothes and hanging them on thornbushes along the bank, we soaked in the river again, the minnows nibbling at our toes. A grey-haired old man with a poncho of tattered goatskin thrown over his shoulders walked his donkey to the water. He gave us a wide, dusty smile and then waved and shouted to us in Setswana, the local language. We shouted and waved back with such bonhomie that he must have thought he'd done us some favor. This was the first human, other than me, that Delia had seen in more than six months.

Our clothes stiff and dry, we drove on to the village and straight for Riley's, two buildings of concrete stucco, with scuffed white paint and green corrugated tin roofs, standing on a sand patch next to the river. Behind the hotel, bar, and bottle store, a long veranda with a red waxed floor was dwarfed by tall, spreading fig trees and the broad Thamalakane River drifting by.

Riley's was the first hotel on the frontier of Northern Botswana. Built as a trading post by settlers who arrived in ox wagons about the turn of the century, for decades it has been a staging point for expeditions north to the Zambesi River, west to Ghanzi, or back to Francistown, more than 300 miles east. Today it is still a popular meeting place, one of three or four in the entire territory of Ngamiland. Riley's has cold beer, meat pies on Saturday mornings—and ice. It was the best place to start looking for an assistant and to visit with friends.

After several months alone in the desert, we had begun to notice subtle signs that told us we needed to see other people, to be part of a social group again. It had become more and more difficult to concentrate on our research without wondering what Lionel and Phyllis were doing or thinking how nice it would be to have a cold beer with someone.

Already smiling with anticipation, we pulled up at Riley's and parked next to a line of trucks, all with rumpled fenders, long bush scratches down the sides, and oil dripping beneath. Behind the low block wall of the veranda, safari hunters in denims and khaki tilted their chairs back from wire tables, each man facing a row of empty beer cans. Weathered ranchers with bushy eyebrows leaned their beefy brown arms on the tables, their dusty, sweat-stained hats perched on the wall beside them. A Botawana tribesman in a red-and-black tunic, wearing a tasseled hat, hurried back and forth with mugs of Lion and Castle beer.

Dolene Paul, an attractive young woman with chopped blonde hair whom we had met on a previous trip, waved to us across the veranda. Born and raised near Maun, she was married to Simon, an Englishman recently trained as a professional hunter. As we moved toward her table, friendly jeers rose from the safari crowd: "Oh Chrrrist! Watch your beer, here come the bloody ecologists!"

Shaking hands, I found myself holding on too long, grasping a friend's hand or lower arm with my left while pumping away with my right. We smiled so much our cheeks ached, and greeted everyone over and over again, repeating their first names several times. Then, suddenly feeling foolish, I quickly sat down at a table and ordered a beer.

It was early afternoon, but no one seemed to have anywhere to go, so we all sipped cold beer and listened to hunting stories. Too anxious to join in, Delia and I made comments that seemed to bring conversation to a temporary halt, and we caught ourselves rambling on, talking much too loudly, and for too long, about subjects that must have bored everyone else. Socially, we were out of practice.

Occasionally the conversation turned to Simon's truck, which needed a new clutch bearing, and how sometime that afternoon they'd have to fix it. Then somebody bought another round of beer.

Since Dolene knew most of the local Africans, we asked if she had heard of anyone who was a good worker and in need of a job, one who might be willing to live with us in Deception Valley. "Can't think of anyone right offhand," she said. "It'll be tough to find a bloke who'll stay in the bush for a long time without other Africans around. You must come round to Dad's *braii* tonight. Maybe one of the other hunters or ranchers will know of someone." A braii—or *braiivlace*— is a southern African barbecue, and though we'd heard of Maun's version, we had never been to one.

Several hours later everyone began to stand and stretch and to talk of driving on over to Dad's. The afternoon was slipping away and nothing more was said about fixing Simon's truck; apparently it could wait until the next day.

Driving over to Dad's, we analyzed the reception we had got. "How do you think Larry acted toward us? Do you think Willy was really glad to see us?" Delia even coached me: "Try not to act so excited when we see everyone at Dad's place."

Dolene's father, "Dad" Riggs, was one of the first white settlers in the area and had for years been a shopkeeper in Sehithwa, a village near Lake Ngami, before moving his family to Maun. Dolene and her brothers spoke Setswana long before they learned English at boarding school in the Republic of South Africa. In later years, Dad had been a stock-taker for the Ngamiland Trading Center, a general store and trading company in Maun.

Dad's place, a pale yellow adobe with a corrugated tin roof, flaking foundation, and sagging screen porch, was hidden behind the trading store. A tack shed, with saddles, blankets, and bridles thrown over a hitching rail, stood near the front corner of the house. Chickens, horses, and goats cropped the sparse grass in the sandy yard while several black children tended the stock. A splintered reed fence enclosed the yard where four or five hunters, ranchers, and their wives lounged next to the front porch on stained mattresses with stuffing peeking through. Dad Riggs strode across the yard to meet us, his weathered face creased with a grin. His wiry blond hair, flecked with grey, curled like wood shavings, and a clipped moustache was crimped over the firm line of his mouth. When Dolene introduced us, he lay a heavy arm around Delia's shoulder and wagged the stump of an index finger, lopped to keep the venom of a snakebite from spreading. "Make no mistake," he said, "*anybody* who lives in the Kalahari is always welcome at my place. Make *no* mistake."

Dad drew us into his circle of friends, and before we could take our places on one of the mattresses, Cecil, his son, a hard-riding, hard-drinking cowboy, pressed cold beer into our hands. Delia sat stroking a goat with long white hair and nervous yellow eyes that was tethered to a water spigot. We listened to more stories of lion, elephant, and buffalo hunts; of clients who'd never held a rifle before coming on safari, about the wounded buffalo that had gored Tony, about the biggest lion, the biggest elephant, and the biggest rifle, the four-five-eight. There was talk of cattle buying and cattle selling, the Rhodesian war, and how Roger had clobbered Richard with a Mopane pole for fooling around with his wife . . . gales of laughter and more beer. Donkeys brayed, dogs barked, and gumba music played from the native huts in the village beyond.

The rusty wire gate squawked and Lionel Palmer and Eustice Wright, magistrate-cum-rancher, strutted into the yard. Eustice's corpulent belly strained at a gap where buttons were missing on his shirt, and he wore great baggy shorts that waddled above legs like knotted walking sticks. His face florid from heat and exertion, he flopped onto a mattress next to Lionel, lit a cigarette and inhaled deeply. "I knew you bloody bastards would be drinking a peasant's brew," he croaked, pouring himself half a water glass of Bell's Scotch whisky from a bottle tucked under his arm. He couldn't understand why he had consented to join such riffraff... "the bloody flotsam and jetsam of Maun!" He downed a generous gulp of whisky, then belched, "Chrrrist! What the bloody hell am *I* doing here anyway!" Everyone cheered.

Lionel and Eustice, the top two social figures in the village, triggered stories of old Maun. "Were you here, Simon, when Lionel and Kenny, here, stole the cash register from Riley's? Yessus, but Ronnie wasn't happy *that* week."

It was Maun's very first cash register and Ronnie Kays, the bartender at Riley's, was very proud of it. While his brother Kenny chatted him up one day, Lionel, Cecil, and Dougie Wright, all professional hunters, grabbed the register, ran to Lionel's truck, threw it in the back and drove away. Ronnie was not amused; when it was finally returned he bolted it to the bar top. Several days later, Lionel and the others walked calmly into Riley's with a winch cable hidden behind them. When Ronnie's back was turned, they threw a loop around the cash register and signaled to someone waiting in the truck beyond the veranda. The cable snapped tight, and the register leaped from the bar top and bounced out the door. It was considered very poor sportsmanship when Ronnie reported the theft to the district commissioner of police.

"I remember one night when Dad got pissed as a lord at Palmer's," Simon led in with his clipped British accent.

"Nothing unusual about that," Cecil laughed.

"Except we took him home and put him to bed with a donkey foal we picked up in front of the butchery."

"He thought it was bloody Christine!" Everybody roared, and slapped Dad on the back.

When I could get Eustice aside, I asked him if he knew of anyone who might work for us. "Who the bloody hell would want to stay in the middle of the bloody Kalahari?" he laughed. "He'd have to be bloody bonkers."

"What about it? Can you help us find someone?" I asked.

"'ell, Mark, that's a tough order. These blokes don't like to be by themselves, you know, and especially not when there are lions crawling about." He took a heavy drag on his cigarette. "Wait a minute, there's a bloke everybody calls Mox—I practically raised him; worked for me for bloody years. A quiet sort, unless he's drinking. Then he terrorizes the whole village. Chr-r-i-i-ist! The guy's a regular Attila the Hun, a real piss-cat. He's got a hell of a reputation with the women . . . they aren't safe from the bugger when he's on a toot. Started getting pissed on *buljalwa* every day, so I sent him to the cattle post to work with Willy. He might go with you, and out there where he couldn't get ahold of any booze he might work out all right. Stop by my place noonish tomorrow, I'll send for him and you can see if you want to take him with . . . and if he'll go."

Dad pushed himself to his feet and announced, to no one in particular, that it was time to get on with the business of the braii. He grabbed the goat by its horns, dragged it to the center of the yard, and slit its throat with a sweep of his knife. It gave a short bleat and sank to its knees spewing blood. I swallowed hard, and glanced at Delia's startled face.

Using a block and tackle hanging from a tree limb, Dad and the Africans hauled the carcass up by its feet, and put half a truck tire beneath its head to catch the drizzling blood. *"Gotsa molelo!"* Dad bellowed, as he knelt rinsing his knife and hands under the faucet. The Africans lit a large pile of mopane logs. The men all gathered round, sawing at the carcass with knives until the goat was quickly reduced to a pile of meat, its skinned head with its bulging eyes perched on top.

Evening shadows grew as tall as the tales, and beer after beer followed yarn after yarn around the bonfire. Daisy, Eustice's Botswana wife, raked coals out of the fire and set a heavy iron pot of water on them. Orange sparks showered into the night. When the water was boiling she stirred in handfuls of mealie-meal with a large wooden spoon. Meanwhile Dad and Cecil shoveled out more coals and laid a big square rack of goat meat over them. By the time the mealie-meal had cooked down to a thick paste, called "pop," the meat was brown and sizzling. We gathered around to eat. Everyone gnawed on goat chops in the firelight, and the grease ran down chins and dripped from glistening fingers.

We were reliving an important part of our evolutionary history as

social carnivores. The hunting stories, the fire, the drink, the camaraderie—all a legacy from the first frail prehominids who descended
from the trees, leaving behind their herbivorous existence in the forests
of Africa to venture onto the savannas. Ill-equipped as they were for
stalking and killing dangerous game, there was a great advantage to
hunting cooperatively, both for the sharing of meat and for the development of language skills for communicating the details of the stalk
and kill. Sharing food while verbally reviewing their techniques reinforced the essential cooperation between hunters, as well as encouraging and teaching the young. Powerful social bonding, together with
the evolution of superior intelligence, made human beings the most
successful carnivores the earth had ever known. As I took part in this
primitive ritual I was reminded that this part of our fundamental nature
hasn't really changed that much in thousands of years.

* * *

After the braii, Dolene and Simon invited us to sleep at Buffalo Cottage. Their bungalow, set on a bulge in the river, was smothered in
flowering bougainvillea, and the head of a cape buffalo glared from
over the door. Our bedroom had a sweeping view of the water, and
clean sheets and towels had been laid out for us on a kaross of jackal
pelts. Before we went to sleep, I ran my fingers through the silky
black and silver hair of the kaross. It had taken thirty Captains to make
it.

Next morning we were awakened by the smell of tea and spice
cookies, as the bare feet of a small native boy padded off down the
hallway. Later we shared a breakfast of toast, orange marmalade, and
more tea with Simon and Dolene, on the veranda overlooking the river.
Several other hunters, bleary-eyed and thick-tongued from the night
before, arrived, and Simon called for another pot of tea.

They all asked us to go fishing the next day, but we felt pressed to get
back to the Kalahari and regretfully declined. "Right then—piss off to
the bloody Kalahari," someone said in jest. Though we knew it hadn't
been a serious jibe, it still hurt, and we worried that perhaps we were
being antisocial. The talk ran to Dad's braii and plans for the fishing trip
and now and then someone mentioned the job to be done on Simon's
truck. We excused ourselves and began shopping for supplies.

The stores in Maun are low concrete block buildings with tin roofs
set along the footpaths and major tracks that run through the village.

We usually had to visit every one in order to find all the food staples and bits of hardware needed for another few months in the Kalahari. Items as basic as truck tire tubes and patches were often unavailable, and we had to arrange with one of the transport drivers to purchase them in Francistown and bring them back with his supplies for the village. Sometimes, when the gravel road was bad, we waited for days to get gasoline and other essentials. Only recently have such perishables as cheese, bread, eggs, and milk become available, with the arrival of the first refrigerators in one or two of the shops.

We made our way from one store to the next along deep sand ruts, jamming on the brakes, banging the sides of the Land Rover, and whistling away the donkeys, dogs, goats, cattle, and children, who sometimes dared us to run over them by dashing in front of the truck. At Maun Wholesalers, also known locally as Spiro's (for the Greek who owned it), a horse was tethered to a rail. A goatskin saddle, a bedroll of plaid wool blankets, and a skin sack of goat-milk curds were tied over its back with rawhide strips. The shop was a single large room lined with shelves of rough-cut lumber; a heavy wooden counter ran nearly all the way around it. One end was stacked to the ceiling with canned foods, bars of Sunlight soap, boxes of Tiger oats, Lion matches, packets of lard, tins of Nespray powdered milk, and other grocery items. There were shirts, pants, cheap tennis shoes, and yards of colorful fabric. Metal bins of bulk flour, mealie-meal, *samp*, and sorghum stood along the front of the shop, and an iron scale with a copper pan and sliding weights sat on the wooden counter above. Saddles, bridles, hosepiping, chains, and kerosene lanterns hung from the rafters.

Two tall, somber-faced Herero women swayed into the shop. In spite of the heat, they were dressed in brightly colored flowing dresses, purple shawls, and red turbans. Puffing on their pipes, each filled a Coke bottle from the spigot of a kerosene drum; a tin bathtub caught the spills. The shop was crowded with men, women, and children lying across the counter, holding out money and shouting their orders to the shop assistants.

Delia began taking cans from the shelves. A young black girl, her dress hanging off one shoulder, tore a strip from a brown paper bag, brushed away the mealie dust on the counter, and began to tally with a pencil stub. I found a hefty axe handle in a tumble of three-legged pots, kettles, tin bathtubs, shovels, and picks leaning against the wall.

Later, when we unloaded in camp, we found that our three months' supply of flour and sugar had been stolen from the truck, along with some other grocery items. We were furious. On our limited budget, there was no going back to Maun until our next regularly scheduled trip, so for three months we were without bread, an important part of our diet. Since most of the door and window locks on the old Land Rover were broken, there seemed to be no defense against being robbed, other than for one of us to watch our goods every minute while in the village. But on our next supply trip, I solved the problem for good.

When we were getting ready to go, the birds in camp suddenly burst into a cacaphony of chirring, twittering alarm calls. We soon spotted two ten-foot mambas winding their way into the trees above our dining tent, apparently intending to have a couple of our feathered friends for lunch. The black mamba is so poisonous that the natives use a term for it that literally means "two step"; according to them, two steps is as far as you ever get if you're bitten by one.

I shot the snakes, took them to Maun with us, and coiled them over the pile of fresh supplies in the back of the Land Rover. It didn't matter that they were dead. The first curious young lad who strolled past our truck, trailing his fingers along the side of the window, suddenly leaped back with a yell and disappeared through the village. The word soon got around that the Grey Goose was not to be molested.

*   *   *

There were two butcheries in Maun, both owned and operated by Greek merchants: the "Maun Butchery" and "Dirty George's," the latter known for its flies, general lack of sanitation, and cheaper meat.

We always shopped at the Maun Butchery, though there was actually little difference between it and Dirty George's. These two places were the only sources of fresh meat for villagers who had once lived by hunting large herds of antelope, most of which have now been displaced around Maun by monocultures of cattle, goats, and sheep. Two towering tribesmen in blood-smeared boots and aprons heaved slabs of stringy, tough meat from blocks to scales to counter, slicing off portions as demanded by customers. I often wondered if the poor quality of the meat was due to the reverence that native ranchers had for cattle on the hoof. A cow is worth much more to them as a display of wealth when it is alive. Many Maun people believed that only the oldest and most infirm and scrawny beasts found their way to the butchers.

When we had finished buying supplies, we drove north on the track to Eustice's small farm at a bend in the Thamalakane. The small frame house stood high above a broad sweep of the river. Behind the house a big yard with tall trees and a vegetable garden fell gently to the water and reed banks below.

As we drove up the long, sandy driveway, Eustice appeared from a side door near the garden. A slender black man in his mid twenties, medium in height and wearing a floppy hat, stood under a jacaranda tree next to his canvas tote-bag. I shook his hand, noticing that he had strong arms and shoulders, but that his legs were those of a gazelle, long and slender.

While Eustice interpreted, I explained that we lived in a camp far away in the Kalahari beyond the Boteti River, and that if he came to work with us, the life would be hard: There was very little water; he would not see other people for months; there would be lions in camp some nights; and we could afford to pay him very little other than his food. The only shelter we could offer was what we could make from a ten-by-twelve tarp. He would be expected to patch tires and generally help maintain the truck. He would help keep camp clean, assist me in hauling water and firewood, and lend a hand with the research when we needed it.

Throughout our one-sided conversation I could see that Mox was extraordinarily shy. He stared at the ground, unmoving, with his hands hanging awkwardly at his sides. Now and then when Eustice asked him something he would utter, "*Ee*"—little more than a whispered assent.

"What does he know how to do?" I asked Eustice. "Can he patch a truck tire or cook?"

"He says he doesn't know how to patch tires, but I taught him to cook a little. He'll try to learn anything you want to teach him."

"Can he spoor—track animals?"

"No, but... 'ell, Mark, any of these blokes can pick that up, but quick."

"Does he speak any English?"

"No."

I glanced at Delia. We were both skeptical. How could we possibly work with someone this shy, who was too embarrassed even to look at us, and who had no skills and was unable to speak our language? According to Eustice, Mox had spent most of his twenty-six years tending cattle for thirty cents a day. He lived with his mother, to whom he gave everything he earned. His father was a skinner for Safari South.

Without skills or the ability to communicate with us, we didn't see how Mox could be of much assistance. If he did agree to come to the Kalahari with us, I felt sure he wouldn't last for more than one three-month stint. But we needed help badly, and for what we could afford to pay, we wouldn't be able to hire a skilled worker.

"Ask Mox if he'll come to the Kalahari for fifty cents a day and food—that's twenty cents more than he's getting now. If he learns well, and if we get another grant, we'll give him a raise."

Eustice rattled off more Setswana, and Mox raised his eyes to me for the first time. They were bloodshot from native beer. Through a ragged cough he whispered, *"Ee,"* and we arranged to meet him at Safari South the next morning.

We had finished packing the Land Rover by the time the sun rose over the river. While we were thanking Dolene and Simon for their hospitality, the others in the fishing party arrived—fishing rods, shot-guns, folding chairs, old mattresses, and cases of cold beer in the back of their trucks. Simon insisted that they should all have tea and biscuits before leaving. We said our goodbyes, and they all settled down around the table. Someone said that if they got back before sunset, they might work on Simon's truck.

We found Mox sitting next to the sandy track on a small bundle of blankets. Inside the bundle was an enamel bowl, a knife, a piece of hardwood for sharpening the knife, a piece of broken comb, a wooden spoon, a splinter of mirror, and a small cloth sack of Springbok tobacco; it was all he owned in the world. Wearing holey blue shorts, an open shirt, and shoes without laces, their tongues hanging out, he climbed to the roofrack and sat on the spare tire.

It was about 9 A.M. when we passed Buffalo Cottage on our way out of Maun. The hunters were still drinking tea on the veranda, and, jacked up on blocks, Simon's truck stood beside the house.

That night we spread our sleeping bags and set our chuck box on the ground near the edge of the Central Kalahari Game Reserve. Mox built a fire, and Delia made a supper of goat meat, mealie-pop, and tea. We sat eating quietly, happy to be back in the bush, yet both of us feeling somehow lonely, missing the warm afterglow that usually follows a visit with good friends. "I feel let down... as though we don't really belong anywhere but Deception Valley," Delia said sadly.

We had gone to Maun not only to get supplies, but also to socialize. Yet, in spite of the generosity of the people in the village, we had come away disappointed and unfulfilled, still feeling that we weren't really a part of any group. We were both overly friendly after long stretches alone in the Kalahari. Our Maun friends did not respond to us with the same *exaggerated* enthusiasm. In contrast to us, they were casual, and we misinterpreted this to mean that we weren't really accepted by them. And since there was no other social circle, it was important that we be accepted. This anxiety grew over the years, and we gradually turned more and more exclusively to each other.

*   *   *

It felt strange to have someone else around the campfire. Mox was silent, totally unobtrusive, yet we felt his presence, as though the shadow of a person was with us but not the person himself. We tried to communicate with him, using what little Setswana we had learned, aided by a phrasebook published by Catholic missionaries. He never spoke unless a question was addressed to him, and then very softly, hardly daring to glance at us. His answers were mostly, "*ee*," and "*nnya*." Still, we were able to understand some of what he knew about the world.

Though he had lived all his life on the edge of the Okavango River Delta and the Kalahari, he had only been into the delta during a few hunting trips with Eustice, and he knew little about the Kalahari. Using a stick to draw a picture of the globe in the sand, we tried to explain that the earth was round and we were from America, across the ocean on the other side of the world. But he half smiled and shook his head, his brow furrowed in embarrassed confusion. He didn't know what "world" or "ocean" meant, in his language or any other. He had never even seen a lake, much less an ocean, and his world consisted of little more than what he could see.

Much later, when the fire had died to embers, I lay on my back looking up at the blue-black star-filled sky. Had we made a mistake? Had Mox? Why had he left the security of the village and his family—his social group—for the unknown of the Kalahari Desert? I turned the question back on myself. Far away to the south, somewhere near Deception, a lion roared.

8

Bones

Mark

A king of shreds and patches
—*William Shakespeare*

SWAYING and ducking tree limbs, his wiry black hair plastered with grass seed and straw, Mox rode on top of the Grey Goose through the woodland of East Dune and into Deception Valley. At the edge of the riverbed, still half a mile from camp, we could see that something was wrong. We raced toward our tree island, the truck shaking and rattling over grass clumps, and found pots, pans, bits of clothing, pieces of hose, sacks, and boxes scattered for hundreds of yards around. Camp was a shambles.

A big dust devil? A heavy storm? What or who could have made such a mess? I began picking through the rubble and found one of our heavy aluminum pots with a large hole, the size of a fifty calibre bullet, in the bottom. Just as I realized the pot had been punctured by a big tooth, nine furry heads emerged to peer at us from a thorny hedge west of camp. We were still standing next to the Land Rover when the lions started strolling toward us in a long single file. Two large lionesses led the way, their bodies swaying with easy power, and five slightly smaller subadult females strode confidently after them. Two yearling male cubs, biting at each other's ears and tail, brought up the rear. This was the same pride I had herded into West Prairie with the Land Rover the night Delia had been trapped in the tent. We had seen them many times since then in this part of the valley, and apparently they were the ones who had raided us.

Like justices arriving in court, they slowly took their places side by side in a half-circle on the perimeter of camp, not more than twelve or fifteen yards away. Licking their paws and washing their faces, the lionesses watched us with idle curiosity and without apparent fear or aggression. Besides the excitement and mild apprehension we felt with them so near, there was also the sad feeling that it would end all too soon, whenever they chose to go away.

Mox felt otherwise. While Delia made a fire and put on a pot of soup, he and I began, with slow, cautious movements, picking up the pieces of camp. But he kept the Land Rover between himself and the lions, and he never really took his eyes off them.

Later, we drove Mox to another stand of trees, 150 yards south of our own, to help him rig a tarp shelter beneath a spreading acacia tree. We lashed together deadwood poles for a frame, and with the canvas fastened over it, a crude but snug hut took shape. To "lion-proof" the camp according to Mox's native custom, we cut wait-a-bit bushes, a shrub with wicked clawlike thorns that snare the flesh and clothing of those passing by, forcing them to "wait a bit" while they unhook the needle-sharp spines. We piled the thornbush in a tight *boma*—a circular enclosure—leaving an entrance for Mox that he could close with a single large bush. When he seemed satisfied with the comfort and security of his quarters, Delia and I went back to our camp, leaving him to make his bed and arrange his belongings.

The lions stood up as we drove in, but then lay down again. Around sunset Delia served up steaming potato soup, and while mealie cakes sizzled in the black skillet, the lions watched everything we did, seldom moving, except to yawn or lick a forepaw.

This was a valuable experience for us, and we took careful note of the way they reacted to what we did: Their widened eyes and tense shoulder muscles expressed fear if we walked too quickly or too directly toward them. Their chins lifted, their ears perked, and their tails twitched with curiosity when I dragged a branch toward the fire. Each posture and expression told us more about how to avoid inciting fear, aggression, or undue curiosity in them.

The cool evening air began to settle from the sandslopes into the valley, and the last colors of sunset were draining from the sky behind West Dune. The forms of the big cats grew dim, lost focus, and finally faded. As it grew dark, more primitive, less scientific feelings settled in on us, and I switched on the light to check the positions of the lions. To our surprise only one large lioness and the two yearling males

were left; the others had quietly disappeared. Despite the protection
his thornbush boma provided Mox, we had to make sure that he was
safely in his shelter.

When I swung the spotlight around, pairs of glowing amber eyes
leaped into view—the lions were roaming all around Mox's camp!
We jumped into the truck, but by the time we got there, three females
had already picked their way across the boma and had their noses to
the canvas. Two others were on the opposite side of the tree, and the
last, and largest, sat on her haunches near the entrance to the circle
of thornbush.

I stopped the Land Rover behind the shelter and put the spotlight
on it. "Mox, are you all right?" I whispered as loudly as I could. No
answer.

"Mox!" I called again, louder. "Are you okay?"

"Ra?" But his voice didn't come from inside the canvas.

"Mox, where are you?" Then I noticed the large lioness near the
boma staring into the tree above her. I followed her gaze with the
spotlight until I found Mox, grinning nervously, perched naked on a
branch not ten feet above the big cat!

I eased the truck between the base of the tree and lioness. She gave
way without a grumble and then came and sat next to my door, peering
at me through the open window. In one smooth motion, as though the
trunk of the tree were greased, Mox slid down, grabbed his shorts,
put them on, and shot safely into the Land Rover.

"Tau—huh-uh." He shook his head, shivering. He mumbled something about Maun as we drove slowly away. We stayed in the truck
until the lions lost interest and moved north up the valley.

In the morning the sweet smell of woodsmoke was drifting through
camp when I opened my eyes. Delia was still asleep beside me. The
muffled clatter of Mox busy with dishes in the kitchen was pleasant,
and it stirred memories of mornings on the farm, my brother, sisters,
and I waking to the smell and sounds of Mother making breakfast
downstairs. It was still early, and Mox's apparent eagerness to get
busy with the details of his new job temporarily eased my doubts about
whether he would last in the Kalahari. I pulled on my cut-offs and
sandals and stepped outside. Clucking hornbills and fluttering flycatchers landed on the branches along the path, begging for their morning's
ration of mealie-meal.

At the kitchen I found Mox sitting on the ground among pieces of litter and garbage scattered about by the hyenas during the night. Carefully picking his way through the mess, he had done the dishes and stacked them neatly away on the table. Now he was innocently cleaning his toenails with the point of our best kitchen knife.

*　　*　　*

During breakfast, lions began roaring in the valley to the north. It was now early May 1975, and although the Kalahari was still receiving scattered rain showers, the dry season would soon be upon us. Within a month or six weeks the lions would be gone, traveling through lands unknown. No one knew how far, or in which direction they migrated, and we wondered if they would return to Deception. If so, would they defend a part of the old river valley as their territory? How could we recognize them if they did come back, especially after such a long time away and at night, when we would be most likely to encounter them?

We had come to realize that at least during the rainy season these predators were the hyenas' chief source of carrion, influencing the diet and range movements of the scavengers to a great degree. If the brown hyenas depended on them for food, it was important to know more about the lions in Deception Valley. Though we still continued our night observations on the brown hyenas, we decided to begin watching local prides more often, to learn what we could about their habits and their ecological relationships to the hyenas.

The best way to be certain of their identities, if and when they came back, was to mark them with ear tags. Perhaps if any of them were shot, the colored plastic discs would be taken to the Department of Wildlife, or they might turn up on some Bushman's necklace. At least it would be a chance to learn how far from the riverbed the prides traveled during the dry-season migration, and how many were killed by man and who was responsible. Finding out which individuals stayed together would reveal something about Kalahari lion social organization, which had never before been studied or described in detail. The immediate problem was to ear-tag as many as possible before they left the riverbed, probably within a night or two, and to keep from alienating them in the process. Above all, we didn't want to permanently alter their natural behavior.

While we were getting our darting equipment ready, we agreed on some ground rules that we hoped would minimize trauma to the lions during immobilization. Whenever possible, we would dart only at night to avoid exposing them, in their unconscious state, to the extreme heat of day. We would tranquilize them only while they were preoccupied with feeding on a kill, and only after we had sat with them long enough to insure their complete habituation to us. We would never chase them with the truck or manipulate them with cowboy tactics (which often result in a great deal of stress, alienation, and even death). A minimal drug dosage would be used, and we would work as quietly and quickly as possible to avoid stressing them unnecessarily. As with the hyenas, our criterion for complete success was an ear-tagged lion who showed no more fear of us or the Land Rover than it had before the procedure.

Time was short, so we would dart as many as possible in one session and Mox would hold the spotlight while we tagged them. But we could cope with no more than three to five in various stages of unconsciousness and recovery at one time, and meanwhile, the others would be roaming about in the darkness nearby. We had never darted lions before, and didn't know how they would react.

Together with Mox, we found the pride lying in Leopard Island, a clump of acacia and ziziphus trees on the west edge of North Pan near Cheetah Hill. Approaching them in a half-circle with the truck, we slowly worked closer until, at about fifteen yards away, they lifted their heads and began to look nervously about for an avenue of escape. Lions and other wild animals are generally much less disturbed if they have the initiative, as when the pride had strolled into our camp. Now, as we moved in on them, they began to feel a bit threatened. I switched off the engine. They immediately relaxed, blinking their eyes and yawning. For the next several hours we sat quietly, letting them get more used to us and hoping that they would make a kill.

We had seen these lions many times and had given each of them a name. The two older females were Blue and Chary. Blue was always preoccupied with chewing the tires of the truck; fortunately they were of a heavy ply and she never punctured one. Chary, a big lioness with a sagging back, was the oldest female in the pride, and for some reason a little wary of us. Even when she put her head on her paws to rest, she never completely closed her eyes.

Of the five subadult females, Sassy was the most bold and curious. She had a broad chest and large frame that promised someday she would fill out to be a very large lioness. She often stalked the truck as if it were prey, creeping slowly up to its rear, ready to spring if it tried to get away. But if our departure didn't happen to coincide with her game-plan, she would find herself jaw to bumper with an animal she didn't know how to tackle. Then she would stand up from her stalk and pat a tire with her paw once or twice or chew on a fender or taillight. Once I forgot that she was still involved with her game and started the motor. Taking a shot of exhaust in the face, she leaped back and then hissed and spat at the tail pipe. She was fascinated with the rotation of the wheels and invariably, when we drove slowly away after watching the pride, she would hurry to the side of the Land Rover and watch them go round and round, her eyes rolling and her chin bobbing in time with each revolution. After that she would trot along behind us, crouching as if looking for a place to deliver a killing bite. Sassy was a favorite with us.

Gypsy could never sit still, and whenever the pride visited camp, she would roam its perimeter or leave to be by herself for a time. Spicy, who once mock-charged me, was the color of cinnamon, and she was pugnacious. Spooky had great round eyes, and Liesa was small, neat, and pretty. The two yearling males, Rascal and Hombre, were constantly irritating one of the adults and getting slapped across the nose.

When we were sitting with the lions that evening, Chary seemed to sense something was being planned, and she moved a few yards farther away to lie under a small bush. While the others slept, she watched us.

Around nine o'clock she lifted her head and began staring alertly into the distance, the muscles of her shoulders tensing. The other lions were immediately still and attentive, looking in the same direction. I raised the spotlight and saw an ostrich picking its way carefully through the bush at the base of North Dune. I doused the light. Chary rose slowly to her feet and began stalking toward the big bird. Gliding like a serpent through the grass, she disappeared into the night. One by one the other females followed; Delia and I were left sitting alone in the moonlight. We didn't want to follow the lions or use the spotlight, for fear of influencing the hunt by confusing them or their prey. The minutes dragged . . . we were anxious to know how the chase was going.

About three-quarters of an hour after they had left us, snarls and throaty rumbles carried from the bushes on the duneslope. The lions were still quarreling over the ostrich when we parked the truck close to where they were feeding in a circle around the carcass. They turned to glare at us, and Chary and Spicy rose to a crouch and laid their ears back, obviously resenting the intrusion. Their muzzles were plastered with blood and feathers and their broad paws were clamped possessively over the bird as they tore out big chunks of red meat. I reached slowly for the ignition key and switched off the truck. They turned back to their kill and settled down on their bellies to feed. We were within fifteen yards of them, so there would be little chance of missing with the old dart gun.

Delia held the spotlight while I fumbled with equipment boxes, syringes, and the bottles containing the drug, trying as best I could to be quiet as I filled the darts. I was all thumbs, and the steering wheel and gear shift kept getting in the way. I felt as if the lions were watching over my shoulder through the low half door beside me; I had taken out the window and frame so that I could swing the rifle around if one of the targets moved past the truck.

Chary's dart was finally filled with its mixture of phencyclidine hydrochloride and xylizene. As she was the most wary, it was important to put her down first so that she couldn't alert the others. Several minutes went by while I sighted down the barrel of the rifle, holding on the line of Chary's back. But she was lying on the opposite side of the carcass behind Hombre, and I couldn't get a clear shot. My hands were sweaty on the gunstock, and it all seemed so unreal, poking a 350-pound lion in the shoulder with a needle from a few yards away.

The lions spat and swiped at each other. Then Chary abruptly rose, looking huge in the spotlight, and turned to step over Blue, who was feeding on her right. I strained to see the sights, lined them up, and squeezed the trigger.

The gun popped and the dart hit the lioness in the shoulder. There was an eruption of snarls and growls. Dust and feathers flew as the pride leaped into the air and across the carcass, their tails lashing like whips. For a tense moment, we froze, half expecting one or more of them to charge the truck. They were milling about, looking from truck to carcass, then off into the night and at each other, trying to place the cause of the disturbance. Suddenly Chary slapped Blue across the nose; a score was settled, the tension broke, and the lions resumed feeding. We settled back in our seats to wait.

Ten minutes after the shot, Chary's eyes began to widen and her pupils enlarged. She left the kill and staggered off to a spot in the thick bush, where we could barely see her. Mox, who seemed to have the eyes of a cat himself, kept watch on her while we darted Blue, Gypsy, and Liesa in succession, allowing time after each shot for the others to relax and resume feeding. Before long, four darted lionesses were lying scattered within a fifty-yard radius of the kill; the other five, young females and cubs, kept feeding.

By now it was almost forty minutes since Chary had gone down. She and the others would begin to recover within an hour after being darted. We quickly drove to the place where Mox had last spotted Chary, and after several minutes of shining the spotlight from the top of the truck, we found her near a stand of bushes; her beautiful amber eyes wide-open. Her ears twitched to the sound of the approaching truck, and she raised her head slightly.

I parked the Land Rover about ten yards behind her, switched off the engine, and stepped out, all the while wondering if I was doing the right thing. Since I wasn't sure how immobilized Chary was, I wasn't keen on walking up to her. My feet rustled the dry grass, and she jerked her head. If she could hear, some of her other faculties must still be functioning, so I clapped my hands twice to test her reflexes. She didn't respond. Stalking carefully up to her, I finally crouched beside her tail, still ready for a flat-out retreat to the Land Rover, and gently nudged her big rump with my toe; she seemed to take little notice.

I signaled okay to Delia, who handed the spotlight to Mox in the back of the Land Rover, and got out to take the equipment to me. The lioness at our feet was the color of dry-season grass, sleek and powerful, her body as solid as the trunk of an oak. We regretted taking advantage of her like this, after having gained her trust. While Delia felt inside her upper foreleg for a pulse, I hurried to squirt salve into her eyes in order to keep the corneas from drying out. Chary was lying on the dart, so, holding her broad paws and using her legs as levers, we rolled her over. Delia dressed the small wound with salve, and I clipped a tag in her ear.

By the time we finished with Blue and Gypsy, more than an hour and a half had passed since we'd begun to dart. Chary and the others were regaining coordination. Furthermore, the undarted lions had sated themselves on the ostrich kill, and they were getting more interested in us, and in what was happening to their pride-mates. There were big cats prowling everywhere, and Liesa still had to be tagged.

When we found her, Liesa was supporting herself unsteadily on her forelegs; she could almost stand up. We would never be able to tag her without first injecting a supplemental dose of tranquilizer. But we didn't want to subject her to the trauma of the darting rifle a second time, especially with all the other lions watching. I went back to the truck and prepared a syringe. *"Go leba de tau, sintle*—watch the lions closely," I said to Mox, and I slipped off my shoes, opened the door, and began crawling toward Liesa.

Both Delia and I knew that this was probably not wise, but we were afraid that if we tried to drive the truck toward the half-drugged lioness, the sound of the engine might terrify her. And if she began charging around the area, we might alienate the entire pride.

Take your time, and don't make any noise. Watch where you put your hands and knees! I thought as I crawled away from the Land Rover, my shadow stretching in front of me, almost to the big swaying form of the lioness, who was sitting on her haunches, looking away. I felt my way quietly through dry leaves and grass tussocks, and twigs that would snap like cap pistols if I put my weight on one. The farther I got from the truck, the more I knew that this was foolish. I was tempted to turn back, but I convinced myself that the lioness, partially drugged as she was, would probably not even feel the injection. I was depending on Mox and Delia to warn me if any of her pride-mates became a more immediate threat. I would bolt for the Land Rover at the first sign of trouble.

When I was about five yards from her my knee crushed some dry leaves with a loud crackle. Liesa swung around and looked directly at me. I froze, waiting for her to turn away, but she held me with her yellow eyes, her ears perked, weaving unsteadily, saliva dripping from her chin whiskers. I was afraid to make even the slightest move; her eyes seemed to narrow as they tried to focus.

"Tau, Morena!" Mox whispered urgently from the Land Rover, warning me of the approach of another lion.

To my right, about twenty yards away, one of the undarted lionesses was stalking me through the bush, crouching, head low and tail twitching. I flattened my belly into the sharp bristles of a grass clump and pressed my cheek to the sand, trying to get out of sight. My pulse hammered in my ears.

It was too far to the Land Rover to run for it; the undarted lioness was too close. I put my arm over the back of my neck and closed my eyes, trying not to breathe, but inhaling sand and ash up my nose. I could see the crooked arms of two hunter friends in Maun who had been foolish with lions, and the broad platter of lumpy scar tissue, from hip to breast, in the side of a little Bushman who had been dragged from his hut. I hugged the ground and waited.

"Mox! Hold the light in her eyes!" I could hear Delia's insistent whisper in the silence. Though he spoke almost no English, Mox understood what to do. He turned the spotlight and fixed it full in the eyes of the stalking lioness. She stopped, half stood up from her crouch, and squinted into the beam.

Liesa must have heard her pride-mate, for she looked in that direction. Seeing a chance, I got slowly to my knees and began backing toward the truck, trying to be quiet in the dry grass. The undarted lioness was raising and lowering her head, still trying to see me through the bright light. She started forward. I flattened myself again. Mox kept the spot on her until she stopped, blinking her eyes, not more than ten yards from me. Feeling like a mouse under the nose of a house cat, I raised my belly from the ground and began crawling in reverse, my arms rubbery with fear. Then the front bumper of the truck was beside me. I lunged through the door and sank back in the seat. I wiped the grit and sweat from my face with a shaking hand.

The stalking lioness finally lost interest and moved back to the ostrich carcass. After what I had just been through, I was somewhat less concerned about frightening Liesa. I drove the truck forward and stopped next to her rump, slipped my arm out the door, and injected her with more drug. About ten minutes later we tagged her. We waited near the ostrich carcass until all the lions had recovered from being immobilized and then went back to camp to get some sleep.

The next evening, the entire pride, except for Chary, came into camp and circled the Land Rover, smelling its tires, bumpers, and grill. They seemed unaware of their blue plastic ear tags, each with a different number. They were the Blue Pride.

*　　*　　*

As much as we hated darting, on one occasion, at least, it allowed us to learn something about the strength of the social bond between male lions. Having grown up with each other, Pappy and Brother had traveled large expanses of the Kalahari together as nomads, without having found a pride of their own to rule. Young males, often brothers, frequently stay together as adults, and these two seemed inseparable.

Immobilizing Pappy was a routine operation. After the shot he sagged to the ground and went to sleep on his side, the dart dangling from its needle just behind the shoulder. Brother, his head raised and his eyes wide, had watched intently as his partner had lost coordination and then consciousness. He looked from Pappy to us, and back to Pappy again, as if trying to understand. Then, ignoring the truck parked just eight yards away, he walked to the downed lion and sniffed along his body until he found the dart. Clamping it between his front teeth, he backed up and pulled. A cone of Pappy's skin clung to the needle and then finally popped free. After chewing the dart up and spitting out the pieces, he walked to his companion's side and licked the small wound made by the needle. He rubbed his head against Pappy's, cooing softly. Then, lowering himself on his forequarters, he took Pappy's neck gently in his mouth and began to lift. But the other's bulk was too unmanageable. After struggling in this way for more than a minute, Brother put his jaws over Pappy's rump and did the same thing; then under his neck again, while cooing. For fifteen minutes he went first to one end, then the other, trying to lift him with his mouth.

Was he trying to stand his companion back on his feet? It certainly looked like it, though we can't be sure. We know that elephants occasionally attempt to lift a fallen family member, and it does not seem unreasonable to suppose that lions might try to do the same thing.

We were greatly moved by this, but Brother was so persistent that we worried that his canines might injure Pappy's neck. I eased the Land Rover to Brother's side and maneuvered him far enough away so that we could ear-tag, weigh, and measure Pappy. Then we rolled the immobile lion onto a tarp, lashed the corners to the back of the truck, and dragged him to a shade tree, where he would be cool during his recovery. Brother followed and lay nearby until Pappy regained consciousness. Then he eagerly rubbed his head and muzzle all over his fallen comrade.

*　　*　　*

"*Tau, Morena!*" It was early morning, only a few days after we had darted the Blue Pride. We had followed hyenas late into the night before and were asleep on the tent floor when Mox roused us. He stood in a patch of sunlight just beyond the flap, pointing to a lion 300 yards east of camp. Through the tent door we could see the male's tottering figure, hunkered over the months-old remains of a gemsbok, as he tugged at the few bones and the brittle skin. Normally, a lion would hardly have noticed such useless carrion—so parched and tough it could not be eaten. But this male was struggling urgently to get the wizened carcass to the shade of Topless Trio, a clump of trees opposite camp on the riverbed. Through field binoculars we could see that he was terribly emaciated and very weak. The gemsbok carcass must have weighed no more than thirty pounds, yet the lion could not move it more than a few feet at a time before stopping to rest, panting heavily. He would straddle the carcass again and again, trying to drag it forward, without much success. Then he would turn, take the paper-dry skin in his mouth, and pull in reverse, with the same result. Each attempt left him weaker, until he finally collapsed, clearly near death from starvation.

We pulled on our clothes, jumped into the truck, and drove slowly toward him. When we got closer, he stared blankly, hardly noticing us. We were appalled by his condition. There was nothing but the shell of a once proud lion. His ribs stuck out sharply, his skin hung in great folds, and I could have encircled his midriff with my two hands. He must have been suffering for weeks, with virtually nothing to eat.

With an enormous effort he stood up and began staggering toward the Topless Trio, and it was then we noticed a dozen or more porcupine quills deeply embedded in his neck, shoulders, and flanks. While trying to get food in his weakened state, he had probably botched a hunt. He made his way to the shade of the tree island, where he slumped to the ground as though his large, bony head and tattered mane were too heavy for him.

We left him there but returned with the darting equipment in the late afternoon. We wanted to examine the lion more closely and to determine his age and his chances for survival. When the dart struck him he didn't even flinch, and he quickly sagged to the ground. We began removing the festering quills, some of them lodged more than six inches deep.

There was one quill that Delia could not pull out. It was sticking out from the inside of the upper right foreleg, and since the lion lay on his side, the left leg kept getting in the way. Mox was watching from a distance, his hand on the fender of the Land Rover. "Mox, *tla kwano*—come hold this," she called, straining to push the heavy leg aside. Mox shuffled over hesitantly, his eyes shifting nervously. We didn't know it at the time, but as a child he had apparently been taught by his tribesmen that if he touched a lion, his arm would rot off. He believed in this taboo, yet he still came forward.

Noticing his halting, tentative movements with the lion, Delia tried to reassure him. "It's *go siami*, Mox—*go siami*; it's okay." She smiled. Mox took hold of the great furry leg as if it would spring to life at any moment and gently pulled it back. After Delia had removed the quill, Mox, still holding the lion paw, spread his palm and fingers against the great calloused pad. He held it there for several seconds and then looked up, the glimmer of a smile in his eyes.

Dusk had fallen. We had nearly finished taking out the quills and treating the wounds with ointment. I was having problems with a broken quill lodged in cartilage just below the knee of the right hind leg. It would not budge, so I got some pliers from the Land Rover, seized it, and yanked several times. No matter how hard I pulled, the pliers just slipped off. It was getting dark, so I asked Mox to switch on the spotlight. When I could see better, I discovered I had not been pulling on a quill, but on the broken end of the lion's tibia. He had a severe compound fracture.

We were faced with a dilemma: According to the dictates of objective scientific research, we should simply allow the lion to die. Even if we tried to help him, neither of us had been trained to deal with such a wound and, furthermore, our efforts would be hampered by darkness. Yet we had already anesthetized him, and we knew from the small amount of wear on his teeth that he was in his prime, no more than five or six years old. So, although he would probably not survive, we decided to do what we could.

The lion would never tolerate a splint. Our only hope would be to open the leg at the fracture, saw off the splintered end of the bone, sew up the torn muscle, and then disinfect and close the wound. If we could somehow entice him to stay off his feet for a few days, the bone might begin to knit.

We drove to camp and assembled some makeshift surgical tools: a broken hacksaw blade, a razor blade for a scalpel, a dish-scrubbing brush to clean out the wound, and ordinary needle and thread for suturing.

It was dark when we got back. Mox held the spotlight while we opened the wound further, scrubbed and disinfected it, and sawed about three-quarters of an inch off the splintered bone. We stitched the muscle and skin back into place and injected a large dose of antibiotic, then clipped an orange ear tag, numbered 001, into his left ear. We stood back and looked at this pitiful wreck of a lion. If he lived, we would call him "Bones."

He would need moisture and food immediately if he was to survive, but he couldn't hunt without stressing his injured leg. Using an old poacher's rifle loaned to us by the Department of Wildlife, I shot a steenbok, and while Bones was still sedated, we placed the twenty-five-pound antelope under his head, where it would be safe from jackals and hyenas until he regained consciousness. Several hours later Bones began to feed on the meat, slowly at first, then gulping down large red chunks. By dawn he had consumed the entire carcass, and he was sleeping deeply when the sun crept over East Dune.

Bones would soon need much more to eat, and he would try to hunt unless we gave him another antelope. Early that same morning I shot a 530-pound gemsbok and pulled it to him at the end of a thirty-foot chain attached to the truck. Lions—Kalahari lions in particular—are prone to drag their kills to the nearest shade. If I unhooked the carcass too far from Bones he would get up and try to move it, perhaps permanently crippling his leg. The problem was how to get his food supply practically up to his nose without frightening him away or provoking a charge. He was fully alert now and, in his weakened and vulnerable condition, undoubtedly more nervous than he would ordinarily have been.

I had the gemsbok no closer than about twenty yards from him when he began to tense. So I slipped out of the Land Rover, loosened the chain, and drove away. Bones got to his feet and limped over to the heavy carcass. He straddled it and, taking the neck in his jaws, began to drag it, putting full weight on his broken leg. The stitches began to separate and blood poured from the wound. The pain must have been intense.

For an hour and a half he struggled to get the gemsbok into the shade, moving it only a very small distance at a time, then pausing, heaving from exhaustion. He managed to drag it only ten yards before his strength finally failed him. He staggered to his shade tree and collapsed, totally drained. It had been a magnificent, though grueling, performance, one we knew he would repeat unless we could get the carcass closer to him.

Over the next hour I inched the Land Rover back to the gemsbok and rechained it to the ball hitch. Slowly we worked it closer to Bones, switching off the motor when he showed alarm, moving in an arc around the tree, until finally the carcass was within four yards of him. He grew more and more agitated as I backed the Land Rover up to the gemsbok, slid out my door, and crept to the rear of the truck. I slowly reached from behind the protection of the rear wheel and began fumbling with the knotted chain, sweat streaming down my face. Meanwhile, Bones, his shoulder muscles tensed and twitching, his eyes wide with fear and aggression, sat watching my nervous movements. I tried not to meet his piercing stare or do anything sudden that might bring on a charge. The chain finally came free, and I scrambled into the Land Rover and drove away.

We watched from a distance as Bones, still not satisfied with the positioning of his carcass, and oblivious to the pain, dragged it safely to the trunk of his tree.

From camp we could see him lying at Topless Trio, and each morning and evening we sat near him in the truck, watching as he gained weight and recovered lost strength. Daily he became more and more accustomed to us, and our hopes grew that somehow he would live. He was treating his leg better now, only getting to his feet to feed on the gemsbok, or to shift position under the tree. But we could not justify shooting more antelope for him to eat, and we believed that, once he had to hunt again, the stress of the chase would surely snap the weakened leg. He would never be able to survive alone.

On the ninth night after his surgery, we were awakened by his bellows flooding through the valley as he moved south down the old riverbed. We doubted that we would ever see him again.

*   *   *

Bones had been gone for ten days. We had seen no sign of him. Early one morning Mox and I were spooring a hyena we had lost in thick bush the night before. While we two tracked, Delia kept pace with us in the Land Rover, taking notes on the brown's route of travel and foraging habits, all encoded in its footprints in the sand. It was slow, tedious work in the heat and thornscrub, and we were depressed by how far we had to go for each bit of information. But it was the only way to learn how far the hyenas roamed from the riverbed and what their activities were in the sandveld, where the bush and grass were too thick for us to keep up with them at night. Mox and I stalked along side by side, stopping often to discuss a spot where the hyena had rested, fed, and socialized with another, or chased a springhare. If we lost the trail for some reason, we often backtracked to learn more information. We weren't trying to find the animal, so it made no difference to Delia and me which direction we went, as long as we learned where it had been and what it had been doing. But when we tried to follow the tracks in reverse, Mox was hopeless; he lost all interest. We often noticed him, standing with his hands clasped behind him, gazing absent-mindedly into the veld. No matter how we tried to inspire him, he could see no sense to "tracking backward." He thought we were a funny lot for trying to spoor hyenas in the first place. To many Africans, and to many other people, hyenas are the scourge of the earth. Why anyone would want to follow their footprints for hours on end was incomprehensible to Mox.

On this particular morning, we spoored the hyena to Leopard Trail. It had been tough going; we were often on our hands and knees, straining to find a single claw mark in the hard-packed soil. The tracks led us northwest into the soft sand of the duneslope, where, near the top, the hyena's spoor cut the fresh tracks of a large male lion. We had had little contact with males in the area, and we were anxious to meet those of the Blue Pride.

Following the lion now, we slowly moved off through the woodlands and into a complex of springhare burrows. Mox and I were abreast, casting around for tracks, when my eye caught the flattened, wedge-shaped head of a very large puff adder. The snake was coiled tightly, and Mox's foot was descending toward it. With no time to warn him, I swung my left arm across his chest, knocking him off balance and backward. In the same instant, the adder hissed loudly and I jumped back. Mox gave me a peculiar grin, but his eyes were wide as we skirted the snake and walked on.

Just beyond the adder, the lion's tracks deepened—he had chased a porcupine. Rain began falling lightly while we spoored ahead, reading the story of the hunt in the sand: The porcupine had run over a low, worn termite mound and made a sharp turn south. Skidding clumsily, the pursuing lion had lost his footing on the greasy clay surface and had fallen. But he must have recovered quickly, for 200 yards beyond we found a pile of quills and a smear of blood.

I felt Mox's hand on my shoulder. *"Tau, kwa!"* he whispered.

Under an acacia bush 100 yards ahead, a big male lion sat looking

through the veil of rain into the open woodland and the valley beyond—
it was a timeless picture of Africa.

Mox and I joined Delia in the Land Rover. We drove toward the
lion and he turned to watch us. Then we saw the orange tag, number
001, in his left ear. It was Bones. He had gained a lot of weight, and
though his leg was not completely healed, the wound was scabbed
over and obviously on the mend. Of course he was full of porcupine
quills; we wondered if his lameness was keeping him from tackling
larger prey.

We sat with him for a long while, glad that this once we had
interfered with the ways of nature. Finally he stood, stretched, and
began to walk away, the only sign of his past ordeal a trace of stiffness
that interrupted the rhythmic roll of his gait. Studying his spoor more
closely, I noticed a slight twist in the track made by the right hind
paw, a trademark that would follow Bones throughout his life. We
would know his spoor anywhere.

*　　*　　*

While counting antelope on the riverbed one morning, we rounded
Acacia Point and discovered Bones on a young bull gemsbok he had
just killed. It had been three weeks since we had seen him in the rain,
and he had filled out remarkably. We were amazed that he had successfully tackled such a powerful and formidable prey little more than
a month after we had taken three-quarters of an inch of shattered bone
from his leg. As the sun rose higher, he began eyeing the shade of
our camp, about 400 yards away. Panting from the heat, he dragged
the carcass toward the trees, while kamakaze jackals circled around
him, snatching meat from his kill. Though he rested every thirty yards
or so, he hadn't the trace of a limp, and we now believed he would
survive. Killing the gemsbok had been the ultimate test, and a testimony to the remarkable recuperative powers of a Kalahari lion.

Bones spent the following two days under a tree twenty yards outside
of camp, feasting on his kill. In the evenings, we sat in the truck on
the riverbed nearby, watching him feed and laughing when he rolled
onto his back and pawed at the sky.

*　　*　　*

We were following Star, our favorite brown hyena, across the riverbed
one night when she suddenly stopped and began to bristle, every hair

standing out from her body. Suddenly she bolted westward: The Blue Pride was on the prowl. Sassy and Blue trotted to the truck and stood peering over the half door at us. At times this made us a little uneasy, wondering if their mood might suddenly become dangerous. But no matter how close they came, they were always playful.

After their initial investigation, Sassy and Blue apparently tired of trying to spook us. Without warning, they launched a mock attack on Spicy, bowling her over and then chasing her in circles around the Land Rover, their big feet drumming on the ground. Their mood was infectious, and the two male cubs, Rascal and Hombre, joined in the fun, all the lions romping in the bright moonlight, except for Chary, who remained aloof as usual.

Abruptly the nine lions stopped their play and lined up shoulder to shoulder, looking north. I swung the spotlight and saw Bones charge into the beam with a powerful stiff-legged trot, his massive head and mane swinging side to side. He strutted to the waiting pride and stood there while each female greeted him in a fluid fusion of her body with his, beginning cheek to cheek, then rubbing along his length until she sidled off his ropey, tufted tail. After their exuberant greetings, the pride lay together quietly, Bones a few yards away. The master of the Blue Pride had come home.

Bones's arrival seemed to have changed the mood of the females. Their playfulness had given way to a calm sense of business as they stared intently into the night, hunting even as they lay there. Sometime later Chary stood and moved off silently, followed soon by the two youngsters, Spicy and Sassy. Then Blue and Gypsy were gone, and finally the entire pride had slipped away into the growing darkness, a long procession, with Rascal, Hombre, and Bones bringing up the rear. The moon was setting toward West Dune.

The pride moved along the riverbed to Last Stop, a small group of trees on the edge of North Pan, where they often scent-marked and rested before leaving the valley. In the early light of dawn they walked slowly toward a herd of seven red hartebeest browsing on silvery catophractes bushes on the west slope of North Dune. An old bull, the tips of his horns worn to shiny nubs, stood a little apart, licking the mineral from a termite mound. Lowering themselves for the stalk, the lionesses fanned out toward the herd, gliding through the brush, ears drawn down beside their heads. Nearly an hour later they were moving abreast, in a line about 100 yards long, still seventy or eighty yards from the hartebeest but moving toward them. Rascal and Hombre stayed far to the rear with Bones. But while the lionesses were stalking north, the hartebeest had turned east; they would miss their chance unless adjustments were made. Chary and Sassy pulled from the line, and slipping behind their pride-mates, they disappeared in the grass to position themselves in front of the antelope. Liesa, Blue, and Gypsy began stalking slowly forward.

Waiting... then moving from bush to grass clump to hedge... then waiting some more, the pride worked its way toward its target. The hartebeest sensed something. Staring back at the lions, they began prancing and blowing their alarm calls. Then the herd cantered away.

The old bull was in the lead. As he dodged an acacia bush, Chary's thick arm flashed out and hooked over his shoulder. He disappeared into the cover, groaning harshly, his feet flailing wildly. The other hartebeest dashed to the top of the dune and stood looking down, snorting and flicking their tails. Within seconds all the lions were tumbling toward the kill. We could hear their throaty rumblings and the tearing of flesh.

Bones heard the commotion too, and trotted past us on his way to join the others, Rascal and Hombre scampering through the tall grass behind him. At the carcass he rushed forward, snarling and scattering the lionesses, and clamping his wide paws over the hartebeest, he began to feed alone. The females, with Rascal and Hombre, watched him from ten yards away.

But Blue began to edge closer, watching Bones and sinking to the ground whenever he shot a glance at her. At about eight yards, she made a slow arc toward the carcass. Bones stopped feeding. A deep rumble grew in his throat and his lips rose to expose his three-inch canines. Blue spat at him. He roared across the carcass, shoveling sand as he charged, and clubbed her across the nose with his paw. The lioness bellowed, her ears pressed to her head as she flattened to the ground again. Bones went back to the carcass, and twenty minutes later the females, followed by Rascal and Hombre, slowly walked away. That night, while their male was occupied with his hartebeest, the lionesses killed and consumed an eighty-pound springbok on South Pan.

*   *   *

It was the end of May 1975 and almost a month since the last rain shower. The skies were pale and cloudless, the cool nights perfumed with the sweet musk of golden grasses, and the morning wind had a cutting edge. All spoke of the coming winter. The heavy clay soil of the riverbed had lost most of its moisture, and the gemsbok and hartebeest herds had fragmented and moved away.

We saw the lions less and less frequently; finally they were gone. We missed the sound of their bellows rolling through the valley on the night wind and wondered where their migration had taken them, and whether we would ever see Bones and the Blue Pride again. We knew it would probably be more than eight months before the rains brought the flush of new grasses and the larger antelope back to the fossil river, late in 1975 or early in 1976. The lions would not be back before then. We began concentrating on our study of brown hyenas, learning as much as we could about every facet of their existence.

9

The Carnivore Rivalry

Mark

Nor heed the rumble of a distant drum.
—*Edward FitzGerald,*
The Rubaiyat of Omar Khayyam

DELIA elbowed me in the ribs. "Did you hear it?" she asked.

"Did I hear what?" I groaned, lifting my sleepy head.

"The drums!"

"Drums?"

"Quick, we've got to answer them!" It was a bright and frosty dawn. She wormed out of her sleeping bag and, clad in nothing but panties, pushed back the flap and hurried outside. Her breath coming in clouds of steam, she huddled against the cold, listening.

"Maybe you've been in the bush too long," I teased her. Then I heard them too. Tum, tum, tum-tum-tumtum—a very low-pitched sound, like someone beating a large bass drum.

"What can I use to answer them?" she asked, searching the kitchen area. I suggested—not seriously—the five-gallon pail we used for our oven and a tent stake. Holding the pail under her arm, she began clobbering the bottom, mimicking the cadence of the drums. After each series, she listened for an answer. But the drums had gone silent. She whacked the pail again and again, and I buried my head inside my sleeping bag to shut out the racket. Finally she gave up and scrambled shivering and subdued back into bed.

For days, at sunrise and sunset, we heard the drums. It must be a Bushman hunting party, we reasoned, since we first heard them south of camp, but later in the west and north. They seemed to be moving up and down the valley but avoiding the part of the riverbed where we could have seen them. Delia kept her pail and tent stake handy. But each time she answered the drums, they fell silent.

Thinking that Delia's clatter had been frightening the hunters, instead of answering one evening when we heard them, we dropped everything and jumped into the Land Rover. Safari hunters had told us that the few truly wild Bushmen left were shy people who avoided contact with modern man. We would probably be lucky even to see them before they ran away.

We drove slowly toward the drums, craning out the windows of the truck and taking a compass bearing each time we heard them. Tense with anticipation, we imagined that at any moment we would round a hedge and see little black men in animal skins, with bows and arrows slung over their backs, gathered around a small campfire, roasting a steenbok for their supper. Or maybe one of the hunters would be beating a drum while the others danced around in a circle. We wondered what they would do when they saw us, and if we should have brought sugar or tobacco to offer them.

We were almost on top of the sound and I was easing the truck around a large clump of bushes, when I stopped abruptly. A few yards ahead of us was a large male kori bustard, feathers puffed out from his swollen neck, strutting through the grass, his beady eyes fixed on us: *Whum, whum, whum-wumwum! Whum, whum, whum-wumwum!* It was his mating call.

We left the kori to his dance and turned toward camp, promising each other that we would never tell another soul about this.

*   *   *

During the nights of the dry season of 1975, we followed Star, Patches, Shadow, or any brown hyena we were lucky enough to find on the open riverbed. If we missed a single night of observation, we felt compelled, no matter what the reason, to enter it in our journal: "Alternator on truck broken, severe wind and sand storm; impossible to follow hyenas tonight," or "had to haul water today; got back too late to look for hyenas." We had to learn as much as we could, and as fast as we could, about brown hyenas, not only for their conservation but also for our own. We still had to prove ourselves as field biologists if we wanted to stay in Deception Valley.

We were especially fascinated with the brown hyenas' relationship to other carnivores, on whom they relied heavily for food. We hadn't yet learned which species they could dominate successfully enough to steal their kills. Leftovers from lions made up the major portion of their diet during the rainy season, but it would be a short-lived brown hyena that tried to take a kill from them. They could only wait until the cats abandoned their carcass. And though the hyenas often appropriated kills from jackals, their interactions with leopards, wild dogs, spotted hyenas, and cheetahs were completely unknown. We planned to investigate these fundamental relationships during the dry season, while the lions were away.

We had been censusing antelope one evening, and it was almost dark when we stopped the truck in camp. Mox's fire glowed weakly from beneath a big, brooding acacia tree. I switched on the spotlight. If we were lucky, one of the hyenas we wanted to follow would be walking by, saving us hours of searching. When I played the light along the riverbed, large yellow eyes winked from the branches of a tree between Mox's camp and ours. The leopard, whom we had named the Pink Panther, was draped over a limb about ten feet above the ground, his tail hanging straight down. He paid no attention to us, apparently absorbed in watching something north toward Cheetah Hill.

I turned the light in that direction, and the shaggy form of a brown hyena came into view. It was Star, moving slowly in our direction, following a zigzag course, with her nose to the ground. In seconds she would be directly below the Pink Panther.

"Mark, he's going to attack her!" Delia whispered. Since one of our objectives was to learn the relationship between brown hyenas and leopards, I didn't think we should interfere. Delia leaned forward in her seat, her hands clenched around the covers of the field journal in her lap. If the leopard did attack Star, I was sure—having seen browns lug off the heavy parts from gemsbok carcasses—that we would see a hell of a scrap. Still, I thought the hyena would surely smell or see the cat and avoid the tree. I was wrong.

Star moved directly under the Pink Panther. Peering down at her, the tip of his tail twitching, the cat carefully drew himself into a crouch. Star began walking circles around the base of the tree, still smelling the ground. The leopard did not move. Half a minute passed. At any moment the attack would come and Star would be torn apart before she could even look up.

When she walked out from under the tree and started south toward Eagle Island, Delia let out a long sigh and settled back in her seat. Star was about 200 yards away when the Pink Panther climbed down and began walking west. The hyena swung to pick up a scent and saw him. Her hackles rose like spikes along her back; she lowered her head and charged. When she was nearly on him, the leopard launched himself toward the acacia tree that he had just left. By the time he was at full pace, his body stretching into another stride, Star's open jaws were inches from the end of his streaming tail. He hit the acacia at full speed, and chips of bark flew from his claws as his momentum swung him around the trunk. Spitting and growling, he reached the safety of the limb just as Star made a last lunge for his tail. Her front paws against the tree, she howl-barked at the leopard again and again as if frustrated, while he glared down at her from his perch. Finally she walked off. The Pink Panther watched until she was at a safe distance. Then he hurried down and slunk into the tall grass of West Prairie.

It must have been a fluke, we told ourselves, a mishap. Surely sawed-off brown hyenas did not usually dominate leopards single-handedly. But the Pink Panther's rivalry with brown hyenas was not over.

Several weeks later, while we sat at the campfire eating supper, a groaning death rattle rose from the darkness just beyond our tree island; a springbok had just been killed. We had started for the truck to go to investigate when the Pink Panther trotted into camp, his muzzle and chest smeared with blood. He stopped within three yards of us, looking quickly over his shoulder, and then hurried up a nearby tree. He had apparently just killed the antelope. But why had he left it?

On the opposite side of camp we found Shadow, the most subordinate brown hyena in the clan, chewing at the belly of the springbok. We waited, and after about twenty minutes, the Pink Panther reappeared from the island, working his way toward her. She paid little attention and continued gnawing on the antelope she had taken from him. He lay in the grass watching her devour his kill, his ears turned back, his tail twitching. Then, as if he could stand it no longer, he jumped up, curled his tail over his back, and took three pounces toward Shadow.

Without hesitation, the chunky hyena launched herself over the carcass directly at him, her hackles bristling and jaws open wide. Again the Pink Panther turned tail, and the two now stormed into camp, where the cat streaked up the tree next to our kitchen boma. Shadow sat below for a few minutes, watching him lick his paws, and then she walked back to the carcass. Ivey, the clan's dominant male, joined her, and the two of them finished off the springbok. The Pink Panther slipped quietly away.

Apart from our having learned a great deal from this interaction, we were gratified to know that neither Shadow nor the Pink Panther had the slightest hesitation about using our camp as a battleground. We had wanted to blend into the Deception Valley scene. This was testimony that we had succeeded.

We had gained a new respect for the brown hyenas. Scavengers they are, but they don't just wait passively for a handout from the predator community. They often steal from quite formidable competitors. Apparently it is too risky for a leopard to fight a brown hyena, whose massive shoulders and neck could absorb many bites and slashes, whereas a single crushing bite from the hyena could break the cat's leg, or even kill it. As for the Pink Panther, losing a carcass was less costly then losing a leg.

The browns are as skillful as they are bold. During the rains, they key on lions to such an extent that the clan's territory almost perfectly overlaps that of the Blue Pride, boundary for boundary. They know the pathways and lying-up places habitually used by lions and leopards, and they keep close tabs on their activities by coming downwind from them once or twice a night to smell if they have made a kill. They even use flocks of circling vultures to help them find carcasses in the early mornings or evenings, and, as we had seen, they find other kills by following the strident calls jackals make when mobbing leopards. If a leopard has not already stashed its carcass safely in a tree by the time a hyena arrives, it soon loses its meal to a brown.

We learned that hyenas not only dominate leopards, but they also chase cheetahs from their kills. Cheetahs are less powerfully built than leopards, and much more timid. In the Kalahari, as distinct from their behavior in East Africa, cheetahs often hunt at night, when the browns are busy foraging. Spotted hyenas, on the other hand, will displace brown hyenas at carcasses, though they wander into Deception Valley so seldom that they rarely compete with their smaller brown cousins.

A pack of wild dogs is apparently too much for one brown hyena. Star appropriated a springbok kill from a cheetah near Acacia Point one night. While she was busy trying to free a leg to cache, the wild dog, Bandit, and two others of his pack rushed to the carcass and chased her off. Two minutes later Star was back, pulling at the leg while the dogs fed at the other end. Without warning, Bandit bounded over the dead springbok and bit Star on the rump. She yelped and galloped away just as the rest of Bandit's pack arrived on the scene. In seven minutes, the wild dogs completely devoured the ninety-pound springbok, except for the horns, skull, spine, and jawbone. Star did not get one more bite of meat and could only finish off the bones after the dogs had left.

Practically speaking, brown hyenas are near the top of the hierarchy in their ability to displace other carnivores at carcasses. The order descends from lions to spotted hyenas, to wild dogs, brown hyenas, leopards, cheetahs, and jackals (the last two about equal in this regard). But since lions are absent during the entire Kalahari dry season, and wild dogs and spotted hyenas are seldom present in any season, brown hyenas are often the most dominant carnivores around. They are not the shy, skulking creatures many people think them to be.

*   *   *

It was late when we swung into camp after our night's observations. Some jumping jacks helped to shake the kinks out of our cramped legs. We then tipped some water from a jerrican into the washbasin, splashed it onto our faces, and headed for the tent to sleep. I was a little indignant when Delia suggested that I leave my shoes outside.

My tennis shoes were more holes than canvas, but they'd carried me a long way in a country where even the best footwear doesn't last long. Each step had widened the holes a little, improving the ventilation and making them more comfortable. But for the sake of domestic harmony—and a fresher atmosphere—I put them on the flysheet over the tent, where the jackals couldn't carry them off, before slipping into bed.

When I got up around dawn, Mox was already on his hands and knees, blowing life into a reluctant fire. Springbok herds were stirring restlessly near camp, whizzing their nasal alarm calls; there was a predator on the riverbed. I parted the flap, slipped into my cold, ragged tennis shoes, and stepped into the frosty morning.

The sun was creeping toward East Dune, the air dead still, crisp, and fresh, one of those special mornings when you have to get moving.

I stuffed some leathery strips of biltong into my pocket and headed for the truck. We had lost track of Bandit and his pack in the bush of North Bay Hill the night before, but maybe now they had come back to hunt the springbok grazing on Mid Pan.

Delia had to transcribe some notes in camp, so I asked Mox to join me; it might be a welcome break from his camp chores. Silent as always, he climbed into the Land Rover beside me, his hands folded on his lap. As we drove along, his sharp eyes missed nothing on the old riverbed before us, but his face was expressionless.

We drifted through the springbok, who were nibbling the drying grasses. It was June, the cold-dry season in the Kalahari. Gemsbok, hartebeest, and herds of other broad-muzzled, nonselective grazers—those that crop away the overburden of straw—had left the valley for the season. It had become more and more difficult for the springbok to find the few remaining green stems. Like the other antelope, they, too, had adjusted their feeding strategies by moving into the sandveld in the evenings. There they grazed greener grasses and browsed leaves, some of which absorbed up to forty percent of their weight in moisture from the humid night air. At dawn they moved back to the open riverbed, where they rested and socialized until evening.

Later, in the hot-dry season, when the relative humidity is at its lowest, fires would sweep the desert again, burning the last moisture from the leaves. To survive, the scattered bands of antelope would eat acacia flowers and wild melons—if there were any—or dig fleshy roots from deep in the sand with their hoofs. There is something pathetic about a handsome bull gemsbok on his knees, his head and shoulders pushing deep into a hole, chewing off woody fiber to get the moisture and nutrients he needs to stay alive. The antelope are remarkably well adapted to this whimsical land. Living and reproducing in large herds during times of plenty, they eke out a near solitary existence by grubbing roots from the barren soil in the severe dry season and drought.

When we were driving through the springbok herd early that morning, something suddenly galvanized their attention. Like iron filings drawn by a magnet, they all turned to the north. I raised my binoculars and saw Bandit and his pack shagging along in rough file, headed toward the dry water hole about a mile away. We caught up with the pack as the dogs wandered over the dried and crusted surface, searching eagerly for water, sniffing with their noses at the clods and cracks in the clay. But it would be more than eight months before the rains would come and they could drink again. Until then, like the other predators, they would subsist only on the moisture found in the fluids of their prey.

Bandit stood on the calcrete rim of the water hole and eyed the herd of springbok across the valley. Then he turned and rushed to the other dogs, touching noses with them, his tail raised with excitement as he incited the hunting mood. The pack crowded into a huddle, pushing muzzle to muzzle, their tails waving like tassels as they welded themselves into a coordinated hunting machine. Bandit raced away, leading the others toward the herd.

Minutes later they had pulled down a springbok, and when Mox and I arrived, it had already been quartered and torn to pieces. Bandit and the other adults stepped back from the kill to let the yearlings feed first, as is the habit of wild dogs. After the young had fed alone for about five minutes, the older dogs rejoined them and finished off the carcass. Then they all pushed their crimson-stained muzzles through the grass and rolled over and over on their backs to clean themselves.

A game of tag began, with several dogs racing around the Land Rover, using a springbok leg for a baton. Mox and I watched the circus: dancing, high-spirited gypsy dogs with rag-tag coats, tattered ears, and broom-sedge tails. Finally the sun grew hotter, and three of the dogs settled into the shade of the truck.

The lower jaw bone of the springbok they had killed was lying about fifteen yards away in the short grass, and if I could get it, we could determine its age. I would have to collect it immediately, however, or one of the dogs would certainly carry it off. Cape hunting dogs had never been known to attack a person on foot, so, gathering up my camera, I eased open the door and stepped out. Mox was shaking his head and muttering, "Uh-uh, uh-uh," while I crept slowly to the front of the truck, ready to retreat if necessary.

I moved ahead several yards, and two dogs raced between me and the Land Rover, one chomping on the ear of the other. Three more streaked in front of me, one of them carrying the springbok's leg jutting at right angles from its jaws. With the pack dancing and dodging around me, I felt a rush of exhilaration, a sense of freedom, almost as if I were one of them.

I began to snap pictures as fast as I could. The wild dogs were running, jumping, and wheeling in hyperanimation, their golden-and-black coats a kaleidoscope in the soft morning light. They seemed totally unconcerned with me. But when I squatted to pick up the springbok jaw bone, the mood of the pack suddenly changed. A young dog turned toward me, first raising his head very high, then lowering it, as if seeing me for the first time. He stalked toward me until he was only ten feet away, his eyes, like black opals, staring me in the face. A loud *Hurraagh!* came from deep in his chest, and immediately the rest of the pack turned on me. In a second they had formed a tight semicircle around me, and shoulder to shoulder, tails raised above their backs, they continued growling as they pressed in on me. Beads of sweat broke out on my face. I had gone too far. A dash for the truck was out of the question; yet unless I did something immediately, they might attack.

I stood up. The effect was immediate and striking: The entire pack suddenly relaxed as if tranquilized. Dropping their tails, looking away, they broke their formation and began wandering about, some returning to play. A couple of them gave me wry looks, as if to say, "Now why did you pull a stunt like that?"

I looked back at Mox in the truck. Poor guy, he'd been treed twice by lions and once by a gemsbok since coming to work for us. He just couldn't understand why anyone would be so foolish as to walk among wild dogs.

I had learned to manipulate the pack. By squatting or sitting I could draw immediate threat; several dogs would dart forward and nip at the camera tripod before springing back. If I thought they were getting too agitated, I stood, and they would back up and relax. After several minutes of this experiment, some of the threat seemed tempered by curiosity. I was interested in their responses to my positions, so I decided to try lying down.

I sank slowly to a sitting position, and again the same young dog gave the alarm. Six of the pack members strutted toward me, tails over their backs, growling and bristling with threat. They were little more than a yard or two away when I stretched out on my back, with the camera on my belly. Strangely enough this posture stimulated more curiosity than threat, and two dogs moved cautiously toward my head, noses near the ground; two others moved in on my feet. The ones on my left seemed content to threaten the tripod. They all smelled rather high, like Limburger cheese.

I didn't worry too much about the dogs at my feet, but it was hard to observe the two coming at my head. Suddenly all four of them began rushing in for quick sniffs of my hair and feet before dancing away. I found that if I wiggled my feet and my head now and then, they were more cautious, content to just stalk in for a whiff and then dart away.

I was waggling my shoes and shaking my head to keep the pack off while I photographed them, and I had taken some great shots of my foot just under the chin of one particular dog. Everything was fine until he touched my toe with his nose a couple of times. He cocked his head, and his face assumed a peculiar look, as if he'd been stunned. Then he turned completely around and began kicking sand over my foot, trying to bury my tennis shoes.

10

Lions in the Rain

Mark

*Deception Valley
January 1976*

Dear Mother & Dad,

We could not have known what the Kalahari had in store for us. All through September, October, November, and December the rains did not come, and at the beginning of January there was not a cloud in the sky. The temperatures soared past 120 degrees in the shade, and the wind blew hot as a blast furnace across the dry, dusty valley. As in the previous dry season, for weeks we were able to do nothing but lie on our cots, dizzy from the heat and covered in wet towels. We tried to conserve our energy so that we could work at night, but by sundown we were always weak from heat fatigue. We ate salt tablets like candy, and our joints ached continuously. We just existed. The sun and wind seemed determined to sear and strip the last vestige of life from the dry Kalahari.

But if the heat was bad for us, it was much worse for the animals. There were no antelope on the old riverbed, and only a few ground squirrels and birds scratched around for food. In the sandveld, gemsbok pawed deep holes in the ground, searching for the fleshy, succulent roots and tubers from which they could get enough moisture and nourishment to stay alive. Giraffe stood spraddle-legged in dry water holes, dragging their heads through the dust in the shimmering heat. The nights were deathly quiet, empty of all sound except the occasional squawk of a korhaan or the cry of a lonely jackal.

Then in mid-January, puffs of snow-white cumulus clouds began to appear each day, softening the harsh glare of the desert sky. But, like apparitions, they disappeared into the great void of heat that gripped the Kalahari. Again and again the clouds challenged the inert high-pressure system that locked the land in drought. Each day they grew, until they stood like great cathedrals with massive columns in the sky. As if in anticipation, small herds of springbok began appearing on the riverbed, their bodies a misshapen illusion in the silent waves of midday heat. They seemed to understand the language of the distant, rumbling clouds. The sky beneath was streaked with rain; we could smell it. Standing at the edge of camp, we willed the storms toward us, but they would not come. And we knew it might not rain at all.

Then, late one afternoon the clouds were back, stacked closely, dark mountains of vapor growing over the valley. A black squall line dropped low and rolled toward the riverbed. The trees seemed to quiver, and we could feel the thunder deep in our chests. Lightning cut across the sky, swirling clouds swept over the dunes, and fingers of sand raced down the slopes with the rushing wind. The sweet fragrance of rain was everywhere, and like an avalanche, the storm broke over the parched desert. We could not contain ourselves. Laughing and singing, we ran from the camp to meet the stinging wall of wind and rain. We danced around, and even rolled in the mud. The storm meant the rebirth of our spirits and new life for the Kalahari. It rained and rained, and that storm ushered in the Kalahari wet season. No wonder that *pula* is the most important word in Setswana. It means "rain" and is both a greeting and the name for a unit of Botswana's currency.

It must truly be one of the wonders of the world to see the Kalahari change from a bleak desert to a verdant paradise. Through eons of time, all the life in the desert has adapted to these extreme conditions and dramatic changes. Animals and plants alike wasted no time getting reproduction into full swing to take advantage of the short and unreliable rainy season. Every living thing from grasshoppers to giraffe, jackals, and gemsbok quickly give birth to their young before the dry season begins all over again. It would be a major challenge for an animal behaviorist to describe the facial expression of a male springbok, who, after standing alone for months on his dusty midden, suddenly looks up to see 2000 females prancing into his territory.

Before dawn one morning another heavy storm charged into the valley. Howling winds drove sheets of rain through camp, and lightning cast the shadows of frenzied trees on the billowing wall of our tent. Before long the legs of our cots stood eight inches deep in water, and we lay listening to the symphony of the thunder accompanied by the wind and rain on canvas. When the storm had passed, the Kalahari stood in soggy silence, as though holding its breath while drinking the life-giving moisture. The only sound was the pok-pok-pok of water dripping on the tent from the trees overhead. Then the deep roar of a lion, the first of the season, rolled through the valley on the still dawn air.

We slogged over to the truck through ankle-deep mud and water and headed north along the riverbed in the direction of the call. North Pan was wreathed in a thin layer of ground fog, and just as the sun appeared over East Dune, a big male lion stepped through a golden curtain of swirling mist. We stopped some distance away, in case he was a stranger and not used to us. Lifting his head, his sides heaving, he came toward us, his bellows punctuated with puffs of vapor. At the truck he stood five feet away, listening for an answer to his calls. And then we saw it—the orange tag, number 001, clipped in his ear. It was Bones!

You cannot know the feeling, and we cannot explain it. He looked at us for several long moments, and then he walked south along the valley, roaring. We wondered where he had been since June, eight months earlier; how far he had traveled, and in which direction. Was he looking for his Blue Pride females? So far we have not seen them, but we hope to anytime now. We followed him to camp, where he sunned himself while we ate breakfast.

Our research is going well, and we are both in good health. Will mail this in a few weeks when we go to Maun for supplies, and we hope to hear from all of you then. We miss you all very much.

Love,

Delia and Mark

*   *   *

A crash, then the sound of splintering wood brought my head up sharply from the pillow. Through the gauze of the tent I could see the full moon settling low above the dunes west of the valley . . . must be near morning. I looked over at Delia, still sound asleep. We had already gotten up three times to coax the brown hyenas from camp. Now they were back again, obviously tearing something apart. Groggy from lack of sleep and thoroughly irritated, I jumped up and, without bothering to dress or light the gas lamp, I stomped down the narrow path in the darkness. This time I was going to make sure they got the message.

I could see a dark form ahead and hear teeth grating on the screen frame I had made for drying lion and hyena scats. Swinging my arms and swearing in a low voice, I strode to within four or five feet of the intruder, stamped my foot, and barked, "Go on now, dammit! Get the hell out of—" I bit off my words as I suddenly realized this was much too large for a brown hyena. With a growl tearing from her throat, the lioness spun around and crouched in front of me, the screen clamped defiantly in her jaws, her ropy tail lashing from side to side.

We had vowed never to put lions in a compromising position, never to threaten them. Half asleep, I had broken our cardinal rule. Bolts of nervous energy shot up my spine as we stared at each other through the darkness. I began to sweat in the chilly night air. It was dead quiet, except for her breathing and the swish of her tail in the grass. We were so close I could have reached out and put my hand on her head; yet I had no idea who this was. "Sassy, you devil, is that you?" I whispered.

The lioness didn't move, and my words fell away in the darkness. Somewhere on the riverbed a plover screamed. I tried not to breathe. Unable to see the lion's face, I wasn't getting any clues. Her only vocalization had been one of surprise and threat when she had crouched down over her hindquarters. She could very well lash out and lay me open from shoulder to waist or send me sprawling like a rag doll into the thorns. If I moved, she might spring at me; if I stood still, she might just turn and walk away.

Delia's voice from the tent behind me sounded small and far away. "Mark, is everything all right?"

Too frightened to answer, I slowly put one foot behind me and began a retreat. With a loud, straining grunt, the lioness leaped into the air, whirled around, and hoisting the screen frame high, she romped out of camp. As I made my way back toward the tent, the drumming of heavy feet and more grunts sounded in the dark around me.

I knelt to light the gas lamp. Delia raised up on her elbow. "Mark, what are you going to do?"

"I can't let them tear up camp."

"Please be careful," she urged, as I started back along the footpath toward the kitchen. I held the lamp low and shielded my eyes with my hand so I could see ahead. The lions seemed to be gone, or perhaps their sounds were being covered by the hissing of the lantern. I moved past our dining tent and stepped around the row of water drums. Three lionesses of the Blue Pride were stalking toward me from only ten yards away; Sassy, as usual, was in the lead. To my right, three others were invading camp along the footpath to the kitchen, and Rascal and Hombre were pushing through the bushes behind the water drums.

There is a great difference in the posture and expressions of mildly curious lions and those bent on destruction. The Blue Pride was keyed up, their ears perked forward, bodies held low, tails thrashing. I had seldom seen them in such a mood—a mixture of curiosity and playful rambunctiousness, with perhaps more than a dash of predatory urge thrown in. They had probably come from hunting along the riverbed.

They had visited us on numerous other occasions in the previous rainy season, and each time had grown less and less afraid of us and the camp's surroundings. Each time it had become more difficult to convince them to leave without damaging anything important. The first time or two I had only to start the truck's motor, raise my voice, or wave my arms slowly to start them moving away. But since then, progressively stronger action had been required.

Now they stared directly at me as they came. It would take more than the usual amount of persuasion to turn them out of camp before they began to ransack it. If they discovered how flimsy the tents were and how much fun to bat around, they might break them down and shred them to pieces.

Sassy, Spicy, and Gypsy were about six feet away. "Okay, that's far enough!" I said in a loud, shaky voice. At the same time I stepped forward and swung the lantern within a foot of their noses. I had used this deterrence successfully before, but this time they quickly dropped to a crouch, their tails whipping up puffs of dust in the path. The other two groups were advancing from each side of me and were now less than twelve feet away.

Unnerved, I took a few steps backward. Then I noticed an aluminum tent pole propped against a tree next to one of the water drums. Confident that this would do the trick, I swung it hard against the empty drum. Wang! Once again they all just crouched.

When they started toward me again, I grabbed a stick of heavy firewood lying near the footpath. Against my better judgment, but seeing no other option, I drew back and threw the block of wood toward Sassy, ten feet in front of me. Turning once in the air before it reached her, it would have struck her cleanly across the snout had it not been for the big paw she raised like a catcher's mitt at the last instant. With astonishing speed she deftly blocked the missile with the flat of her pad and grounded it at her feet. She then looked at me for a second before seizing the chunk of wood in her jaws and strutting out of camp. It was as if my rash action had broken the tension; the rest of the lions sprinted after her.

Swinging the lantern from side to side, trying to see in the underbrush along the path, I made my way quickly back to the tent, where Delia was waiting anxiously. As I pulled back the flap to step inside, the lantern reflected the amber eyes of lions standing all around the Land Rover, which was parked just off the back corner of the tent.

"These lions are in a hell of a funny mood," I whispered. "We'd better get into the truck. I don't know exactly how we're going to do it, though."

Delia pulled on jeans and a shirt while I watched the big cats playing around the Land Rover. One was chewing a tire. Bones stood near the left front fender, his head taller than the hood, and as he turned to the side, I could see the heavy scar over his right hind knee.

We waited, crouching near the corner of the tent; some of the lions were now lying around the truck. Meanwhile, one of the others stole the spade from near the campfire, and another romped out of the reed kitchen with a large tin of powdered milk.

About half an hour later, Bones began roaring, and the entire pride joined in the chorus. Continuing to bellow, the two near the door on the driver's side of the truck moved to the rear. We crawled along the wall of the tent and slipped quietly into the cab.

When the morning sun crested East Dune, I sat dozing with my forehead against the steering wheel and Delia was slumped against my side, her coat pulled up snug around her neck. The dull thonk of rubber in trouble and a movement of the steering wheel brought my head up sharply. I leaned out the window to see Sassy lying on her side next to the front wheel, her long canines poking into the tire. Having spent themselves during the raid on camp, Gypsy, Liesa, Spicy, Spooky, Blue, Chary, Rascal, Hombre, and Bones lay sprawled in the pool of warm sun around our truck. The Blue Pride had come back to Deception Valley.

Rascal and Hombre had grown up considerably, despite the rigors of the long dry season, and each sported a fringe of patchy, untidy mane. The young females had lost most of their adolescent spots and their forelegs, chests, and necks had thickened. They were adults now, but obviously still youngsters at heart.

* * *

It was urgent that we learn as much about the Blue Pride as we could during the short rainy season: the size of their territory; what prey they ate, how much, and how often; how their kills influenced the movements and feeding habits of the brown hyenas. We were also interested in finding out how their social system compared with that of lions in the more moderate climate of the Serengeti Plains. Within two to four months, depending on how long the rains lasted and how late the large antelope prey stayed in the valley, the lions would migrate away again.

But even when the lions were near Deception Valley, they spent most of their time in duneslope woodlands and bush habitats, where it was very hard to follow and observe them, especially at night, when they were most active. Unless we stumbled onto them while following brown hyenas, homing on their roars was our only way of finding them.

Typically, we would have just gone to sleep after hours of night work, and then hearing a lion's bellow, we would jump out of bed swearing and fumbling for the flashlight. Whoever found it first dashed for the truck to take a compass bearing on the sound. We had no more than about forty seconds before the first series of roars died away. If we didn't get a fix then, invariably, it seemed, the lion would not call again. We would be left standing nude in the dark, often with skinned or rope-burned shins and toes from running the gauntlet of thorns, sharp tent stakes, and tie-down lines that lay along the path out of the trees to the truck. As soon as we would crawl back under the covers, another bellow would echo through the valley.

If we managed to get a bearing, we would pull on our clothes and climb into the truck. Then Delia would hold the compass on her lap, directing me as we drove along. We were able to find the lion about half the time, unless it was moving when it bellowed, as lions often do. Crude as this technique was, we began to get good rainy-season information on the Blue Pride's movements in the valley and what antelope they were eating.

Nearly even evening, the roars of the Blue Pride were answered by lions farther south in Deception Valley. We grew more and more curious about these neighbors, especially since observations on only one pride would never give us a reliable picture of Kalahari lion ecology. We would have to head south to locate and observe as many of the other valley prides as we could.

The idea was a little intimidating at first, since it was an expedition that we had never made and for which we were not equipped. We would have to find our way deeper into the Kalahari along the shallow, meandering riverbed, which would be totally obliterated in places, blocked with sand dunes. Alone in our battered old truck, without a back-up vehicle or any radio communication, and with only the food and water we could carry, we might lose the river course and wander around for days trying to find our camp again.

Nevertheless, we decided to do it. We packed the Land Rover with water, cooking pots, fuel, spare parts, and essential food and bedding. I wrapped our only tube of tire patch solution in a piece of plastic fertilizer bag to keep it from evaporating and from being punctured in the toolbox. The thornbush would be heavy in places and a flat tire or two was inevitable. The plastic fertilizer bag would seal small leaks in the radiator if stuffed into the grill and then set alight—according to an old Bushman tracker who didn't know much except how to come back from the Kalahari.

Early one morning, we set off south along the riverbed. We left Mox standing at the edge of camp, a piece of paper in his hand. Our note read:

To Whom It May Concern:
On April 6, 1976, we left camp to explore Deception Valley south from this point. If it has been more than two weeks since our departure when you read this, please go to Maun and ask someone to send a search plane out to fly along the valley.

Thank you,
Mark and Delia Owens

It was highly unlikely that anyone other than Mox would ever see our scrawl, but we felt better having left it just the same. Mox had instructions to walk along our truck spoor, east out of the reserve, to a cattle post if we were not back after the sun had risen and set fourteen times.

As we drove south, the familiar line of West Dune with its picturesque acacia woodlands followed us for a mile, and then fell easily behind. A stranger took its place: The riverbed grew narrow and tentative, less distinct. Soon all that was familiar about the Kalahari disappeared, and we were headed toward a flat horizon of thornscrub, grass, and sand.

Several miles later a bottleneck in the riverbed spilled out into a

generous open plain, or pan, where hundreds of gemsbok and hartebeest and thousands of springbok grazed on lush grasses. "Springbok
Pan," we wrote in our log for the first time. Other antelope sipped at
shallow water holes a few feet across, where hottentot teals dabbled
in the mud. White storks, migrants from chimneys in Europe, and
their white-bellied cousins from North Africa strolled along picking
up grasshoppers. Black-shouldered and yellow-billed kites, tawny eagles, lappet-faced vultures, and kestrels hovered and turned through
the sky, while jackals and bat-eared foxes trotted over the savanna
pouncing on mice and snatching grasshoppers from grass stems.

We drove slowly through the herds and across the pans, then found
our way back into the narrow part of the rivercourse. Giraffes craned
their necks curiously at us from low, shrub-covered dunes, close along
either side. Never had we seen so many antelope—herd after herd
cantered aside as we passed.

Later we rounded a bend, and a large conical sand dune with a lop-
sided cap of woodland loomed ahead, blocking the river channel; there
seemed no way around. We drove straight up the side to the top and
stood there in the wind, feeling minute against the endless savanna.
The river channel beyond splayed in several directions, like the un-
braided ends of a rope; it was not altogether obvious which tributary
we should take.

From the truck's storage box I pulled out a tattered photograph, a
composite of tiny aerial pictures taken by the British Royal Air Force
years ago. In printing this collage of photos, the geographic features
along the edges of the smaller prints had not been carefully matched
up by technicians in the Department of Surveys and Lands: they were
scattered about like the pieces of a jigsaw puzzle. As a navigational
tool, this enlarged mosaic was fuzzy and inaccurate, but it was all we
had. From the picture it looked as if the middle fork was the channel
most likely to be the continuation of Deception Valley, so we set
course along that one, stopping to look for lion tracks in the mud at
water holes, to collect scats from tree islands, and to study old kill
sites.

We tried to keep track of our position, but in many places the old
channel was shallow and covered with the same vegetation found in
the bordering sandveld. Every now and then we would stop, worried
because we had lost touch with Deception. Then, by standing on top
of the Land Rover to get above the flat terrain, we would find the

narrow trough again, faintly visible as it wandered away to the north or south of us through the waving grass-heads. At each temporary campsite I took star shots with an old Royal Air Force bubble sextant from a World War II bomber. But it wasn't much help without an accurate map.

Looking back, those nights far from base camp seem as if they were of another world. We lay on our backs beneath stars and planets set like diamonds in the inky black of space and undimmed by any lights of human civilization. Meteors left blue-white trails across the sky, and manmade satellites hurried along on their journeys through space. No one on earth knew where we were; we barely knew ourselves.

* * *

The roll of RAF photographs fluttered in the wind as I tried to flatten them on the hood of the truck. Squinting in the bright sun, Delia and I studied a large lightly shaded area that appeared to be about fifteen miles south of our position.

"It's huge! It must be several miles across." In the aerial photo the pans looked much larger than any others we had seen in the Kalahari.

"There must be stacks of game there." Delia added. "Hyenas, too—and lions."

Because of our limited food, water, and gasoline, we were hesitant to leave the riverbed in search of the pans; it was our only landmark and navigational aid. But we needed to know to what extent the wildlife was using the pans, and if we drove due south and recorded mileage readings from the odometer as we went, it should be easy to find our way back—especially since we could follow our truck tracks through the grass. After double-checking our supplies, we turned straight south, toward the center of the big circular depressions we had seen on the map.

It was slow going. The ground was studded with grass clumps, pocked with holes, and spiked with dry bushes. Pitching and rolling in the truck, we could manage only two to three miles per hour. Every few hundred yards I stood in front of the truck and sighted along the compass to pick a tree, dune, or some other feature in the distance to aim for as we drove. We slowly made headway, but, battling the soft sand and tough thornscrub, the Rover had begun to use much more gasoline. Travel on the hard-packed soil and through the shorter grasses of the riverbed had been much faster. Even more worrisome was our water consumption; we had to stop every quarter of a mile or so to clean grass seed out of the radiator and pour several cups of water over it to cool the motor. As we churned along I was hoping that the pans would be where they were shown on the aerial photo. I was beginning to have doubts about having left the river channel.

Hours later, we stopped—hot, irritable, itching from the grass seed and dust. The spot where the large pan should have been had come and gone. After another look at the photos we drove farther south, then east, then west, getting more and more confused about where we were, relative to the pans. And by now we had lost our north-south spoor from the riverbed. I clawed my way to the top of a thorn tree; swaying in the wind and straining through the binoculars, I could see nothing but rolling sandveld in every direction. Every ridge, every stand of trees or clump of bushes, looked bewilderingly familiar and unfamiliar at the same time.

I climbed down, my legs and arms scratched and bleeding, my clothes torn. We glared once more at the RAF photographs, and I finally noticed that the edges of the pan were fuzzy and unclear, unlike the sharply defined features of those familiar to us near camp. I tried to think what had gone wrong.

"Unbelievable . . . *Unbelievable*!" I moaned. "You know what this is? It's a piece of dust! We've been driving for hours toward a piece of damned dust!"

Decades before, an RAF aerial reconnaissance crew had become careless, and a fleck of dust, charged with static electricity, had invaded their camera and left an impression of itself on the film. Enlargement had made the impression much bigger, so that it looked almost identical to the images of Kalahari pans. We had been searching for a phantom.

Going back was not just a simple matter of heading north for the riverbed. The valley was so indistinct in many places that, unless we found our tire tracks, we could easily drive right across Deception without ever knowing it. During all the driving, we had stopped keeping notes on mileages and directions; neither of us could remember whether we had last driven east of our spoor or west. In fact, we had not seen our north-south tracks in the four- or five-square mile area we had searched.

With Delia perched on the hood of the truck, I began driving slowly west, searching for the tire tracks that would take us safely back to Deception Valley. But after a few minutes of staring at the sea of waving grasses, our vision began to swim so much that we probably wouldn't have seen the spoor to the north if we had parked right across it. Forty minutes and two miles later, we turned back east, still looking. But it was hopeless and we were using too much precious gasoline and water. We turned and headed north toward Deception Valley.

Delia rode in the spare tire on the roof, where she should be able to see the channel of the riverbed. She *had* to see it. Bits of chaff, grass straw, and grasshoppers drifted into my lap from the windows and vents. My mouth was dry from the heat of the motor and the desert. I reached around behind the seat, had a swig of hot water from the plastic bottle, and passed it up to Delia.

I couldn't help wondering just how far a person could walk from where we were—wherever that was. Lionel Palmer had been hunting lions near the border of the game reserve one day when he and his native tracker saw ahead of them what appeared to be a man's head set down upon the sand. What they found was a fourteen-year-old native boy near death. After they had revived him with water, they learned that it was the morning of the third day since he had set off on foot from one cattle post to another. He had lost his way and had soon drunk all the water he carried in his goatskin. He walked only at night, covering himself with sand during the day in order to stay cool and help his body hold moisture. After two nights he had buried himself for what would have been the last time had Lionel not come along. I doubted that we could last for more than two days, either.

We passed through several shallow depressions ringed with catophractes, the brittle silver-leafed bush that fringes Kalahari pans and fossil rivercourses. We hoped the shrub was a sign that we had found the channel. I stopped and we both got up on the roof. Shielding our eyes from the glare, we tried to follow the nuances of the slopes around the shallow bowl; but none of them ran into the flat, open channel of the wandering riverbed.

The longer I drove, the more convinced I became that we had crossed Deception at one of its indistinct points and were driving to nowhere. We stopped, talked it over, and decided to go only three more miles. If we still hadn't found the river valley, we would turn back at an angle and shoot for another intercept, hoping to hit a spot where the old channel might be deeper and better defined.

Hunched forward over the steering wheel, my shoulders tight with fatigue and tension, I looked back at our last half-empty jerrican of water. Suddenly Delia shouted and banged on the roof. "Mark, I see our spoor! Off to the left!" Her eye had caught the faint line of tire tracks running through the shorter grasses of a small pan. I grabbed the water bottle and passed it up to her for a well-deserved drink. We were so relieved at the sight of those two arrows pointing the way back to the riverbed that we camped right there. The next morning we followed our tracks to the valley.

We had been gone from camp for only five days, but we had used a lot of water looking for the nonexistent pan. Logically, we should have headed straight back to camp and at least made it to the water hole on Springbok Pan. But there was still more of this end of the valley to see, and so we drove on, looking for a place to top up the jerricans. There had been no rain in this area for quite some time, and the several pans we found were filled only with mud and animal tracks. The Kalahari was drying up.

By noon the following day, the riverbed had become shallow, intermittently obscured by ridges of bush-covered sand, and more and more difficult to follow. We came to a grove of trees at a calcrete pan. Less than an inch of water covered the grey, muddy bottom, and antelope droppings floated on the surface. Never mind all that—it looked like an oasis to us. I shoveled out a deeper hole, and while we waited for the water to clear, we sat under a shade tree drinking tea and chewing strips of biltong. Later we scooped up water with pots and strained it through my shirt into our jerricans. When we had finished, I dug out another hole, and after stripping, we sat on the slimy bottom and bathed. After drying off in the wind, we smeared lard on our faces and arms to ease our burning skin.

The next day the ancient riverbed faded into the desert, so we turned back for home. Several days later we crossed the large conical dune at the turn in the valley and entered Springbok Pan. "Lions!" Delia pointed to an open stand of acacias: Two males and five lionesses were sleeping in the canopy of a fallen tree, beside a giraffe they had killed. The males, who had dark coats and thick, jet-black manes with halos of golden hair around their faces, raised their heads to look at us, yawning deeply.

We named the males Satan and Morena (which in Setswana means "a respected man"). The largest female we called Happy, and the others Dixie, Muzzy, Taco, and Sunny. Stonewall, a scraggly male adolescent, completed the pride. We set up our four-by-six nylon pack tent under the trees nearby, and early the following night we were able to ear-tag some of the lions. They all recovered well and, after sleeping off their hangovers, began to feed on their giraffe kill again. Later that night we scoured the riverbed for a couple of hours, searching for brown hyenas before heading back to the tent. I was tired from the darting, but Delia was determined to find a hyena, so she drove away to continue looking. I crawled into the pup tent for some sleep.

But I was too keyed up to doze off. I lit a lantern and set it just outside the gauze flap to keep the insects from crawling in, and propped up on one elbow, I began writing in my journal. Sometime later I heard a sound, rather like someone slapping his leg. It took a moment before I realized that it was a lion shaking his head. I slowly reached out and doused the lantern. I felt a little uneasy about my visitor, for we did not know these lions the way we did the Blue Pride. The moon was nearly full, but suddenly a great black shadow blocked it out. Satan stood within inches of where I lay.

At twelve feet four inches long and more than four feet tall, he could have squashed the tent like a bubble with just one paw. His shadow moved; a twang sounded and the tent sides shook. He had stumbled over a tie-down line.

Satan was very still for a few long moments, the shaggy silhouette of his mane against the side of the tent. His feet made a crisp rustle in the grass as he moved around the tent toward the flap. A second later he set one of his forepaws directly in front of me: I was looking right under his sagging belly at the riverbed. The belly tensed, he lifted his head, and his roar carried away into the valley. *Aaoouu-ah aaooouu-ah aaaooouuah-ah aaaooooouuah-huh-huh-huh-huh.* When he had finished he stood perfectly still, ears perked, listening to two lions answer from not far away. Then he walked over to them and joined their chorus of bellows, all three lying together in the moonlight.

In a few moments I heard the truck coming. "I came as soon as I heard them roaring," Delia said, unzipping the gauze and slipping in beside me. I was still stirred by my encounter with Satan.

"Incredible—*Incredible!*" was all I could say. It was not until after dawn that they moved away to the west, still bellowing at Bones and the Blue Pride, who answered from the valley six miles to the north.

* * *

We found the Springbok Pan and Blue prides as often as we could, knowing they would leave at the end of the rains. The Blue Pride was not difficult to observe, since our camp was a favorite point of interest along their route through the valley.

Our relationship with these lions had gradually changed. As we had learned to recognize facial expressions and postures that indicated their moods and intentions, and as they had become less curious about us, we found we had little to fear from them, as long as we did not create a setting that they might interpret as compromising or threatening. This is not to say that they had become house cats; we realized that they were still wild and potentially dangerous predators. Yet, even when we blundered into them during the many times they had come into camp, they had never done us any harm. We no longer hurried to the truck when they wandered into camp, but sat quietly under the ziziphus tree or at the fire while they moved around us. Because we no longer felt threatened, we could more fully appreciate and enjoy them in our close encounters. We were not just observing them, we were knowing them in a way that few people have ever known truly wild lions in their natural state, and this was a unique privilege.

When we began our research, most of the information about lions in the wild had come from studies by Dr. George Schaller on East African prides, particularly those in the Serengeti. Our observations were beginning to reveal that lions in different parts of Africa do not necessarily behave the same.

The portion of real estate used by a pride is termed its *area*, and it may overlap others.[1] The territory, a smaller portion within the area, is defended against intruders—lions from other prides and nomads. In the Serengeti, a pride may move its territory around within its area to take advantage of seasonal changes in the densities of prey. However, they still defend the territory against foreigners.

The rainy-season behavior and ecology of Kalahari lions was similar to the year-round behavior and ecology of the Serengeti populations. By spooring the Blue Pride, we had learned that during the rains their area is comparable in size to that of some Serengeti prides, roughly 130 square miles. However, because the prey communities are different, the diets of the two populations of lions are quite dissimilar: Serengeti lions feed mostly on wildebeest and zebra, whereas gemsbok, springbok, hartebeest, kudu, and giraffe make up most of a Kalahari lion's prey. Wildebeest are included when they are available.

Each East Aican lion pride has a nucleus of related adult females (grandmothers, mothers, sisters, and daughters), their young, and from one to three dominant males, who are unrelated to the older lionesses. The females usually remain in the same pride until they die, although a few may be forced to become nomadic if the pride gets too large. But when they are about three years old, young males are expelled by the dominant adult males. They become nomads, wandering widely and without territory, until they reach their full size and have well-developed manes, at five to six years of age. From two to five of these prime males form an alliance, or "coalition," which often includes brothers or half-brothers, and after collaborating in driving established older males from a pride area, they assume possession of its harem of resident females.

During the rainy season, Kalahari prides, too, are made up of several females who associate together. The difference—as we were to learn later—is that, unlike the lionesses of Serengeti social groups, they are often not closely related.

The behavior of the two groups is, however, very similar: In prides of both the Serengeti and the Kalahari there is a great deal of touching and camaraderie. While sleeping during the day, Sassy often rolled over and placed her paw on Blue's shoulder; Blue nuzzled Chary's flank; Chary's tail dropped over Spicy's ear; and so on throughout the pride. Everybody was in contact with someone else, except for Bones, who usually lay a few yards apart. The females hunted cooperatively, as well. In the evening and at sunrise, when they weren't sleeping, hunting, or feeding, they licked one another's faces and romped in play.

One of the most striking differences between Kalahari Desert lions and those in the Serengeti is related to the amount of rainfall in each area. Because the Serengeti normally receives more than twice as much rain, it has a greater number of large prey animals that are permanent residents. Furthermore, there are usually places for lions to get water year round. But in the Kalahari, as we have described, when antelope herds drift away from the fossil river valley, the lions disappear for months and stop defending their riverbed territories. The questions of how much their ranges expanded, what they ate, and where they found water to drink were intriguing. But we were especially curious about how their social behavior changed in response to diminished prey resources and other ecological constraints. It was our desire to answer these questions that eventually led us to new and exciting discoveries about desert lions, and lions in general.

*   *   *

In the meantime we began to study more about how lions communicate. When pride members are together at close range, they signal their moods and intentions with a combination of ear, eyebrow, lip, tail, and general body postures. Even the pupils of the eye have expressive value.

Blue was resting with the Blue Pride in Easter Island one morning when she noticed a lone gemsbok, an old bull, entering the riverbed at South Pan. Her ears cocked forward, her eyes widened, she lifted her head, and the tip of her tail began to twitch. Seconds later Sassy and Gypsy had picked up her cues and were looking in the same direction. Blue had as much as said to them, "I see something interesting over there."

After they had killed the gemsbok, Bones arrived, intending, as usual, to take the carcass from his females. Sassy faced him, her eyes little more than slits. With her mouth three-quarters open, she bared her teeth, wrinkled her nose, spat, and growled. She was expressing defensive threat, saying in effect, "I'm not going to attack you first, but you had better not try to take my carcass." Unfortunately for her, Bones took the carcass in spite of her threats.

After having snarled, growled, and cuffed one another during feeding, lions make up by engaging in an elaborate face-licking and head-rubbing ritual. By the time they have washed all the gore from one another's faces, peace has been restored to the group.

A lion often locates others and advertises its claim to a territory by roaring or bellowing. To roar, a lion draws air deep into its chest, tightens its abdomen with great force to compress the air, and then releases it through its vocal cords, the sound erupting from the throat with such energy that it carries great distances. Occasionally, when the Blue Pride assembled around the truck, roaring in unison, the metal floor buzzed in sympathetic resonance.

A lion's roar consists of three parts: The first one or two sounds are low moans; these build in volume and duration to a series of four to six full bellows, followed by a number of grunts. Both males and females usually roar while standing, their muzzles pointed forward, parallel to the ground, or slightly lifted. But they may also roar while lying on their sides or while trotting.

We noticed that Kalahari lions roared most often when the air was still, moist, and at its most efficient as a conductor of sound. They almost always roared after a rainstorm and during that part of the night when the relative humidity was highest, from about 4:00 A.M. to half an hour after sunrise. In the valley, and under the conditions described, the sound carries the farthest, up to eight miles, to our comparatively unspecialized human ear. Sometimes the Blue Pride also roared in apparent response to the morning or evening calls of jackals, who also vocalized soon after storms.

In the dry season, however, we could hear a lion no more than a mile and a half to two miles away. Actually, in fact, they rarely roared at that time of the year, possibly because large antelope prey were so scattered that it was not economical for them to spend the energy to advertise and defend territories; or perhaps it was a waste of energy to try to communicate through the dry air. It may also have been that subgroups of prides were so spread out, looking for food, that they would not likely have heard one another even if they had roared.

Whether or not pride-mates are successful at locating each other by roaring depends on whether the recipient of the call chooses to answer. Bones regularly became separated from his lionesses, especially if he had taken their kill from them. The females would move on until they made another one, often miles away from him. One, two, or three days later, when he had finished his carcass, Bones was faced with the problem of finding his pride. He would walk in the general direction they had taken, roaring as he went and listening for answers. In the rains, when the territory size was comparatively small, his roar carried the full length of his domain, and he could reach the females wherever they happened to be. Usually they answered him, and the pride would reunite.

But sometimes the lionesses seemed less than anxious to get in touch with Bones. On several occasions he walked down the riverbed roaring and passed within several hundred yards of where the females lay silently in the bushes. He called repeatedly, smelling the ground and looking in every direction as he continued on down the valley. But for some reason, perhaps to protect their kill, his pride would not answer him. When in estrus, however, it was often the females who first put out a call for Bones. With them it was, at times, a matter of "Don't call us; we'll call you."

Lions can also coo as gently as a baby. This *aaouu* sound is tossed softly back and forth among them when they are moving through thick cover. It apparently helps them keep track of each other, as well as providing mutual reassurance in uncertain situations. Sometimes at night we were able to find and follow a pride through the bush by stopping the truck and listening for this genial call. In the early years, before the Blue Pride lions felt completely at home in our camp, we were often awakened by their coos as they moved through our tree island investigating the tents, water drums, and other pieces of equipment that were strange to them.

A third way lions communicate is through olfaction—by scent-marking and smelling. The Blue Pride walked through the valley at night along scent-paths, often coincident with trails made by antelope or our truck. In most areas the trail was defined only by scent-marking; there was no visible path. Taking one of these routes, Bones would often stop at a bush or small tree, raise his head into the lower branches, close his eyes, and rub his face and mane against the leaves, as if reveling in the scent from a previous mark, and perhaps also applying it to himself. Then he would turn, raise his tail, and spray urine, mixed with secretions from two anal glands, into the branches. Certain bushes and small trees along his route were irresistible favorites, including the acacia bush next to the window of our tent. He would never pass by without giving it a squirt or two. To our unsophisticated noses his odor never lasted for more than several minutes after he had gone. The females also marked bushes, but only occasionally.

Sometimes these bushes became visual signposts as well. Bones never failed to spray a seven-foot albizzia tree on North Pan when he walked by. Its bark had been shredded by the Blue Pride lionesses in sharpening their claws, and its limbs had been twisted and broken because they couldn't resist playing in it—all of them at once. After three or four of the lionesses had managed to get into the canopy, another would try to climb up the trunk. One of them would end up hanging beneath a limb while the newcomer stood on top. Rumps and tails poked from every quarter of the poor tree, until the inevitable happened: A limb broke, dropping the lionesses to the ground. In the end, the albizzia was reduced to a tangle of woody rubble that Bones, nevertheless, continued to spray every time he passed.

A scrape mark is another type of olfactory and visual signpost used by male and female lions. The sign is made when the individual hunches its back, lowers its rump, and rakes its back feet over the ground, tearing up the turf with its claws while dribbling urine into the soil. Lions mark territory in this way, and scrapes are often made while roaring to foreign prides. Two young males who had recently taken over a new territory, scraped twenty-six times in three weeks along a 400-yard stretch of our truck spoor on the riverbed. By comparison, the older male they had replaced usually scraped once or twice along the same route in a similar period of time. The youngsters also jetted the same bushes he had marked. They were making sure that every lion in the valley knew they now owned this piece of real estate.

Besides marking territory, scent probably identifies the lion that left the mark and indicates how long ago it passed the spot. It also communicates the condition of females in estrus. George Schaller reported that Serengeti lions can locate each other with scent, and he observed one male track two others for a kilometer by smelling their trail. Kalahari lions appear to be less successful at this, especially in the dry season, possibly because the scent denatures more quickly in the arid desert heat. We once watched Bones circle, his nose to the ground like a bloodhound, searching for Sassy, who had left him only thirty minutes earlier for better shade; she was a mere 200 yards away. He kept losing her scent and circling back to their former resting place, but he could have seen her if he had only looked in the right direction. When he finally did stumble upon her, he turned back his ears, squinted, and looked away. If I hadn't known better, I could have sworn he was embarrassed.

Whenever he was smelling a female's scent, Bones would lift his head and raise his lips to expose his teeth. Then, as air passed through his pharynx, he would wrinkle his nose in a grimace. This behavior, termed *flehmen*, is a way of "tasting" the scent, or better discriminating its chemical message by passing it over a special pouch, filled with sensory cells, that is located in the roof of the mouth. A lion showing flehmen reminds me of a wine connoisseur who draws air into his mouth and breathes it out his nose to better experience the bouquet and flavor of his selection.

* * *

Lions generally kill large antelope by suffocation. First they knock or pull it down, and then seize and hold its throat, or occasionally clamp their jaws over its muzzle. I had always been curious about how they could accomplish this with a giraffe, who may weigh up to 2600 pounds and whose throat may be seventeen feet above the ground. Late one afternoon the Blue Pride showed us their giraffe-hunting technique. They had eaten little more than a gemsbok calf and a springbok fawn for several days—not much for over 3000 pounds of hungry lions. After spending the day in Tree Island on South Pan, they began to hunt through the open woodlands of West Dune. A light rain began to fall, and they lay down along either side of a heavy-game trail used by antelope to cross from the fossil river to the bush savanna. Their heads were raised and their ears were perked to catch any sound, each one of them looking in a slightly different direction. For nearly two hours they had scarcely moved, lying like statues. Instead of stalking, Kalahari lions often hunt by waiting along game trails, especially where there is little cover.

But now all the females drew themselves to their haunches, leaning forward, their muscles bunching. Near the foot of the dune a large bull giraffe walked into view, browsing the green leaves from the tops of the acacia trees. Chary and Sassy were closest to him; they slowly rose to a low crouch and each began a divergent course around the unsuspecting giraffe. Liesa, Gypsy, Spicy, Spooky, and Blue spread out in an arc across the trail. Over the next hour they stalked slowly toward their prey, using grass, bushes, and trees to cover their approach. At the same time, Chary and Sassy managed to skirt the giraffe and hide in the grass beyond it, along the same trail but farther west.

The five lionesses who were working together got to within thirty yards of the giraffe. Suddenly he wheeled and went thundering down the trail toward the dune, his tail curled tightly over his rump, flinging chunks of sod from platter-sized hooves. When it seemed they were about to be trampled by the 2000-pound bull, Chary and Sassy sprung the ambush. The giraffe dug in his feet, trying to stop and sidestep the lions charging from both front and rear. But his hooves failed him in the wet sand. Like a collapsing tower, he slewed forward out of control, right into Chary and Sassy. Instantly the other lions were at his flanks, raking and tearing at his belly and sides. The giraffe bolted forward again, trying to outrun the lions, but Blue locked her jaws around his right hind leg just above the hoof, set her own legs stiffly in reverse, and hung on.

For twenty-five yards the giraffe staggered forward, his eyes white and his breath ragged, dragging the lion clamped to his leg. Refusing to release her hold, Blue's claws plowed up clumps of grass and left deep furrows in the sand. The others ran along beside the bull, slashing at him until his entrails burst from his body. Finally he collapsed, flailing weakly at the gang of predators.

There was no way Bones could drive his hungry females from this mountain of meat; there were too many of them and too much of it. But during the week that the Blue Pride spent at this giraffe kill, we noticed that the relationship between Bones and the two young males, Rascal and Hombre, had changed dramatically. The youngsters were now nearly three years old, and shaggy ruffs showed where their manes were coming in. Their very presence seemed to incense Bones. At first he wouldn't let them feed at all, driving them off with snarls whenever they ventured too near him at the carcass. Only after he had sated himself did they manage to snatch a few bites.

By restricting their food supply Bones was forcing independence on Rascal and Hombre; before long they would leave the pride to become nomads. The next two or three years would be a critical period for them, without females to help them hunt. In the coming dry season, prey would be scarce and there would be little cover for hunting, and what was more serious, their predatory skills would still be dangerously underdeveloped. They could easily starve to death—many young inexperienced males do—before they were big enough and aggressive enough to acquire a pride of females and a territory. Somehow they had to survive together until the rains, when hunting would be easier.

In the Kalahari Desert it may be more important that young male lions learn how to hunt on their own than it is in a more moderate climate, such as that of East Africa. As adults they will be separated from their pride females more often, and for longer periods, than males in areas where prey is more readily available and lion pride territories are generally much smaller. When a Kalahari male appropriates a kill from his females and they move on, it frequently takes several days for him to find them again. During this time he may have to hunt alone, killing somewhat smaller prey like springbok, young gemsbok, and steenbok.

Rascal and Hombre were growing up fast, and as the weeks passed, they were less and less inclined to back down in confrontations with Bones. They would often seize pieces of carcass, snarling and threatening him—muzzle to muzzle—before he cuffed them into retreat. They were developing the aggression they would someday need to take over and hold a pride area and its females.

* * *

During these early years we learned a great deal about the wet-season diet of Kalahari lions by watching the Blue Pride and Springbok Pan Pride hunt. To supplement this information we collected, dried, crushed, screened, sorted, weighed, and identified bits of horns, hooves, bone chips, and hair in dozens of lion scats. One day I called Mox to join us at the edge of camp, where we sat with bandanas tied over our faces, a smelly cloud of white dust rising around us as we smashed lion feces with a hammer. He arrived just as I was pouring the powdered remains of a scat to be weighed into an extra dinner plate. When he saw what we were about, he clamped his hand over his mouth— "Ow!"—shaking his head and staring in slack-jawed disbelief.

But before long Mox—though a little reluctant at first—had disappeared in his own white cloud, hammering and grinding away at a pile of feces. A day or two later, however, I noticed that he was no longer bringing his enamel plate to our camp to be washed with our dishes.

11

The van der Westhuizen Story

Delia

It is not easy to remember
that in the fading light of day...
the shadows always point toward the dawn.
— *Winston O. Abbott*

WITH A LONG sweeping motion Mark stripped the silver-grey leaves from a thin catophractes branch. He dipped the stick very slowly into the drum of gasoline, pulled it out again, and pinched the spot where the coating of liquid ended. "This has to last us for eight more weeks."

It was May 1976, twenty-one months since we had received the $3800 grant from National Geographic. Once again our money was nearly gone. Without another grant very soon, we would have to abandon our research and earn the funds to get home. We also desperately needed money for radio-tracking the brown hyenas and lions, who were difficult to follow in the thick bush savanna, where they spent most of the dry season. In this habitat we could usually only follow the hyenas for an hour or so before losing them. And so far, we had no idea where the lions traveled in the hot months. We had done just about all the research on them we could do, without more sophisticated equipment.

A few days after Mark had checked our gasoline supply, a bush plane zoomed down the valley just above the treetops and buzzed camp, circling and dive-bombing the island like a mobbing bird. We ran out just in time to see a small bundle tumble out the window and the aircraft waggle its wings in salute and speed away. Our mail from Maun, tied up with string, lay in the grass. We never found out who had done us this favor.

We opened the package and found a handwritten message from Richard Flattery, Maun's new bank manager, telling us that a Mr. van der Westhuizen would be in the village soon with some money for our project. Van der Westhuizen was the name of the director of the South African Nature Foundation, to which we had applied for a $20,000 grant. Declaring the night a holiday, we celebrated with pancakes and homemade syrup.

Next day we packed the Land Rover and started for the village before the sun had reached East Dune. Night had fallen when we wound our way through the jumble of earthen huts, each softly lighted by a flickering cook fire and shrouded in a drifting haze of smoke. The Flattery's house was a flat-topped stucco standing opposite the reed fence of Dad Rigg's place. Through the screen door, patched and repatched with an assortment of mesh, we could see Richard cleaning fish over a bucket. His wife, Nellie, was frying fresh bream over a gas stove.

"Glad to meet you...heard all about you...yes, Mr. van der Westhuizen has money for you. We'll tell you all about it—stay and have some food and a cold beer."

We sat down to a meal of fried fish, potatoes, and fresh bread in a small raftered dining room that might have been in an English cottage, except for an active termite mound protruding through the floor.

As it turned out, Richard knew little about the grant—only that Mr. van der Westhuizen would be arriving in Maun the very next morning. Later on, after a pleasant evening, we asked Richard to invite our prospective sponsor to join us for lunch the following noon at the "Riviera," a ramshackle retreat set on the banks of the Thamalakane River.

The owner of the Riviera, an innkeeper from Selebi Phikwe, had given us permission to use the camp on our supply trips to Maun. The complex consisted of five dilapidated reed-and-straw huts that clung to the steep banks of the river like abandoned birds' nests. The largest hut, which we used, had a partially caved-in roof and leaned heavily toward the river, straining against guy wires that tied it to a massive fig tree. The shaggy encampment, a welcome refuge from the desert, was all but hidden in tall grass. We pulled two rusted camp beds from under the fallen section, swept off the spotted mattresses, and hung a mosquito net—more mends than mesh—from a rafter over our sleeping bags.

Mark Muller, a young bush pilot, also stayed at the camp, in one of the smaller thatched huts. The next morning we were awakened by a grinding clatter before dawn. Muller was starting his ancient Land Rover, a roofless relic resembling a World War II German staff car. He left it ticking over idly at the top of the bank while he went back to his hut for something. The next thing we knew, the nose of the truck crashed through the wall of our house with a splintering of reeds, stopping six feet from our bed. The hut swayed dramatically around us as thatch, reed, and poles rained down. We both jumped up, afraid the house might collapse any second, but it slowly steadied itself. Muller ran down the hill after his runaway truck, muttering incoherently to himself. "Sorry," he said and backed out of our bedroom and drove away.

We began immediately preparing a special lunch for our meeting with Mr. van der Westhuizen. Mark stoked up a rust-eaten pot-bellied wood stove that squatted beneath the fig tree, and, tears streaming down my face from the cloud of smoke that belched from the stovepipe, I baked a loaf of orange bread while Mark went for supplies. About noon we laid out a lunch of cold sliced mutton, fresh fruits, and hot bread—the most extravagant meal we had prepared since coming to Botswana.

Sitting on tin trunks on the reed house veranda, we ate our lunch with Mr. van der Westhuizen, a soft-spoken man with greying hair and a slight limp. In the wide, lazy river, a few feet away, coots splashed among the reeds, and on the opposite bank, a group of baboons moved toward the water's edge to drink.

As Mr. van der Westhuizen quizzed us about our research we grew more and more puzzled. He seemed to know almost nothing about us or the nature of our work.

Finally Mark asked, "Haven't you read our proposal?"

"Proposal?"

"The one we submitted to the South African Nature Foundation."

"I don't understand. Oh, . . . I'm afraid there has been some mistake. I'm not from the Nature Foundation." He went on to explain that he was an architect from Johannesburg who had heard about our research and wanted to donate $200 of his own money to our project.

Two hundred dollars would barely fill up our auxiliary gas tank and pay for the trip to Maun. We tried to conceal our dismay, saying, "We really do appreciate your contribution, it couldn't have come at a better time." But, it was no use. We heard little more of what he said, and after an eternity Mr. van der Westhuizen drove away in his shiny new truck. We stared silently at the river.

* * *

A vise crushed both sides of my head, and a sharp wedge pressed down from above, splitting my brain. The pain of resting my head on the pillow was unbearable. I tried to sit up, but a wave of nausea swept over me. Under the soft mesh of the mosquito net Mark slept restlessly beside me. Without moving my head I nudged him, "Mark...some pills...I must have malaria."

He felt my forehead, then eased from the bed, and brought me six bitter chloroquin tablets from our first-aid kit. I swallowed them with great difficulty. He carried me to a mattress on the floor of one of the smaller huts, which had no holes in its walls. There was no reason to take me to the mission clinic in Maun, which had nothing better for malaria than chloroquin and where there was a good chance of picking up tuberculosis or something worse. In the rainy season Maun was rife with malaria. According to the hunters, "You either take the pills, sweat out the fever, and get better, or you die."

The hut was dank, dark. I was buried under heavy blankets of scratchy wool, but I was still stone-cold, my skin clammy. Mark lay next to me, trying to keep me from shivering, but I could feel no warmth. The blood in my head pounded against my skull, and a brilliant light from one tiny window stabbed at my eyes.

Then my body began to burn. With all my strength I shoved Mark away and threw back the covers. The sheets were damp and a putrid odor smothered me. For a long time my mind floated in darkness, and then there was a kind of peace. I saw home, live oaks and Spanish moss, the red-brick house where I grew up, and Fort Log, built with pine logs as a fortress against some imaginary neighborhood Indians. But when my thoughts tried to focus, I thrashed in the bed and cried out. Home was far away. *Clickety-clack, clickety-clack, you can't get off and you'll never get back. Clickety-clack.*

After a long time, the light from the hut's window grew softer and my mind began to clear. Tap-tap-tap-tap. We would stay in Africa somehow, and make it work. Tap-tap-tap. Mark was working on a borrowed typewriter set up on a tin trunk near my mattress. He came over to me. Clean sheets and warmth, a snug fresh feeling caressed me. His familiar smile, a kiss, hot soup, and cold, cold water welcomed me back. I tried to get up, but a firm hand gripped my shoulder and pushed me back . . . rest.

During the days that I had been delirious with fever, Mark had stayed at my side, writing proposals to conservation organizations all over the world, describing our progress and needs. When I was much better he drove into Maun one morning to mail the stack of thick envelopes. I propped myself up on pillows and waited for him to come back. Though still a bit woozy, it felt good to sit up. I watched two scimitar-billed hoopoes flitting about in the fig trees just outside the window. An hour later I could hear the Land Rover growling its way back through the sand.

"Hi, Boo. Glad to see you sitting up," Mark said quietly. He sat on the edge of the bed. "Feeling better?"

"Yeah—I think we can get back to the desert soon." I smiled at him.

"Well, we can't rush it," he said. He walked to the small window.

"Didn't we get any mail, any news from home?" I asked.

"Uh . . . no." He went on staring blankly at the river beyond the trees.

"But isn't that a letter from Helen?" I had recognized one of my sister's personalized envelopes tucked in the back pocket of his cut-offs.

His hand shot to his hip. He turned and came to the bed, his face full of pain. "God, love, I didn't want to tell you until you were stronger. There's some bad news. It's your dad. He died of a heart attack about six weeks ago."

I sank numbly back in the bed. "My mother—what about my mother?" I heard myself ask. "And we don't even have the money to go home."

My father had been one of our staunchest supporters, writing letters of encouragement, sending addresses and reference books, not to mention the newspaper clippings about football games that piled up in our post box at Safari South over the months.

Mark lay down beside me. One of the hardest things to bear during our seven years in Africa was being away from home at such times. While we were gone, Mark's mother passed away, and his grand-

THE VAN DER WESTHUIZEN STORY / 169

mother. And besides my father, I lost my grandmother. And I missed the marriage of my twin brother. We struggled with feelings of guilt because we were not at home to help our families through the difficult times, or to celebrate the good ones.

"If you want to go home, I'll get the money from somewhere, Boo," Mark told me.

"Let's make our project succeed, that's the best thing we can do." I whispered.

When I was finally strong enough to go into the village to see the doctor, he warned that I had not only malaria, but also hepatitis, mononucleosis, and anemia. "You must not try to go back to the Kalahari for at least a month," he said sternly in a thick Swedish accent, as he peered over his spectacles. "You must rest, or run the risk of a relapse. If that happened out there, you would be in serious trouble."

But I could rest in camp just as easily as in the dank hut at the river, and we had to do as much research as possible before our money was gone. So I didn't pass on the doctor's comments to Mark and, instead, pretended to be feeling better than I really did. Three days later we were ready to leave for Deception.

On our way out of the village, our friends at Safari South, always ready to help in one way or another, loaned us a high-frequency, long-range radio. This meant that at noon every day, at least during the safari season, we could be in contact with the hunters in the field, or with someone in their Maun office. For the first time since our project began, we would be able to reach the outside world. But unless we received a grant soon, this would be our last trip to the Kalahari.

Back at camp we rationed our gasoline, food, and water more strictly than ever. Using only one point three gallons of gas each time we followed hyenas at night, and one gallon of water per day, we could last three months. By then we should have received word from our new grant applications. In the meantime we would get some solid data on lions and brown hyenas. At first I was too weak to stand the pounding in the truck, so I rested in camp while Mark followed the hyenas or lions by himself. But I slowly recovered, and for eight exhausting weeks we worked with mad enthusiasm, knowing we would soon have to leave Deception Valley.

*　　*　　*

"Zero, zero, nine, do you read me?" came the garbled voice of Phyllis Palmer on the radio.

"Roger, Phyllis, go ahead."

"Delia, Hans Veit, the director of the Okavango Wildlife Society, is in Maun. He would like to meet with you to discuss a possible grant for your project. Can you come in? Over."

We looked at each other and rolled our eyes. It might turn out to be another van der Westhuizen story, but what choice did we have? "Roger, Phil. We'll be in touch as soon as we get in. Thanks."

In Maun, two days later, we were relieved to find that Hans Veit really was the director of the Okavango Wildlife Society and that a grant for our research was very likely. But we would have to go to Johannesburg for further discussions with the society's Research Committee before a final decision could be made.

Once in the city, we negotiated a grant with the society for two years of research in the Kalahari. The funds would allow us to get a better second-hand truck, a tent, and, most important, to make a round trip to the United States to see our families, consult with American researchers, and buy much-needed radio-tracking gear for the lions and brown hyenas. To be able to follow the predators consistently in the long dry season would mark a major turning point in our project.

But the first thing we did in Johannesburg was to walk to a bakery. Standing in front of glass cases filled with small pink and yellow iced cakes, chocolates bulging with nuts, cookies covered with cherries, and puffy cream pastries, Mark and I ordered two of everything in the shop. Carrying our stack of neat white boxes tied up with string, we walked to a green park and sat in the sun. After inhaling the sweet, warm aroma of the freshly baked goods, we took a bite from every one and finished off our favorites. Laughing and talking, our lips covered with powdered sugar, we lay on our backs to rest our aching stomachs.

12

Return to Deception

Mark

Though the sky be dark, and the voyage be long,
Yet we never can think we were rash or wrong,
While round in our Sieve we spin.
—*Edward Lear*

IN OCTOBER 1976, we flew back from New York to Johannesburg and found the city buzzing with ugly reports of the terrorist war in Rhodesia. The conflict was spilling across the Botswana border near Francistown. Farther south, along the only main road to the east and north, terrorists were beating and shooting travelers at roadblocks set up along the 500-mile route to Maun.

It had been four hectic weeks since we had left Deception Valley, and we were anxious to get back and to begin radio-tracking lions and hyenas. But it would be risky to try to enter Botswana at this time. The Soweto riots had still been smoldering when we had left Johannesburg for the United States, and now there was the threat of total anarchy to the north, perhaps all over southern Africa.

For some months Botswana had resisted getting embroiled in the conflicts on her borders. But now it was rumored that terrorist training camps had been established near Francistown and the village of Selebi Phikwe. In the past months, Angolan refugees had begun to appear in Maun, many of them suspected terrorists, and the mood of the native people in the village had begun to sour toward whites. On one tense occasion, Delia had been heckled by a group of men while shopping at the Ngamiland Trading Center. Such a thing would have been unheard of even two years before. An atmosphere of fear and suspicion had begun to drift through the settlement, like smoke from the evening cooking fires. Remote as it was, Maun had become tainted by world politics.

In response to the conceived threat from Rhodesia, Botswana had hurriedly formed the Botswana Defense Force, or BDF. Together with mobile police units, this ill-equipped instant army was to roam the countryside seeking out the insurgents who were supposedly infiltrating from Rhodesia and South Africa. We had heard a number of reports about innocent people having been injured or killed either by terrorists, the BDF, or the mobile police—no one seemed to know for sure who was responsible.

We had promised ourselves and our families that if the political turmoil became too threatening, we would delay our departure for the desert, or even leave the country altogether. But as we packed our gear, we rationalized that we might wait forever and never hear more than rumors, and that we would be safe once we got to our remote Kalahari camp. The worst terrorist incidents were mostly confined to Francistown, on the border with Rhodesia, but when we passed through there on our way north to Maun, we planned to make it during midday hours, when we thought it least likely there would be trouble. We bought a second-hand Toyota Land Cruiser, loaded it with a ton of supplies, and began the long trip north.

Early in the afternoon of the first day we reached the Botswana border, where the road abruptly changed from macadam to gravel, with the usual corrugations and deep ruts. Other traffic disappeared. We drove on alone, a dust trail rising behind us. Our tires crunching on loose stone, we hurried past meager mealie patches and occasional mud huts enclosed by bomas of thornbush. No one waved, and if they looked at us at all, they seemed to scowl.

On a lonely curve in the road, a wooden pole, freshly cut and still covered with bark, blocked the way. Ten or fifteen black men stood to one side—were these police, terrorists, or soldiers? No uniforms; but that didn't mean much. Several of them carried stubby olive-green submachine guns slung at waist level; others held rifles. My skin prickled with fear and my hands wrung the steering wheel. I wanted to keep going, but they strolled across the road and stood facing us, their weapons ready. We had no choice but to stop.

We locked the doors and I rolled down my window, leaving the Land Cruiser in gear, and its motor running, my feet poised on the clutch and accelerator. I would break through the roadblock if they ordered us to get out. A young black with blood-shot eyes strolled up unsteadily, his machine gun leveled at the door. He pushed his face to the open window, and I could smell *bujalwa*, the native beer, on his breath. The others peered into the back, lifting the canvas, pointing to the cases of canned food, the new tent, and the other supplies, while they chattered among themselves. The youth at my window began firing questions at me, his finger on the trigger: Who were we? Where were we going? Why? Whose truck was this? Why was it registered in South Africa? How could just the two of us need so many cans of powdered milk and so many bags of sugar?

After a while, the men at the back started conferring at the side of the road, talking quickly in a language we didn't understand, and the youth joined them. I stifled an urge to let out the clutch and make a run for it. They had no obvious vehicles and could not pursue us, but I was afraid they would open fire.

"Don't get out of the truck...Get down on the floor if I tell you to!" I whispered to Delia.

The men looked our way as the hard-eyed teen-ager strutted toward the truck, his gun still leveled at my door. He leaned through the window glaring at me, silent. My stomach turned over as I remembered one of the stories we had heard in Johannesburg. A young teacher from Europe, on his way to a school in northern Botswana, had been dragged from a bus and savagely beaten in the face with rifle butts — they hadn't liked his beard. Before we had left the city, Delia had pleaded with me to shave mine off. "They're just rumors," I had tried to console her. Now I wasn't so sure.

"You — go." The words came thick and slow. I wasn't sure I'd heard him right.

"*Go siami*—okay?" I asked. Saying nothing, he stood back from the truck. Without taking my eyes off him, I slowly let out the clutch. The other men stared at us as we began to pull away. I pushed the gas pedal flat to the floor, accelerating as fast as I could. In the rearview mirror I could see them watching us, and the skin on my back crawled. A young girl had been shot in the back just days before as she and her parents were driving away from a similar roadblock. "Keep down!" I shouted to Delia and hunched forward over the wheel, urging the truck toward a bend in the road.

Several miles later I pulled over and held Delia close for a moment.

We were both wrung out from the encounter. When we had regained some composure I unfolded a map of Botswana. "We've got to get off the main road as soon as we can," I said, looking at the great white void that was the Kalahari, in the middle of the chart. "There has to be another way to get there, even if we have to cut cross-country." But from that point it was more than 200 miles overland to Deception Valley. Even with the extra fifty gallons of fuel in our new reserve tank and the ten gallons of water in our jerricans, I didn't see how we could make it.

"Mark, what about the old spoor Bergie mentioned that time, the one that comes up from the southeast?"

"That's an idea... If we can find it, we'll take it."

We drove off the road and about 200 yards into the bushes, where we slept the night fitfully, rolled up in our new yellow tent, afraid it would be seen from the road if we set it up. The next day, in a small village on the edge of the Kalahari, we crossed a shoal of white calcrete and drove through a kraal and around the corner of a trading store. Suddenly we found ourselves heading into the desert on what looked like a cattle trail. It worried its way in several different directions around dry washouts and pans, but according to our compass, we were headed in the general direction of Deception Valley, still 190 miles away across the desert.

By nightfall we were miles from the last road, the last village, the last contact with humans. From here on we would reckon distance in time, and time by the positions of the sun, moon, and stars in the sky. I slipped off my watch and laid it in the ashtray; it spoke the language of another world.

No longer afraid to be seen, we built a fire that night and sat talking quietly about how good it was to be back. A lion roared from the west, close. Slowly the tightness drained from my chest, and I relaxed again for the first time in weeks. The trappings and anxieties of man's artificial world—the airport crowds, the city traffic, the wars and Watergates—were behind us. Primitive, unscarred Africa embraced us again. As we settled into our bedrolls, we wondered whether any rain had fallen in the valley, and how Mox had handled village life while we were gone.

Thick thornscrub, hot, heavy sand—the going was tough beneath a broiling midmorning sun the next day. We drove farther and farther into the desert, and the track began to lose itself in the grass and thornbush. Would it disappear altogether? Giddy from the 120-degree heat, we wiped our faces and necks with a damp rag whenever we stopped to cool the steaming radiator. Delia rode with her feet on a cardboard box; we could have fried a steak on the metal flooring.

"I smell gas!" I jammed on the brakes, but by the time we jumped out, gasoline was pouring into the sand from every corner of the truck bed. The big auxiliary tank had shifted, snapping off its out-flow line. The precious fuel that we needed to get us to camp was quickly running away.

"Hurry! Get something to catch it in!" We both ran to the back of the truck and began rummaging through the load. But nothing would hold more than a gallon. I dove under the Toyota and rammed my finger into the hole I had drilled to bring the fuel line through the truck bed. But the tank outlet had moved away from it and had come down flat against the steel; there was no way to reach the broken nipple to plug it.

I grabbed a packet of putty from the tool kit in the cab, and sprawling in the sand again, my shirt and shorts soaked with drizzling gasoline, I frantically tried to force a gob into the small space. But I couldn't get to the break, and the putty wouldn't have sealed it if I had—not with gasoline spraying out.

The tank was bolted to the bed of the truck, with the fuel line beneath; above were the truck's iron gate and a ton of supplies. In a fit of panic, we began throwing cases of tinned food, jerricans of water, tools, and other equipment from the back, while our gasoline continued to stream away. I grabbed a wrench from the toolbox, and while Delia hurled things off the truck, I sat astride the tank, working with clumsy, frenzied movements to unbolt the iron gate and steel straps that held it to the truck bed.

Minutes passed; the sound of the gasoline dripping into the sand was maddening. Working feverishly over the tank, I kept saying to myself, There must be something I've overlooked... *think!*

Finally the tank was free, but no matter how hard I strained, with my hands around its wet, slippery bottom, I could not lift it. I jammed the spade under it and tried to lever it upward—still it wouldn't budge. My tennis shoes were soaked with gasoline, and for the first time the thought of an explosion and fire shot through me.

I finally got control of myself, stopped fighting with the tank and began to think more clearly.

"Pour the water out of one of the jerricans—we'll have to fill it with gas!" I shouted.

I grabbed a roll of siphon hose from behind the front seat, rammed one end into the tank, and sucked on the other. Gasoline gushed into my mouth. Choking and spitting, I stuffed the hose into the jerrican. After we had filled the second can, I hopped into the truck, seized one end of the long tank, and heaved. Now that it was lighter, we finally managed to pick it up and stand it on end. The fuel ran away from the broken line and, after it had dried, we plugged the hole with putty. The rest of our gas was secure.

We sat on the truck bed sick with exhaustion, our mouths like cotton, and I was still trying to spit out the raw taste of gasoline. Then I noticed that the lids were off *both* jerricans. In the confusion to rescue our fuel, we had poured out all of our water.

Even so, I wasn't sure we had managed to save enough gas to take us to Deception Valley. We had no way of knowing how deep the sand ahead would be, no way to predict how fast the truck would use fuel. If we ran out of gasoline, we would have to walk, and even if we moved only at night, when it was cool, we wouldn't get much farther than about twenty miles on foot without water. I ran my hand up and down the side of the tank, trying to feel the coolness that would indicate the level of the liquid inside. If we could at least be sure we were on the right track...

We sat in the shade of the truck and pushed our bare feet deep into the cool sand below the surface. Our choice was to go ahead with no water and limited gasoline or to turn back and travel the main road through Francistown, risking confrontation with terrorists or the military. Suddenly Delia hurried over to the cab and tilted the front seat forward. She had filled an extra two-quart water bottle and put it there before leaving Johannesburg. Small consolation—except for a sip or two, it would have to be saved for the radiator.

While we were picking up our scattered tins of fruit, I thought of the syrup in the cans; we could drink it on our way to Deception Valley. So, having decided to go on, we tied the upended auxiliary tank onto the bed of the truck, reloaded our supplies, and sat in the shade waiting for nightfall and coolness. At sunset we stashed some cans of fruit behind the front seat, and I poked small holes in the tops of two others with my pocket knife. We set off through the desert evening, sipping the cloyingly sweet juice to quench our thirst.

For hours we followed the yellow patch of our headlights, mesmerized by the continuous sea of grass that swept toward us. Finally, nodding and too sleepy to go on, we stopped. We stood beside the track and breathed deeply of the cool night air. The sweet smell of the grasses and the bush was refreshing. Checking our course again and again with the compass, we drove until sunrise and then slept in the tire tracks beneath the truck, until the heat of the day grew intolerable.

The next night, each time we stopped to siphon more gasoline from the reserve tank to the main tank, I grew more worried. By morning its level was very low.

At sunrise we pushed on, and as I drove, I thought about trying to collect water that night by spreading our canvas on the ground to catch condensed moisture. If worse came to worse, we had a mirror to use for signaling an airplane—should one fly over—and food in the back of the Toyota. But the going was slower than ever, and without a detailed map, we could not be sure how much farther it was to Deception.

*   *   *

Midafternoon on a late November day, we finally sat on the crest of East Dune, shading our eyes against the sun and squinting down at Deception Valley. We should have been elated, but we had been through too much, and we were stunned by what we saw. Below us, shrouded in heat haze, the ancient river sprawled between the shimmering dunes. Not an ant, not a blade of grass, not a living thing graced its bleak and scorching surface—a slab of grey earth with bleached bones and bits of white calcrete scattered about.

We had somehow hoped that in Deception we would find relief from the heat, drought, and searing wind. Instead, it was more of the same. Neither of us spoke.

A dozen dust devils skipped across the parched plains, as if the ground was too hot to touch. Sunburned, with cracked lips and reddened eyes, we slowly crossed the riverbed and stopped at what was left of camp: A heap of twisted poles, some shredded, sun-bleached canvas, and a scattering of rusted tin cans all lay under a layer of broken twigs and sand. The hanging shelf dangled from its tree limb by a frayed length of rope, and the shelter we had built for shade had collapsed to a stack of reeds. Crescents of sand had climbed the windward sides of the water drums at the edge of camp. Except for the wind whistling through the trees, a ghostly silence greeted us.

We tried to come to grips with feelings of utter despair. Typically blithe and overoptimistic, we had hoped for rain. There had been none. There were no antelope, and no lions or hyenas. Only wind, thornscrub, sand, and heat. A tattered strip of canvas still holding to a corner of the buckled tent frame flapped harshly in the wind. We tied rags over our faces against the stinging sand carried by the gale and slowly picked up a pot here, a tin can there. What are we doing here...and why? we wondered silently.

Sunset...the heat broke and the wind fell. The desert stood in numbed silence. The red sun, looking bloated and lop-sided in the blowing dust and sand, sagged ponderously behind West Dune. From the woodlands in Captain's territory, the stirring cry of a jackal lifted across the valley. We knew why we had come back.

*   *   *

From here on, our entire study would depend on making the radio-tracking equipment work, but as had been the case with the darting, neither of us had had any previous experience with it. Radio telemetry was still in its infancy, and other field researchers had found the system temperamental and, in many cases, more trouble than it was worth.

"Test the equipment under conditions that resemble as closely as possible the actual field situation," read the instructions. Delia paced off 400 yards from camp, strapped on one of the radio collars, and began crawling on her hands and knees across the riverbed, simulating the movements of a foraging hyena. I was spinning dials and flicking switches on the new receiver, pointing the antenna in her direction, when I heard the soft scraping of Mox's feet behind me. We had driven to Maun and brought him back to camp several days after returning to Deception.

I turned to where he stood, broom in hand, about to sweep out the tent. A curious look came over his face as his sharp eyes focused first on Delia in the distance, and then at the receiver and antenna in my hands. I put one of the large collars around my neck, "Radio—wire-less—for peri—hyena—inside here." I showed him the lump of pink dental acrylic that molded the transmitter to the heavy belting.

Mox looked from the collar to Delia, then back at me. "Wireless!" I said urgently, holding up the receiver and pulling the whip antenna out from the collar.

"Ow! Missus...*peri? Peri*...music?" he asked quietly. He shot another look at Delia, and his hand went to his throat. The corners of his mouth struggled against the faint beginnings of a smile; he sniffed and turned. Despite the times Mox must have wanted to laugh at us, he had always managed to control himself—afraid, I think, that we might take it as a sign of disrespect. "Huh," he sniffed again. Shaking his head, he walked into the tent. I was reminded of his views on backtracking hyenas.

No matter how I held the antenna or turned the receiver, at distances of more than 400 yards—far less than the mile-and-a-half range we had been led to expect by the manufacturer—I could not hear a sound from the transmitter on Delia's neck. We sat on the riverbed, our heads on our knees. This was a critical defeat for our research. Though it meant a special trip to Maun and months of delay, all we could do was send the radio equipment off for tuning and readjustment. We boxed it up and gave it to a bush pilot, who flew it to South Africa.

Meanwhile, we turned again to our old method of searching for brown hyenas. But knowing how limited our chances were of finding one that way, it wasn't easy to motivate ourselves for the long hours of driving over rough ground, watching the light sweep back and forth across an empty riverbed. We sang and recited poetry to stay awake.

* * *

The beginning of the rains in early 1977 brought welcome relief from the heat. The antelope herds filtered back to the valley, and our clucking flock of hornbills arrived from the West Dune woodland early one morning, landing on the tea table and begging for bread crumbs. Delia was out of her chair and getting a bowlful of yellow mealie-meal almost before they had landed.

A lion roar awakened us one morning, and from our bed we could see a large male sauntering down the riverbed toward camp. Propped up on our elbows, we watched, through the mesh window of the tent, a herd of 1500 springbok split neatly down the middle, the antelope ambling aside to let him pass. They knew he was not hunting.

When he was thirty yards from camp we could see the orange ear tag, number 001. Once more Bones had come back to Deception Valley after the dry season. Pausing at the acacia bush beside our window, he gave us a casual glance and, lifting his long tufted tail, shot urine and scent into its lower branches. He roared again, then listened—head high, ears perked—looking far up the valley to the north, where a chorus of lions answered. He walked quickly in that direction, and we followed in the truck.

At the water hole on Mid Pan he stood looking across the valley at the approach of a long single file of lions. Delia raised her field glasses. "Mark, it's the Blue Pride!" Bones walked a short distance toward them, before casually lying down. His pride trotted up, each rubbing cheek to cheek with him and then sliding along his body in greeting. Afterward Sassy, Spooky, Gypsy, Spicy, and Blue headed straight for the truck, and after smelling it thoroughly, they chewed on the tires until I bumped the starter to make them quit. Chary, her back sagging lower than ever, remained aloof from all this tomfoolery and watched from a safe distance.

We sat with the lions for a while, then headed back. When we pulled away from the pride, Sassy, still as fascinated with the turning of the wheels as she had been as a cub, trotted along with her nose close to the rear bumper, while the others followed in a long line behind us. Wearing one of his rare smiles, Mox came to the edge of camp and stood there drying a plate. We must have looked like Pied Pipers.

With Sassy leading, the lionesses invaded camp. Bones lay down near the fireplace. As always, Mox slipped out the back of the island, circled around, and joined us in the truck. By now he was quite used to these surprise visits by lions, leopards, and hyenas, and thoroughly enjoyed the Blue Pride's traveling circus. Sassy grabbed the hosepipe from the water drum. Holding her head high as if she had killed a prize snake, she pranced out of camp with her trophy. Grunting, sprinting, dodging, and turning, their claws tearing at the grass, the others gave chase in a great game of keep-away. Blue pounced on the trailing end but Sassy kept pulling, and the hose snapped in two. Spicy and Gypsy grabbed at one of the pieces and soon the thing was reduced to mere bits of green plastic. Leaving us to figure another way to conveniently get water from our drums, the lions marched off to sleep the day in "Lions' Rest," the bush hedge 200 yards to the west.

We observed the lions whenever we could find them near the river-bed, but early one morning in late May 1977, we followed the Blue Pride when they passed camp and went hunting, north up the valley. It was the last we saw of them that year, justifying our fears that they would leave the valley before our radio telemetry equipment came back. An entire season of lion research was lost.

* * *

When our radio equipment was returned for the third time, it worked no better than before. There was nothing to do but go ahead and improvise with it on the brown hyenas, since the lions were already gone. We had neither the money nor the time to purchase another system.

The only way to increase the range of the collar transmitters was to increase the height of the receiver's loop antenna. Holding it at arm's length from the window was not enough, so we tied it to a tent pole section, and by adding other lengths beneath, we could raise the antenna to twenty or twenty-five feet above the truck. Delia and I stood in the back of the Land Cruiser, developing and practicing a technique for getting the antenna up and down quickly in order to keep up with a moving hyena. Meanwhile, Mox noted every detail of the operation from wherever he happened to be working.

One night we immobilized Star, bolted a radio collar around her neck, and laid her gently beneath a ziziphus tree in Bush Island, a thick clump of bushes close to camp, where we could keep an eye on her recovery. We watched her until dawn, anxious for her to regain her senses and move off into the bush so that we could try out our radio-tracking technique.

After a quick bite to eat at camp, we went back to find that Star was gone from Bush Island. We didn't panic. She couldn't have gone far, and we should be able to find her easily with the radio collar. Delia climbed into the back of the truck, and began swinging the loop antenna, while I switched the receiver to Star's frequency. I immediately heard a beep-beep-beep in my earphones. "I've got a signal! Null—back left—now right—a little more—peak! That's it. Get a compass bearing and let's go."

We drove west into the thick bush of the sandveld, scanning ahead for any sign of Star. A couple of minutes later we still had not seen her, so I stopped and held up the loop antenna. "She must really be moving out. I can barely hear her signal now." I climbed into the back of the truck, stumbling over the clanking pile of tent poles, and began putting segments together. When the antenna was up about fifteen feet, the winds, which begin with clockwork regularity on dry-season mornings, suddenly sprang to life and set the antenna mast in dramatic motion, its middle bending and swaying. I struggled to hold it up while Delia, a tangle of guy ropes in her hands and her face red and rigid with determination, began running back and forth through the thorn bushes, shredding her clothes and scratching herself on the wicked briers as she tried to guess which way the pole would veer next.

A gust of wind set the rickety affair careering toward the west like a piece of spaghetti. "Get on this side—hurry, I can't hold it!" I growled.

"I am—I *AM!*"

"Hold it. Okay, I've got a fix. Let's go." We unhooked the bits and pieces of our jury-rigged antenna mast and drove through the bush in the direction of Star's radio signal. But after several hundred yards we could hear nothing but thick static on the receiver. "We've got to get this damned antenna higher." I began adding more poles to the mast while Delia tried to control the reeling head of the antenna with the guyropes. By now she was nearly in tears and I was mad as a snake— not at her, but from the frustration of it all.

The antenna was about twenty-five feet above the truck and barely under control, and the wind was getting stronger by the minute. Suddenly the worst happened. With an almighty wiggle, the mast crumpled and the antenna soared off into a thornbush. Delia stood holding the limp guy ropes, tears welling in her eyes, while I glared at the doubled-over, soda-straw mast.

"Let's get to the top of the dune!" I spat. "If we can't get her signal from there, we never will!" We threw the antenna pieces into the truck again. Bouncing and bashing through the bushes, we drove to the top of West Dune, more than 120 feet higher, and stuck the unbent sections of the antenna pole together. Still no signal. We were absolutely deflated. We had waited for this equipment for months, and it was useless.

A stony silence filled the cab as we drove toward the riverbed. When we neared camp, Mox was standing next to the tent, waving his arms and pointing east, an ill-concealed grin on his face. Not 100 yards away, Star was walking across the riverbed in full view, headed opposite the direction we had taken.

We gave up on raising the antenna. Instead I would hold it out of the truck's window and monitor the signal while trying to stay within range of the hyena's transmitter. Without any idea where Star slept during the day, we tried unsuccessfully to locate her with the radio gear, driving long grids for several miles east and west over the dunes from the river valley. When that didn't work, we went back to our old method of searching the riverbed for hours every night. Once we had found her with the spotlight, we were able to use the radio receiver to follow her through the sandveld bush savanna. So long as we stayed within 200 or 300 yards of her, we could hear her signal and, in spite of the limitations of the radio equipment, we were able to get some useful service from it.

The night after she was collared, we found Star on North Bay Hill. We were determined to stay with her wherever she went, though we had no idea how far she would take us from camp, or in which direction. We had packed the truck with extra food, water, and other camping gear. For all we knew, we might end up fifty miles from camp by morning.

She turned east and disappeared in the tall grass and bush of the sandveld. It was the last we saw of her for the next twelve hours. We followed her weak signal through a nightmare of thorn thickets and dense woodlands to East Dune, then north along its crest through heavy underbrush. We drove over logs, around stumps, and through walls of thornbush often ten feet high and so tough and impregnable that the front wheels of the truck were actually lifted off the ground at times. After two or three nights of this, the Toyota's electrical wiring, exhaust pipe, and brake lines were torn away. We drove without them for a week or two, until I found the time to encase them in heavy rubber hose and wire them solidly against the chassis. Twigs, bark, and whole branches would shower onto the hood when the brush screeched along the sides, clawing at the truck. Whenever I could, I held out the spotlight or the radio antenna, to see what was ahead or get a new fix on Star's signal. Delia took compass bearings, mileage readings, and notes on the habitat and behavior of the hyena. I never could understand how she handled the flashlight and compass and wrote legibly, all at the same time, in the pitching truck.

Since there were no lion kills to scavenge, Star sought out the thickest cover, where her chances of surprising a leopard, civet, serval, or jackal with a fresh kill were much better than in more open habitat. She never stopped to rest, so neither could we. The scratches and bruises from a night's follow would last for days after. But for the first time since our study had begun, more than three years earlier, we were getting the details of how a brown hyena lived away from the riverbed during the dry season.

It was just after dawn one morning that Star came out of a woodland and into an open glade of tall grass on a duneslope. We could see her from the crest above. She was smelling her way through the grass when two tall, thin forms loomed directly ahead, like lampposts rising out of the grass and scrub. Star froze, then lowered her head and stalked forward. The necks grew longer, until two ostriches stood up, ruffling their wings and peering around alertly. Suddenly the female dashed away, her feathers flickering. But the large black male fanned his wings and swept toward Star, his big horny feet clipping through the grass and stomping the ground. She bristled and rushed forward to meet him. When they were still a few yards apart, the ostrich broke to his left and, dropping his wing, dragged it along as if it had come adrift from his body. Then he fell in a tumble of black and white feathers. When Star didn't fall for the ruse, he stood up and started turning in circles, his "broken" wing hanging limp. It was a spectacular deception, but Star had been around too long; she wasn't fooled. She roamed over the area, her nose to the ground, until she finally found the nest. It was a brown hyena's dry season bonanza.

She stood among the eggs—cream-colored globes the size of summer melons—and opened her jaws wide, trying to get her mouth around one to pick it up. Her teeth lost their purchase on the smooth shell, and the egg popped out of her mouth, her jaws chopping together. She tried again, standing over her muzzle, driving her weight down upon her canines until the shell gave way and the succulent nutriment was released. She lapped up three eggs at the nest, and then carried off and cached eight others in separate hiding places for future meals.

The sun found us sitting on a dune top miles from camp, munching biltong rolls and sipping cold coffee—strained through a cloth to get out bits of our broken Thermos. I patched two flat tires while Delia reviewed her log of compass and odometer readings to work out a course to camp. Star had led us over a twenty-two mile zigzag trek. We had no idea how long it would take us to get home.

We learned so much from our first follows of Star that we soon put radio collars on Shadow, Patches, and Ivey. After each night's tracking, we would rest as best we could in the heat and then try it again the next night. Several days of this and we were ready to feed each other to the lions. Two nights of rest usually quelled these atavistic desires, but because the heat made sleep during the day impossible, we were constantly fatigued.

From the beginning, we had been fascinated by the many mysteries surrounding brown hyena ecology. The bits of information about their range movements, social behavior, and feeding habits that we had managed to get by following them with a spotlight during the rainy season had only raised new questions that whetted our appetites for more. Inadequate as it was, our radio equipment gave us an inside look at the dry-season world of the hyenas. By following their signals, we quickly gained a great admiration for the ways these tough, adaptable scavengers manage to grub out a living in such a harsh and unpredictable environment.

We were continually amazed at how Star survived the rigors of drought. Since carrion items are so widely scattered in the dry months, and since brown hyenas scavenge most of their food, they must expand their ranges to nearly twice their wet-season areas and walk formidable distances to find enough scraps to eat. Including the zigs and zags she made while foraging, many nights Star covered distances of more than thirty miles while searching for food. The energy she needed to push through thornscrub and to muddle through loose sand during these nocturnal marathons must have been considerable, yet she often ate very little or nothing at all. Some nights she fed on nothing more than a single horn, a hoof, a flap of parched skin, or a few sun-bleached splinters of bone—perhaps a pound or two of carrion—all from carcasses picked over and discarded months before by lions, jackals, vultures, and other hyenas.

Brown hyenas are also remarkable for their ability to go for months—even years, in times of drought—with nothing to drink. They actually kill little more than about sixteen percent of what they eat, but they do hunt occasional small prey in the dry season, digging springhares and other rodents from their burrows and occasionally stealing bird or antelope kills from jackals, leopards, or cheetahs. The tissues of such prey and carrion are needed as much for moisture as for food, for though a hyena can make a meal of bone, it provides little fluid, even if it is fresh. If the rains have been good enough to produce a crop of wild melons, the browns eat these for moisture.

At the same time that we were studying brown hyenas, Gus Mills was conducting a thorough study of the hyenas' foraging behavior in the southern Kalahari. But there were still many unanswered questions about this species: for example, the type of social organization they maintained. Mills and his colleagues were then describing the brown hyena as a solitary species,[2] but we often saw as many as five of them meet and socialize while scavenging on large kills left by lions. We knew that, at least during the rainy season, they lived in clans with a social hierarchy. Perhaps, after the lions and their large antelope prey had left the valley in the dry season, such group contact between the browns would dissolve, since there would be fewer large carcasses to bring them together; perhaps individuals would then have their own separate ranges.

By radio-tracking them, we did indeed learn that clan members met less frequently in the dry months and that individuals foraged alone. But though they were often separated by several miles, they stayed in touch with one another by leaving scent marks along common pathways over a shared range. Furthermore, the group maintained its social hierarchy throughout the long dry months. But we still did not know why the hyenas bothered with all these social conventions. Since they had to separate and travel all over the landscape to find enough to eat, why did they even try to keep contact with one another? The answer to this question continued to be one of the main objectives of our brown hyena research.

One day Ivey, the dominant male of the clan, was sloping through the bush and woodland along the crest of East Dune. The zigzag course he followed improved his chances of picking up the scent of a kill made by a predator. By a scavenger's standards it hadn't been a bad night; he'd already run a jackal off a freshly killed guinea fowl and fed on a kudu leg he had cached some days earlier. No longer hungry, he only had to scent-mark the eastern boundary of the clan's territory to finish his night's tour.

A stranger's scent suddenly filled his nostrils. He froze in mid stride, hackles bristling, as McDuff strode from a stand of acacia bushes not fifteen yards ahead. A large male with a cape of blond hair over his thick neck and broad shoulders, he carried his big, brutish head high. When he saw Ivey he seemed to grow even larger. Both males squared off, pawing the ground with their front feet. They measured each other for several seconds, and then Ivey lowered his head and attacked. McDuff stood fast, waiting. They went for each other's neck with unearthly barks and shrieks, shouldering each other and muzzle-wrestling through churning dust and splintering bushes. The stranger scored first. He knocked Ivey off balance and seized him by the side of his head, shaking him violently. Blood dripped from Ivey's face. He screamed and thrust back and upward, struggling to break the vise-grip, tearing his face as he tried to wrench free. Like a warder leading a prisoner, McDuff marched Ivey in circles, shaking him until he flung him to the ground.

McDuff tried to shift his grip, but suddenly Ivey spun around and grabbed his neck. Turning and stumbling like sumo wrestlers, they chewed at each other's neck and face, pausing only to catch their breath. Finally, in a great show of strength, McDuff turned around and stripped Ivey's teeth from his neck. Then he began to run, with Ivey charging at his hind legs and nipping at his heels. But instead of leaving the territory, McDuff circled back, and fighting on the run, the two males disappeared into the bush.

Several nights later we found Ivey again. His neck and face had been severely mauled. But the thickened skin on the neck of a brown hyena can take such punishment, and it would heal quickly. There were more battles between Ivey and McDuff, who usually won, and we saw Ivey less and less frequently; then not at all. McDuff seemed to be everywhere, feeding at kills with Patches and Star, and patrolling the clan's territory as he scent-marked up and down the valley. The Deception Clan had a new dominant male.

There was something about brown hyenas that greatly puzzled us: In nearly two and a half years of following individuals at night, we had learned very little about their reproduction and had not even seen any of their young. Where were they hiding them? We had seen several adult females with Pogo and Hawkins, but there was no way to tell who had been their mother. Our study would never be complete without knowing how often the hyenas bred, how many young survived, and how they were raised.

Shadow appeared at an oryx carcass one night in the dry season of 1977, and we noticed immediately, and with great excitement, that her udders were heavy with milk. It was our first opportunity to follow a hyena mother to her den. Carrying a leg from the carcass into the sandveld, she quickly disappeared into the brush while we followed close behind her. At the base of West Dune, near Leopard Trail, her signal changed pitch and then suddenly vanished. "She's gone into a den!" I said, thinking she must have moved underground, where her radio would not transmit well. Three hours later neither she nor her signal had reappeared. Confused by this development, and disappointed that we had not confirmed the location of a den, we strung some toilet paper in a bush to mark the spot and drove to camp.

At dawn we were back, listening to the radio and watching for some sign of Shadow. Just after sunrise her signal came on, getting closer and louder: She was coming to the den again. When it stopped moving we took bearings on it from three points around the area. Then, like a phantom, it was gone. Two hours later we had seen or heard nothing more of the hyena.

After two and a half years of waiting, we were determined to find this den. I drove slowly ahead through the bush while Delia observed from the roof, to make certain that we didn't get close enough to frighten the mother. When she tapped on the roof, I stopped. In front of us was the entrance to a small den dug from the center of a springhare colony. On the mound of earth were the tiny tracks of hyena cubs. I leaned out my window, hung more tissue on a bush, and backed quietly away.

We sat near the den for the next ten days, and as far as we could tell, Shadow never came back. We were mystified, and we wondered if she had eaten or abandoned her cubs. Perhaps we had stressed her too much in her nervous maternal state; or maybe our scent had driven her off.

David Macdonald[3] found that dominant female red foxes harass low-ranking females with cubs and cause them to abandon their litters. Shadow had the lowest status of any member of the Deception Pan Clan, and Patches or Star may have been intimidating her more than usual. For whatever reason, the chance to find a hyena den had slipped by us.

Months later we had an almost identical experience with Patches, the highest ranking female in the clan, who also led us to a den. But, just as Shadow had done, she, too, stopped coming to the area before we could get a glimpse of her young, leaving not a trace of them behind. We were baffled about what the female brown hyenas were doing with their cubs. We had no way of knowing then that the answer to this question was closely related to the mystery of why they are social.

13

Gone from the Valley

Mark

I look down now. It is all changed.
Whatever it was I lost, whatever I wept for
Was a wild gentle thing, the small dark eyes
Loving me in secret.
                                    —*James Wright*

IT MUST HAVE BEEN the rustle of his feet in the grass that awakened me. I opened my eyes to find Bones standing a few feet away, spraying his acacia tree beside the gauze window of the tent. "Good morning, Mr. Bones," I said. "Fine morning today. What are you doing here so late in the season?" He swung his face toward the window and watched us for a few seconds, his tail still snaked to the limbs. Then he padded down the footpath through camp. We followed in our bare feet as he smelled the flap of the dining tent and headed for the fireplace. Mox was washing dishes with his back to Bones when he strolled past the hyena table. Suddenly the lion's 450-pound bulk filled the kitchen.

I whistled softly. Mox looked over his shoulder, dropped a tin plate and towel back into the water, and shot around the reed wall into the bushes. A minute later he came up behind us. "*Tau*, huh-uh," he chuckled softly. He had grown to love the lions as much as we, and even reported their locations when he heard them roar in the night. "*Msadi* Blue—*huuooah*—*kwa, kgakala ya bosigo*," he would say in the morning, pointing to the dunes ("Last night Missus Blue roared that way, far away").

Bones went up to the utility table and took a large tin of powdered milk in his mouth. His canines punctured the can and a white plume shot past his nose. He sneezed, shook his head, and sneezed again. The water kettle was steaming on the fire grate, and when he touched the hot handle with his nose he jerked back. Then he walked down the path and into the reed bath boma. His tall backside filling the narrow entrance, he lifted his head to the wash table and found the pink plastic tub filled with the leftover water from my sponge-bath. My arms had been smeared with grease when I washed the night before, and it had taken lots of powdered detergent to get them clean. Bones began to drink the black, sudsy water, his muzzle filling the tub and his immense pink tongue lapping the water to a froth. The more he drank, the more the water foamed, until white suds covered his nose. When he finally finished, he looked up, gave a deep sigh, and belched, blowing a large bubble that hung on the end of his nose. He sneezed again, the bubble exploded, and he shook the suds from his muzzle.

With one end of the tub jutting from his mouth like a great pink bill, he strutted out of camp. Chewing on the tub, he walked north along the riverbed, dropping bits and pieces of pink plastic from his mouth as he went. A long trek to North Tree, then east over the dunes, and finally he lay in tall butter-colored grasses, warming himself in the autumn sun, his mane and the grass straw part of the same pattern.

Later he started walking eastward again. "Sure wish we knew where you were headed," I said. But this was in June 1977, before our radio gear had been returned, and since we could not collar him, there was no way to follow his migration. The dry season was well under way, and there was no water anywhere in the Kalahari. We wondered if he was headed for the Boteti River, and if he would join the rest of the Blue Pride there. It would be months before the rains and the shifting herds of antelope brought him home to Deception Valley again.

"Here's to you, my rambling boy," Delia said softly, as Bones turned and walked away through the savanna.

*   *   *

By September 1977 it was hot again and we were exhausted from several months of radio-tracking the brown hyenas. We had not seen the lions since Bones's last visit to camp in June, so we had not been able to collar them either. Short on supplies, we went to Maun to restock and rejuvenate ourselves. We were driving along the northeast corner of the reserve, on our way back to camp, when we met Lionel

Palmer and two of his hunting clients, a druggist and his wife, from Illinois. Hot and tired, we gladly accepted their invitation to stay the night in the safari camp, a mile east of the reserve boundary.

The camp was set in a clearing on the edge of a belt of acacia woodland that followed the Deception rivercourse for miles west into the Central Kalahari Reserve. Five large bedroom tents of heavy canvas sat beneath shade trees. Deck chairs and small cocktail tables were arranged around a campfire in the center of a sandy clearing. The mess tent stood a few yards away, trim and tight beneath a big acacia tree. Inside there was a long dining table, a gas-powered freezer, and a refrigerator.

In the kitchen area, surrounded by a reed windbreak, native Africans were busy cooking and baking bread in a large metal box half-buried in the ground and smothered in hot coals. One young man sat thumbing a hand piano, a palm-sized board with metal strips of different lengths, each producing a different pitch. Another was weaving a basket from strips of grass. The cupboards were filled with imported Swedish hams, American mayonnaise, and tinned seafood.

The skinning and salting of the trophies were done at the end of the encampment, 100 yards from the dining tent. There, dozens of hides were draped about, and vultures sat waiting eagerly in the trees. Piles of horned skulls were wired with metal tags, each with a hunting client's name and address.

As we rolled to a stop, several black waiters with red jackets and tasseled hats shouted, *"Dumella!"* and clapped their hands in greeting. One of them led us to our quarters, a twelve-by-fifteen, dark green manyara tent that had large mesh windows and a fly sheet for shade. At either end of the stoop was a canvas washstand, and in the middle, a table with a mirror, a can of insect spray, a flashlight, a new bar of soap, and a washcloth and towel, all neatly arranged in a row. Inside were two tall iron beds with thick mattresses made up with clean sheets and heavy blankets. Two chairs, another table with more insect spray, and a lantern stood at the far end.

Delia ran her hand wistfully along the sides of the tent. "Imagine having a camp like this in Deception."

"Umm. But I'll bet they've never had a lion spray the bushes right outside their window."

"Right," Delia replied, "and I wouldn't trade that for anything. Anyway, they've cut down all the bushes and cleared away the grass."

"Tisa de metse!" Lionel called from his tent to Syanda, a tall, laughing Kenyan with greying hair who was in charge of the service staff. Syanda relayed the order, and soon a young tribesman, treading on the cuffs of his baggy blue coveralls, carried two five-gallon pails of slopping hot water to a reed shower stall. An empty pail with a shower head welded to the bottom dangled from a tree limb. The bucket was lowered, filled with steaming water, and hoisted back over the stall, where there was a wooden slatboard to keep the sand off our feet while we washed.

Freshly dressed, we met the others at the fireplace. Rows of glasses, an ice bucket, and bottles of Chivas Regal whisky, South African wine, and soft drinks were neatly arranged on a cloth-draped table. Nandi, one of the camp attendants, crushed together the smoldering butts of three branches left from the morning's fire and the flames jumped to life.

Wes, the druggist, was a middle-aged man with a fleshy face, thick black hair streaked with grey, and delicate, almost feminine hands. His wife, Anne, a school teacher, was small, neat, and pleasant. They were a khaki couple: khaki jackets with dozens of pockets, khaki hats, khaki shirts, and pants, khaki cartridge belts, and boots. They had brought to the camp bags and suitcases full of insect spray, tubes of block-out, and bottles of lotion. Like most of the clients on a hunting safari they looked straight out of an L. L. Bean catalogue. But, they were friendly, and we genuinely liked them.

Syanda, a white linen cloth draped over his arm, brought a stainless-steel tray with smoked oysters, mussels poached in wine, and cubes of fried springbok liver. We drank, ate hors d'oeuvres, and dissected the day's hartebeest hunt. To the west, beyond the trees, a jackal called.

Sometime later, Syanda announced dinner. The chinaware and wine glasses on the long mess table reflected the yellow light of tall gas lamps. Two waiters, who had been standing smartly against the wall when we entered, passed a tureen of steaming gemsbok-tail soup. The main course was eland steaks, french fried onions, stuffed baked potatoes, asparagus, and freshly baked bread with butter, accompanied by glasses of clear, cold wine. Coffee, cheese, and gooseberry pudding completed the meal. With a little help from the rest of us, Lionel finished the store of expensive wines his clients had brought with them on safari.

The last flame had spent itself on the reddened embers of the fire when, prompted by Lionel, the woman asked us to tell the story of Bones, who had by this time become somewhat of a legend in northern Botswana.

Anne's face grew concerned when she heard about Bones's near death, filled with festering porcupine quills, the splintered shank of bone stabbing through the skin of his leg. She and Wes sat forward on their chairs, watching our faces closely across the fire. They were absorbed by the details of the surgery that cleaned the wound and removed the shattered bone, and of sewing the muscle and skin back into place. Bones's miraculous recovery and his return to dominate the Blue Pride, as well as his special relationship with us, made their eyes sparkle with tears. When we had finished there was a long silence, and then Anne said, "That's the most beautiful story I've ever heard. Thank you for telling us."

The sszzzz of the heavy zipper on the flap awakened us the next morning. "*Dumella!*" Nandi greeted us, setting a tray of tea with cream and sugar on the nightstand between the beds. We were sipping from our cups when another servant trudged to the stoop, filled the canvas washstands with piping hot water, and laid out towels and washcloths. Breakfast, served in the mess tent, was fresh fruit, sausage, bacon, eggs, toast, cheese, jam, and coffee. All this luxury in the bush costs clients from $750 to $1000 a day.

It was still early when we left the hunters on a track near the boundary of the game reserve. Wes was sitting on a special seat across the back of the truck, his rifle clipped at the ready in a rack in front of him. We waved goodbye and turned southeast into the game reserve toward camp.

Just before noon, a few minutes ahead of our scheduled radio contact with the Safari South Office in Maun, we crested East Dune above Deception Valley. In camp, I set the HF radio on the fender and clipped the wires to the battery terminals. It crackled to life, and Delia stood waiting for the call. I went to the dining tent and started transcribing field notes.

"Zero-zero-nine; zero-zero-nine, this is four-three-two, do you read, over?" It was Dougie Wright, another hunter, calling us from Lionel's camp, where he had just arrived with more clients.

"Four-three-two, this is zero-zero-nine. Good afternoon, Dougie. How are you? Over." Delia answered.

"I'm afraid I have a bit of bad news for you, Delia, over."

"Oh . . . Well, okay, Dougie . . . what is it? Over."

"Lionel and Wes shot one of your lions this morning."

" . . . Oh . . . I see." I could barely hear Delia as she asked, "Do you know the color and number of the ear tag, Dougie?"

"Uh . . . he had an orange ear tag in his left ear . . . number zero-zero-one."

"Mark! My God! It's Bones—they've shot Bones!" She choked and dropped the microphone. I started from the dining tent, but when I got to the truck she was gone, running across the riverbed.

"No . . . no . . . no . . . no . . . no!" Her sobs drifted back to me on the wind.

14

The Trophy Shed

Mark

When I look behind, as I am compelled to look before I can gather strength to proceed on my journey, I see the milestones dwindling toward the horizon and the slow fires trailing from the abandoned camp-sites, over which scavenger angels wheel on heavy wings.

—*Stanley Kunitz*

THE SHED, old, dark, and musty, was filled with animal skins, stiff, salted skins with shrunken ears and hairy rawhide strips for tails. Each had a bullet hole, some several.

Against the bamboo walls, shelves held stacks of bleached white skulls: There were wildebeest, zebra, buffalo, impala, kudu, leopard, jackal, and dozens of others. And lions. Each had a red metal tag wired through a socket where once a clear, bright eye had been.

We found his skin in a bale of others. The orange tag, 001, was scarcely visible inside his wizened ear. Heartsick, I pried at the folds of cartilage with a screwdriver, but they would not give up the tag. Pebbles of rock salt showered our feet when we pulled his hide, flat and stiff, out of the bale, its hair like so much steel wool. The scar from his broken leg and our crude surgery was still there over his knee. We made some hurried, clumsy measurements for the sake of science, wrote them in a notebook, and walked outside into the bright sunlight. Tears were brimming in Delia's eyes, and it was some time before I could speak.

Bones was killed in the dry season, when there was no water in the

Kalahari. More than a thousand gemsbok had passed through the woodlands along the Deception rivercourse and east out of the reserve into the Safari South hunting concession area. He had apparently been following these antelope, for the Kalahari was ablaze with grass fires and prey was extremely scarce.

They had found him resting under a bush with Rascal and one of the Blue Pride lionesses, just a few yards outside the reserve boundary. When he heard the truck he raised his head from his paws. Wes and Lionel drove to within fifty yards, stopped, looked at him through binoculars, and shot him through the heart. If they had seen him sleeping next to our camp, they would have known that they could practically walk up to him and put the muzzle of the rifle against his head. Hadn't they seen the orange ear tag? It wouldn't have mattered if they had. Unfortunately, lions don't understand the laws of man; he and the others had become fair game when they left the game reserve to find food.

At the shot, the lioness dashed away into the bush, but Rascal stood at Bones's side, snarling defiantly and charging whenever the hunters tried to collect their trophy. They were able to frighten him away only by firing their rifles into the air and driving at him with their truck.

The news that Bones had been shot crushed us. We stood beneath the ziziphus tree, cursing and crying and holding each other as the hurt slowly seeped in and fed on his memory. For days we were swamped with depression. If they had seen his ear tag, it was incomprehensible to us how they could have shot him since, only hours before, they had been so touched by his story. Finally, there was despair: Bones had been the symbol of what we had hoped for, and believed could exist, between man and other animals. When he was shot, everything we had been trying to accomplish for the conservation of Kalahari wildlife seemed lost. He had first been our patient, then our friend and mascot. One friend had killed another.

As trained biologists, we knew that we could not fault anyone personally for the death of Bones. He had been a legal trophy, and it was not the hunters' fault that he had left the game reserve. Besides, carefully regulated hunting can be a useful tool in the conservation and management of some animal populations. Unfortunately, many governments insist that wildlife is worth conserving only if it can pay for itself through hunting, tourism, or some other means. Knowing this was true of Botswana, we tried to overrule our emotions and deal with Bones's death on a more rational level.

Though we had never objected to hunting per se, a few of the hunters we knew openly admitted that they were consistently disobeying Botswana's hunting regulations and other recognized sporting codes. They described chasing animals with their trucks, letting clients shoot several antelope until they got the trophy they wanted, setting grass fires to make tracking easier and to burn lions out of thickets, hunting inside game reserves, and shooting game in areas where the quotas had already been exhausted. We had no way of knowing whether they were exaggerating, but this began to strain our friendships.

Because of our responsibilities as ecologists, we encouraged better enforcement of hunting regulations by the Wildlife Department and objected strongly whenever we discovered that one of the hunters had shot animals inside the game reserve. We also recommended that desert lion quotas be reduced and that license fees be increased. These actions were difficult for some hunters to understand, and it was easy for them to believe we were working against their interests, especially since they had helped us so much through the years. But not all the hunters or clients shared these attitudes or participated in illegal practices, and some remained our good friends until the day we left Botswana.

Botswana's Department of Wildlife has next to the smallest budget of any of its government agencies. It is hopelessly understaffed and cannot hope to effectively patrol large, remote areas. Officials told us that in one year alone, more than 600 lions, most of them males, had been *legally* shot by ranchers, safari hunters, and tribal hunters. Additionally, a large, undetermined number, again mostly males, were shot by poachers for the black market trade in skins.

Unfortunately, the Botswana government has encouraged the eradication of all predators outside parks and reserves by enacting a sweeping predator control law. It permits the shooting of predators on ranchland if they are deemed a threat to livestock, crops, water installations, or fences, whether or not they have actually molested domestic animals. This is reason enough for native people to kill every predator they see outside game reserves and national parks. Another section of the new law permits a rancher to keep the skin of a predator that has killed his stock. A lion skin brought about three hundred pula (about $300) on the market in 1978. This law cites as predators two endangered species, cheetahs and brown hyenas, as well as lions, leopards, crocodiles, spotted hyenas, baboons, monkeys, and jackals.

Safari hunters told us that "shootable" lions—those with full manes—were quickly becoming scarce in most desert hunting areas and that they had been virtually eradicated from others. Some of their clients were shooting young males that had mere fringes for manes, simply because they had bought a license and could find no older lions.

This startled us. We were concerned about how long Kalahari lions could withstand such a high mortality rate, especially since it was primarily among the males. Surely the welfare of the population would be undermined. Studies of Serengeti lions, by Brian Bertram,[1] showed that the reproduction of pride females who have lost their males suffer from a reduced fecundity for a considerable period after new males join the pride. A strange lion may even kill cubs that are not his own, so that the females will come into estrus sooner and bear his cubs. If, whenever their males were shot, Kalahari pride lionesses experienced a reproductive depression similar to that found in Serengeti prides, the population might very well be threatened. We had to do something to find out.

Since no long-term wildlife studies had ever been conducted in the Central Kalahari before, the most fundamental knowledge of the lion population was totally lacking. No one, including the Wildlife Department, had even the vaguest idea how many lions there were. And though we had been studying them whenever we could, lions had only been on or near the riverbed, where we could find and observe them, for two or three months each year. In these brief periods, we had been able to learn very little that would benefit their conservation. Out of our feelings for Bones grew a compulsion—perhaps an obsession—to do more. We were determined to find out how many lions roamed the Central Kalahari, what they ate and whether there was enough of it, what habitats were needed by them and their prey, and what, if anything, was threatening the survival of the population. It was essential to know how many were being shot and trapped each year, how many were dying naturally, and how many cubs were surviving to offset this mortality rate.

Since the Kalahari is a desert, one of the most important and intriguing considerations was how lions, and other predators, were satisfying their need for moisture. Large as it is, the reserve contains no water except during the brief rains. But did lions *have* to drink? The longest anyone had seen wild lions go without drinking was nine days; perhaps they could survive even longer than that. But even if they could, presumably they would be forced to leave the protection of the reserve for months each year to find some place to get water. Maybe Bones had been traveling to the Boteti River when he was shot. If this was true, then, large as it was, the Central Kalahari reserve was too small to offer adequate habitat during dry seasons and drought.

Even if we could learn all these things about lions, in order for any of it to benefit the Kalahari population, we would have to sell the Botswana government on the idea that predators are a valuable resource, one that could bring in much more money if they were conserved. At the time the attitude of many officials was that since predators prey on cattle, they are vermin that must be eradicated.

A large scale study of lions in thousands of square miles of untracked wilderness would be impossible unless we could somehow maintain daily contact with our research animals year round. The use of an airplane and radio-tracking equipment was the only way this could be done. But the very idea that we could get a plane for our research seemed preposterous. Neither of us knew how to fly, and I had only been in a small plane a few times in my life. Furthermore, airplanes are exorbitantly expensive to own and operate in Africa. With just a Land Rover to keep going, we had practically starved while trying to keep our research funded. It was ludicrous to think we could raise enough money for an airplane. But we had to try.

15

Echo Whisky Golf

Mark

Only when we pause to wonder
do we go beyond the limits
of our little lives.
          —*Rod McKuen*

ON A HOT AFTERNOON in late October 1977, we stood on a dusty track in Maun reading a letter from Dr. Richard Faust, director of the Frankfurt Zoological Society in West Germany. I was electrified by the news that the society was seriously reviewing our request for an airplane. But they wanted to know my pilot's license number and how many hours of flying experience I had. Somehow I would have to learn to fly before answering that letter.

Leaving Mox in the village, we raced back to camp, threw our best clothes into the truck, and set off for Johannesburg. At four o'clock in the morning several days later, covered with dust and grime, we quietly slipped inside the gate at Roy and Marianne Liebenberg's home in Benoni, a Johannesburg suburb. We had met Roy, a captain for South African Airways, in Maun about a year earlier, when he had flown some tourists to the Okavango River delta. He had been interested in our research and had offered to teach me how to fly if I ever needed to learn. Spreading our sleeping bags on the ground to get a little sleep before dawn, I hoped he would remember his offer.

At 5:30 A.M. the milkman stepped over us, carrying a wire basket of tinkling bottles, and a couple of hours later Roy and Marianne came to investigate the truck and the two lumps parked in their front yard. Captain Liebenberg, a middle-aged man, neat, soft-spoken, and precise, pulled at his sharp nose; a grin cut through the black stubble on his round face. Almost before we could get out of our sleeping bags, he offered us the use of their guest cottage while he taught me to fly.

Six weeks later, after numerous delays from bad weather, I had almost finished my training. We wrote Dr. Faust that I was about to get my license and that I had accumulated forty-one hours of flying experience. I assured him that Roy had given me adequate preparation for bush flying.

We could hardly believe it when the grant was approved and the money sent. That someone we had never met could have such faith in us and our abilities was very gratifying. After shopping around, we bought a ten-year-old blue-and-white Cessna tail-dragger with EWG— Echo Whisky Golf—painted under the wings.

Our first heady reactions quickly gave way to some serious contemplation. We had been so determined to get an airplane and learn to fly, that we had not given much thought to the next phase of the project: getting the plane to camp. With a mixture of anxiety and anticipation, I realized that soon I would have to fly into the Kalahari, an area so remote and featureless that Botswana law prohibited pilots with fewer than 500 hours of experience from flying over it. For the next year, until I could get the required hours, we would have to avoid flying near Gaborone. If civil aviation officials there learned that we were operating a plane in the desert, they would probably ground us, and that would be the end of our project at Deception Valley.

Other problems seemed even harder to solve: Once Echo Whisky Golf was at camp we would not only have to maintain her, but also find a way to haul the thousands of gallons of fuel needed to keep her running. Besides the logistical problems, just learning to fly around the desert without getting lost would be a major challenge.

At dawn the morning after I received my pilot's license, I got ready to take off on my first flight across the Kalahari. It was only my third solo cross-country, and Roy seemed more nervous than I. "Now don't forget, after crossing the Gaborone-Francistown road there will be no more landmarks to help you navigate. Make sure of your position at the railway line before going on." He double-checked the pencil I had tied around my neck and made sure that the black plastic sheet for making emergency water was stowed in the tail.

I kissed Delia goodbye and shook hands with Roy. They would drive the Land Cruiser and pull a trailer back to camp, leading a heavy truck from the Botswana Wildlife Department loaded with drums of aviation fuel. They watched anxiously as I climbed into Echo Whisky

Golf and taxied across the field to the grass airstrip. Then, when I turned and began revving the engine for takeoff, Roy came running toward me, waving his arms frantically and pointing toward the wind-sock. I was headed the wrong way. Waving back—and smiling sheepishly—I spun the plane around and accelerated down the runway. With a roar and a rush of cold air, I slipped into the smooth morning sky. The sense of freedom and the exhiliration were narcotic.

Euphoria didn't last long. At 300 feet the plane began to fly sideways—or so it seemed. I had climbed into a strong crosswind. I eyeballed a drift correction on the mountain peak ahead, then leveled off just below a layer of cloud at 1500 feet above the ground. On the radio, Jan Smuts International Airport advised that the stratus would be lifting and that the weather looked okay for a flight to Botswana. The charts showed the ground elevation dropping away from Johannesburg, and I relaxed a little, knowing that I would have more room between cloud and ground as I neared the Kalahari.

Half an hour out, the radio chatter died away and there was nothing but the droning engine and whistling wind. A gap between two peaks let me through the Waterberg Mountains, and soon the last traces of civilization faded as the Kalahari began to unfold below. Four hours from now, I would have to find a tiny tree island with two tents in the middle of this incredibly vast wilderness. Without navigational aids or any way to check my position on the featureless map, it would be a little like trying to push a piece of frayed thread through the eye of a needle. I could only hold my compass heading and hope that I had made an adequate correction for the crosswind that was trying to push me off course.

Instead of lifting, the clouds began to drop and rain began to fall. To stay beneath the clouds, I began a low descent. "Stay high so you can see Real Deception Pan," Roy had cautioned. "Remember, it's the only landmark you have to show you where the camp is." But the clouds forced me lower and lower until I could see grasses bending in the wind and rolling, bush-covered sand ridges flashing by a few feet below. Because I could not hold my altitude, I might fly past Deception Valley and never see it. Flying for hours with no way to fix my position, I felt suspended in time and motion—lost.

I had been in the air about three hours when av-gas fumes suddenly flooded the cabin. A thin stream of fuel was trailing along the underside of the port wing root and over the rear window. My stomach went hollow. The previous owner had agreed to replace the rotten rubber fuel bladders inside the wings, and I had foolishly accepted his word that this had been done.

The stream seemed to be growing now, spreading from just outside my window farther out along the wing, flowing over the flaps and spraying into the slipstream. The port-side fuel gauge was flickering lower. I fumbled for the fuel selector and switched it to left, hoping to burn as much of the escaping gasoline as I could before it was gone.

I didn't know if the fuel in the starboard wing would carry me to camp; I did know that the slightest static spark could turn EWG into a fireball. I opened the windows to let more air circulate through the cabin.

At the same instant the leaking fuel bladder split. Green gasoline gushed from the wing and ran along the fuselage and off the tail wheel. I quickly switched the fuel selector to the starboard tank and banked the plane right and left, looking for a place to land. There was nothing but thornbrush and small trees below. I tried to broadcast an emergency call on the radio, but all I heard was static. I was too low for my transmission to carry very far, and it was unlikely there was anyone within hundreds of miles to hear my call, anyway.

The fumes were getting stronger and my head had begun to ache. I trimmed the airplane for level flight and rummaged around inside, trying to see if fuel was leaking into the cabin. The carpet inside the rear cargo hatch was wet. High-octane gasoline was seeping through the hatch and into the *battery hold* behind the jump seat. Though I had already turned off the master switch, the risk of an explosion and fire were magnified. Worse, there was nothing more I could do about it.

I took the plane down to just above the savanna, so I would have a chance to crash-land and get out if a fire did occur—assuming EWG didn't just blow up. As the minutes passed, the fuel gauge needle for the damaged tank quickly dropped into the red, then finally came to rest against its peg. The stream of gasoline under the wing narrowed as the ruptured tank finished emptying its volatile load. With the danger of an explosion reduced, my main worry was whether the other tank had enough fuel to get to camp. I figured and refigured until my pencil lead was a nub. With almost half an hour to my estimated time of arrival (ETA), the gauge began to flicker just above EMPTY. Finally it settled onto the mark and stayed there. Each minute seemed like an hour, and I kept waggling my wings to see if there was enough fuel left sloshing in the tank to make the gauge flicker.

Make sure all electrical and fuel systems are switched off, unlatch the door, and attempt the landing with the control column all the way back... The procedures for an emergency landing kept running through my mind, over and over.

I gripped the control wheel hard. My neck was stiff from craning to recognize some feature in the flat, homogenous bush savanna below. A dozen times I imagined that the engine was changing pitch and that I could feel a strange vibration. There was still no good place to land if it quit.

Just before my ETA, I squinted past the whirling propeller. I was sure my imagination was playing tricks again—a round, slate-grey depression with a frazzle of white cloud hanging low above it, took form in the mist. It was just to the right of my course. I held my heading, afraid to make a mistake and waste precious fuel. But suddenly the shallow riverbed of Deception Valley flashed underneath.

I throttled back and sailed over the wet, waxy surface of Real Deception Pan. The fuel gauge was solid red. Let the engine quit—now I could land and walk to camp!

There had been a lot of rain, and several dozen giraffe stood in the pan, peering curiously at what must have looked to them like a huge bird sailing past. I drifted slowly up the valley above herds of springbok, gemsbok, and hartebeest grazing on the lush green grasses of the riverbed. Then I was over Cheetah Pan, the Midway Islands, Jackal Island, Tree Island, Bush Island, and finally camp. All the water holes along the riverbed were full, but the surface of the airstrip we had made—months before we even knew we were going to get a plane—seemed firm enough for a landing. The stall warning bawled, my wheels bumped, and I taxied to camp. Echo Whisky Golf was home in Deception.

The tents had been flattened by a heavy storm. Water and mud stood everywhere; I needed Mox. After propping up one end of the tent and chasing a six-foot banded cobra from under the bed, I took a short nap on the soggy mattress. Then I refueled the good tank from a drum of av-gas we had previously hauled to camp; I would fly on that one fuel bladder until I could replace the damaged one. Because I didn't know my way around the Kalahari from the air yet, I felt a bit uneasy about this, but I couldn't wait around for a new fuel tank to be ordered from the United States. I climbed back into the plane and flew to Maun.

While we were in Johannesburg, Mox had enjoyed village life to its fullest. I found him sitting beside his *rondavel*, his head sagging between his knees, trying to recover from his latest drinking bout. His eyes were bloodshot and bleary, and he had a bad cough from smoking the harsh stemmy tobacco he rolled in strips of brown paper bag. Walking on unsteady legs, he fetched his kit and threw it into the back of the truck I had borrowed.

When we arrived at the village airstrip, he must have realized he was about to be my first passenger, because he sobered up within seconds. He tried to explain that he had never been in a *fa-ly* before, and I tried not to understand, knowing that if he refused to go I would have no other way to get him to camp. He bumped his head on the wing, and before he could focus on what was happening, I had him belted in and we were taxiing for the takeoff. I shoved the throttle forward, and Echo Whisky Golf lunged ahead, gobbling up runway. No bush plane is quiet, especially not on takeoff, and this one roared like the very devil. Mox's eyes bugged out of his head and he clutched at the seat, the door, and the dash. I kept shouting, "It's *go siami! Go siami!*—It's okay! Okay!" and suddenly we were airborne and climbing away.

I throttled back to a cruise climb and banked to establish my heading. When Mox saw his village and the river quickly growing smaller, a wide grin broke across his face and he began pointing to the huts of friends along the water. I showed him the controls and demonstrated their effects, every movement of my hands and feet producing a different sensation. With each new maneuver, a laugh caught in Mox's throat. In fact, he took to flying so completely, and was so proud of the status it gave him among his tribesmen—all of whom depended on walking or donkeys—that some of the hunters nicknamed him "Neil Armstrong."

Mox and I spent the next three days cleaning camp. At 1:30 A.M. of the third night, I was awakened by the sound of the Land Cruiser's engine. By the time I had pulled on my clothes, Delia and Roy were coasting to a stop near the tent. At first I didn't recognize our truck—with its lumps of dried mud and grass, it looked like a poorly formed adobe brick.

Roy and Delia crawled slowly out of the cab and stood in the glare of the headlights, their hair matted with mud and grass-seed, eyes sunken with exhaustion. After our happy, if grubby, reunion, they explained that the four-ton fuel truck from the Wildlife Department was hopelessly bogged in mud sixty miles east. It would have to be dug out if we ever hoped to get all fifteen drums of fuel to camp. Furthermore, they had had to offload all of our groceries to lighten the Land Cruiser enough to get through the muck. They had left behind bags of flour, mealie-meal, and sugar, cases of canned food, and the new tents and the equipment for the plane, all of it stacked on the wet ground near the trailer they had been pulling.

For the next five days, dawn found Mox and me bouncing through tall grass and bush savanna, headed for the mired fuel truck. Together with the driver of the four-ton Bedford and his assistant, we shoveled tons of mud away from the undercarriage and hauled in rocks to put under the big wheels, but the soft ground swallowed up everything, and whenever we had the truck ready to back out, a rain shower made the ground like grease again. Each night we rolled a drum or two of fuel across the bog, loaded it into our Toyota, and headed back to camp. The drivers had made a camp near the truck, and we left extra food for them.

On the fifth day, we arrived at the marsh to find the fuel truck gone and most of our aviation fuel lying on the ground. We never saw the truck again, and that was the last time the Wildlife Department ever offered to help us transport av-gas to camp.

Trying to get the rest of the fuel to camp on one last trip, Mox and I loaded ten of the last eleven drums into the back of the Toyota and onto the trailer and chained the last one onto the front bumper. Then we set off for Deception Valley.

We had just crossed into the game reserve, twenty-eight miles from camp, when, with a shriek of tearing metal, the truck leaped into the air and began to roll onto its right side. Mox was thrown to the floor, and my head slammed against the roof as I fought to regain control. I managed to keep us upright, until the truck smashed through a grove of thick brush, slewed heavily around, and heeled over onto its left wheels. The fuel drums shifted at the same instant, and I thought we were going over. I cut the steering wheel hard left and jammed on the brakes. We came to a stop in a shower of sand, leaves, and broken branches.

Dazed and bruised, I stuck my head out the window. At the end of a long, deep furrow a mangled fuel drum spewed high-octane fumes and gasoline into the air. Muttering in Setswana, Mox pulled himself up from the floor. Eyes white and round, hands shaking, he reached for his pouch of Springbok tobacco and tore a strip from a brown paper bag. *"Nnya!* Petrol—*mellelo*—fire!" I shouted and grabbed his hand.

The drum on the front bumper had come unchained and we had run over it. Luckily it hadn't exploded. I slapped some putty into the fuming hole, rolled the drum off the spoor, and picked up the battered exhaust line that had been ripped off the Toyota. After straightening the truck's buckled shock spring with a sledgehammer, we limped off toward camp, Mox drawing deeply on a wrinkled fag.

I would have to fly Roy back to South Africa the next morning, so early that night, Delia and I collapsed onto our cots, and Roy bunked down on the dry spot he had been using on the floor of the supply tent. Mox had a brand-new tent of his own that we had brought from Johannesburg.

We were not quite asleep when lions began roaring from the airstrip. We jumped out of bed. Maybe it was the Blue Pride! But we had not seen them since Bones was shot, and since Rascal and one of the females had watched the shooting, we worried that they might no longer accept us.

At the airstrip, instead of the Blue Pride, two young males we had never seen before lay in the middle of the runway squinting at the spotlight, their blond, patchy manes like the peach-fuzz beard on a fifteen-year-old boy. They were carbon copies of each other, and we speculated that they were brothers. One had a distinct J-shaped scar on his right hip. Totally unconcerned with us, they began bellowing again, making quite a respectable racket.

We drove back to camp somewhat disappointed. As I flopped onto my cot I thought to myself, surely these two youngsters didn't intend to take over the Blue Pride territory. They could never take the place of Bones.

16

Kalahari Gypsies

Mark

> Like the river, we were free to wander.
> —*Aldo Leopold*

IT WAS one thing to write the proposal, but quite another to actually find lions, put collars on them, and then track them over tens of thousands of square miles of Kalahari wilderness. January 1978 had come and gone by the time we had the plane and its fuel in camp. We had no idea how long it would take us to collar the lions, but we were racing against the coming dry season, when they would disappear into the vast bush savanna away from the fossil rivercourses, for eight months or more. Major international conservation interests had invested in our project, and now it was up to us to demonstrate that what we had set out to do was, in fact, possible. If we failed, we were not likely to get more financing.

Our plan was to immobilize and radio-collar lions and brown hyenas along the full length of Deception Valley, as well as some lion prides in the Passarge and Hidden Valley fossil river systems to the north. The immediate problem was how to find these roving predators in such a remote area and then get to them on the ground for the collaring. Our only chance of spotting them would be to catch them on the open grasslands of the riverbeds in the early morning.

Each dawn for the next six weeks, we roused ourselves for a quick bowl of uncooked oatmeal and powdered milk. Then, our pockets stuffed with sticks of biltong, we hurried to Echo Whisky Golf, standing cold and wet with dew in the pale dawn.

"Switches on, master on, throttle set." Delia shivered behind the foggy windshield.

"Contact!" I spun the propeller and stepped back. A puff of white smoke, a sputter, and a roar—Echo Whisky Golf came to life.

Boeing, a springbok who used the airstrip as the center of his territory, pawed a spot on the ground, urinated on it to freshen his midden, and trotted casually to one side as we taxied into takeoff position. He had become so tame that we had to be careful not to run into him during takeoffs and landings.

On takeoff, camp flashed by and we could see Mox stoking the fire, smoke curling through the trees. We cruised slowly north above Deception Valley, our foreheads pressed against the side windows as we looked for lions on the fossil riverbed.

With Echo Whisky Golf, the days were gone when we groveled like turtles, slowly dragging the metal shell of our Land Cruiser laboratory and home wherever we went. Now that we could see beyond the next rise, our view of the Kalahari was no longer limited to a few square miles of savanna. We soared above the sinuous channel of the ancient river, casting our long shadow across the browns and greens of the sandveld at dawn and dusk. From the air we discovered and named new pans and oxbows—segments of fossil river channels cut off ages ago by ridges of shifting sands. Hidden Valley, Paradise Pan, Crocodile Pan—they had been there all along, lost beyond the dunes. Recently there had been heavy rains, and now the riverbed was dressed in green velvet, a sparkling necklace of water holes along her length.

From the beginning, Echo Whisky Golf made it plain that she did not like life in the bush. Her alternator burned out, and for more than two months, until we could get parts to fix it, I had to start the plane by hand. The engine would run on the current from its magnetos, but without electricity from the battery, we could not use the radio and our compass was out.

The Kalahari looked featureless and there was only one tiny area around camp we could recognize. We followed the fossil river valleys to keep from getting lost, but they were shallow and indistinct in places; at times dark cloud shadows, like inkblots, disguised them in the desert topography. As for the compass, we found out just how badly it was behaving on our first flight to Maun together: We arrived over Makalamabedi, instead, a village more than forty miles east of our course.

When flying far away from Deception, we always carried food and survival gear stowed in the plane's aft baggage compartment. We seldom knew which direction from camp we would be, at any particular time, or how long we would be gone, and there was no one, other than Mox, with whom we could leave a flight plan. I was irritable much of the time, in those early days with Echo Whisky Golf, but now I realize how much stress I felt from flying an aircraft with so many mechanical difficulties in such a remote area. If we had gone down somewhere, our chances of rescue would not have been very good. And there was always the danger of holes. We often had to land and take off in tall grass, where a badger's excavation, a fox's den, or a springhare's burrow could easily have broken off one of the plane's wheels. Later, however, I learned a technique for spotting holes, by flying low while running my wheels lightly over the ground—"feeling" the surface before landing.

*　　*　　*

"Lions—there, in that tree island!" Delia shouted above the engine. I banked the plane steeply and we swept low over the flat-topped trees. Below us a pride of lions was sprawled near a hartebeest carcass about half a mile off Tau Pan on Hidden Valley. I throttled back and pulled flap. We skimmed the top of a nearby tree, and after Delia had dropped some toilet paper into its branches to mark the spot, I took a compass bearing on a big forked acacia on the pan, and we set course for camp.

We packed the truck with camping gear, food, water, darting equipment, and cameras. Delia then drove off in the direction of the lions. I watched the truck until it was lost in the wavering heat mirage; it was the first time she had driven into the Kalahari alone.

Several hours later, when she should have reached the general area of the lions, I took off and flew along the course she had taken, searching the savanna for the white speck of the truck below. I finally spotted it, crawling like a beetle through the bush. Though she wasn't far from our rendezvous at the forked tree on the riverbed, Delia was off course. Unless she changed direction, she would miss the valley altogether. I dropped low and flew just above the truck, directly toward the tree. She stopped, took a compass bearing on the plane, changed course, and drove on. Satisfied she was going the right way, I flew to the trees where we had found the lions earlier that morning. The hartebeest remains were still there, but the lions had gone. I started making slow turns over the area, hoping to find the pride.

Delia arrived at the tree and began driving the truck back and forth along a relatively smooth section of the pan to make a landing strip. Then she walked up and down with a spade, filling in holes and knocking down the worst nodes of clay and the taller grass bases, which could damage the plane. When she was 300 yards from the truck and satisfied that the strip was safe, she turned to walk back to the Land Cruiser. Glancing up, she found the lions: They had moved from their carcass to the trees along the edge of the riverbed. She had been so busy filling in holes that she hadn't seen them coming. Now, strung out in a line, they were headed straight toward her, the nearest one not fifty yards away and moving between her and the truck. With no other cover for miles around, she stood rooted to the ground. We could usually trust the Blue Pride, but these were strange lions, and this was almost certainly their first encounter with a human.

The lionesses began to walk more slowly and deliberately toward Delia, staring intently, raising and lowering their heads, watching her every move. She could hear Echo Whisky Golf turning lazy circles in the sky not more than half a mile away, but she had no way to signal me.

Slowly she began to back away, trying to read the lions' expressions and postures. But suddenly she realized that, by retreating, she was inviting their pursuit, so she forced herself to stand still. The lions kept pressing forward, and when they closed to thirty yards, her fear reached a primal level. She raised the spade, wielding it like a club, and from deep inside her came a sound so primitive it could have come from a Neanderthal woman. "HAARRAUGGH!"

As if on command, the lions stopped and slowly sat down on their haunches in a long line, their heads and necks craned forward, watching the primate that stood before them brandishing her weapon.

Delia held her ground, terrified that if she moved, the lions would again follow. Yet she had to get past them to the safety of the truck. The longer she stayed where she was, the greater the chances that they would come for her. Slowly she took one step, then two, then began moving obliquely past the lionesses, holding her spade at waist level, her eyes fixed on the pride. They tracked her like radar, their heads slowly turning as she worked her way past them, waving her spade and beginning a long arc toward the truck.

She had flanked the pride and begun backing away when one of the lionesses abruptly stood up and stalked quickly toward her, her head low. Resisting an overpowering urge to run, she stomped the ground, screamed, and waved the spade high above her head. The lioness stopped, one forepaw poised above the ground. Delia stood still. The lioness sat down.

Again Delia backed toward the truck, and again the lioness followed. She yelled and slammed the spade on the ground, and the lioness sat. Once more the predator and her ape-prey played the game to the same conclusion. But now Delia was nearing the truck. When she was about ten yards away, she threw the spade toward the lioness and ran for the Land Cruiser. The lioness leaped for the spade and was sniffing it when Delia jerked open the door and scrambled to safety. For several minutes she lay on the seat, trembling.

The sound of the plane grew louder, and Echo Whisky Golf glided in for a landing, the lions watching it intently from nearby. I taxied next to the truck and cut the engine. "Great! You've found the lions," I said cheerfully. Then I noticed her face, pale and wide-eyed, her chin resting on the window frame. I jumped from the plane into the Land Cruiser and held her tight.

That evening we darted and collared three of the lions, naming the group the Tau Pride. The next day, flying low over our section of the valley, we spent several hours searching for the Blue Pride in all of their favorite lying-up places. We had not seen them since Bones had been shot, and we found no trace of them now. Perhaps, after their encounter with hunters, they would not come back to Deception Valley.

Instead, we found the Springbok Pan Pride, who held the territory south of camp. Early in the evening we put transmitter collars on Satan, the dominant male, and Happy, one of the females. While they were recovering, we set up a small camp under the wing of the plane, just 100 yards or so away from them. Delia hung a mosquito net from the wing strut and unrolled our bedding and I set out the jerrican of water and the chuck box and built a small campfire. Before long the kettle was steaming and rattling over the coals and a hash of dried meat, potatoes, and onions was sizzling in the skillet.

The fire died to red coals and the moon peered over the dunes, flooding the valley with silvery light. Sitting under Echo Whisky Golf's broad wing, we could see springbok herds grazing along the riverbed. The lions were beginning to roar when we slid into our bedrolls.

Sometime later I woke up. The moon had set and a layer of cirrus clouds obscured the stars. I fumbled for our flashlight. Its batteries were nearly spent, as usual, and the weak yellow light hardly penetrated the darkness. But as I brought the beam slowly around, I caught the faint glow of nine pairs of large eyes in an arc around the plane. The entire Springbok Pan Pride was peering at us from no more than twenty-five yards away.

Echo Whisky Golf was still an object of curiosity to the lions, and I had the feeling they would love to sink their teeth into her tail and tires. She must have looked to them like an enormous powdered milk can with wings. Delia and I sat up for an hour or so, talking quietly and occasionally switching on the flashlight to check where the lions were and what they were doing. Eventually, one by one, they slowly disappeared into the night.

Later, a large herd of springbok stampeded past us, grunting and "whizzing" their nasal alarm calls. Then a deep, rattling groan was followed by slurping, cracking, tearing sounds, and the throaty rumbles of feeding lions. With our flashlight, we could just make out the Springbok Pan Pride quarreling over the spoils of their kill, about thirty yards off the wingtip of the plane. We had trouble falling asleep until they had finished feeding.

The morning sun was streaming through the netting that hung from the wing above us when we were awakened by Satan's feet swishing through the wet grass nearby. His thick mane rolled over his massive shoulders as he sauntered along, his radio collar scarcely visible in the tumble of jet-black hair. He lay down under a small tree and watched sleepily while we brewed coffee and toasted slivers of biltong on the fire.

We spent the day lolling under the wing of the plane and watching the sleeping lions. They seemed totally unconcerned about their collars, wearing them like light necklaces. About four o'clock that afternoon we took off in the plane to check the operation of the transmitters from the air and to search for more lions farther south along the valley. We were delighted to find that even at a distance of forty miles, we could still pick up Satan's radio signal.

As we turned back toward Springbok Pan, I was alarmed by a wall of black clouds rolling in. We had been preoccupied with the radio equipment and hadn't noticed the storm gathering behind us. We had to get back on the ground and secure the plane before the squall line hit the valley, so I gave Echo Whisky Golf as much throttle as I dared and pushed her nose down in a race for the riverbed.

Suddenly we were bouncing and rolling violently in the severe drafts that ran before the gale. The winnowing grass below indicated a wind of at least forty miles per hour, and when we reached the valley it would be blowing directly across the narrow rivercourse. We would have to make a crosswind landing and keep the plane from weather-cocking, ground-looping, and digging its wing into the ground. Little of my flying experience had prepared me for this, and I vaguely remembered the plane's flight manual warning against landing in cross-winds of more than twenty miles per hour.

We made the valley as the first raindrops splattered fat against the windshield. "Push your seat back, tighten your safety belt, and put your head on your lap," I shouted to Delia over the crackle of hail that pelted the fuselage. I could just make out the tire tracks that marked the hole-free landing zone on the riverbed. We banked at a dizzy angle and dropped toward it. I crabbed Echo Whisky Golf into the sheets of hail and rain blown by the incredibly strong wind. I held full left rudder and rolled the starboard wing hard down into the storm.

Finally we were on a semblance of a glide-slope toward our touch-down point. But it took a balancing act to keep the plane lined up in the gusts. First too little bank and not enough rudder, then too much—we were slewing away from our airstrip and the riverbed. I crabbed even more to hold us in line. It seemed as if we were landing sideways!

I would have to straighten the plane and drop the right wing heavily into the wind at the same instant the right wheel touched the ground. If I straightened out too soon the gale would blow us sideways, and either shear off the landing gear or flip us onto our backs. As soon as the left wheel hit the ground and we had slowed enough, I would stand on the brakes and turn into the wind.

We slid sideways over an island of acacia trees, the ground rushing toward us. When we were gliding just above the grass, I pulled back the throttle and tried to line up for the landing. The stall warning horn bawled. Then suddenly the wind held its breath; without its force, all my maneuvers were wrong. The right wheel slammed into the ground and there was a loud cracking sound. Before I could react, we bounced out of control.

I rammed the throttle to full power and tried to recover flying speed and gain some height. But a gust slapped a wing and the plane reeled across the riverbed toward the dunes. I held full throttle and at the last instant it recovered, turned, and climbed away. I glanced at Delia; her head was still resting on her knees.

A thunderstorm ends the hot, dry season and brings life-giving rain to Deception Valley. Occasionally these storms blew down our tents and wrecked our camp.

Top: *After the rains the grasses attract herds of springbok and other desert antelope back to the valley.*
Bottom: *Our camp within a "tree island." There were no other people for thousands of square miles.*

Top: *Captain, the black-backed jackal, narrowly escapes the jaws of the brown hyena, Star.* Bottom: *Captain fights off vultures to protect a scrap of food.* Overleaf: *A dry-season veld fire more than fifty miles wide sweeps across the dunes at night, headed for our camp . . .*

The Pink Panther often lay in the tree above our camp and once slept an arm's length from us near the open flap of our tent.

Chief, one of the hornbills, gets a handout from Delia.

Top: *Muffin, Moffet, and the Blue Pride often rambled into camp to drink our dishwater, play with our gear.* Bottom: *Star bites Pogo to assert her dominance in the hyena clan's hierarchy.*

Bones in the duneveld woodlands near the riverbed.

Top: *Wild dogs have enormous ranges in the Kalahari and settle only while they have young pups.*
Bottom: *A wild dog from Bandit's pack smells Mark's feet. Most of the animals there had never seen humans before.*

Top: *Bandit and another wild dog play-fight after feeding.* Bottom: *Bones, still weak from the surgery on his broken leg, struggles to drag the gemsbok we gave him into the shade.*

Top: *Spicy and Spooky, two sisters of the Blue Pride, with an early-morning springbok kill.* Bottom: *We place a radio collar on Happy of the Springbok Pan Pride.*

Top: *Bones is challenged by Blue as she tries to reclaim a part of her springbok kill.* Bottom: *Bones drags a gemsbok he has killed, putting his full weight on his injured leg.*

Bones mates with Blue.

Top: *Blue with Bimbo and Sandy, her infant cubs. In drought, when prey is scarce, lion prides disband; Blue and her family were alone for months.* Bottom: *Bimbo takes cover in the long grass while his mother is hunting.*

Top: *Bimbo, now two years old, peers through the branches at Mark.* Bottom: *Delia puts a radio collar on the Pink Panther.*

Pepper, a brown hyena cub, ventures away from the protection of the communal den at sunset. At such times she became more vulnerable to predators such as leopards.

On her first foraging expedition away from the communal den, Pepper surprises Delia as she emerges from the bath hut.

Top: *Brown hyenas like McDuff are among the rarest and least-known large carnivores on earth.*
Bottom: *We carry Ivey into the shade to recover from immobilization (Photo: Bob Ivey).*

Top: *Diablo, Happy, and Spicy feed on a gemsbok they have killed.* Bottom: *Captain fights to maintain his status in the jackals' social hierarchy.*

Top: *Star picks up her cub Cocoa to carry him to the communal den over three miles away.* Bottom: *Pepper tries to get Toffee to play. All the cubs of the clan were reared at the one communal den.*

Top: *Springbok pronking in alarm at the approach of predators.* Bottom: *Starbuck, Happy, and Dixie of the Springbok Pan Pride teamed up with Spicy of the Blue Pride for a springbok hunt and share the kill.*

Top: *Pepper and Cocoa carry a steenbok carcass toward the den.* Bottom: *Dusty tries to bite open an ostrich egg, an important source of nutrition in the dry season.*

Top: *Mark returns at sunset from a day's aerial radio-tracking.* Bottom: *Pepper greets Delia during one of her observation periods.*

After prolonged drought, the second-largest wildebeest population in Africa nears the end of its 350-mile trek to water and forage.

Top: *At last the drought is broken by a spectacular night thunderstorm.* Bottom: *A springbok and gemsbok stand in the rain.*

When it rains, even the giraffe leave the dune woodlands for the mineral-rich grasses on the riverbed. Overleaf: Chary's and Sassy's cubs have one of their first drinks of water ever. During the drought, for more than nine months, the lions' only liquid came from the fluids of their prey.

Top: *Mate with the remains of a springbok killed by the Blue Pride after the rains.* Bottom: *Wild dogs, too, find food more easily.*

Chary of the Blue Pride rests beneath a ziziphus tree in the rainy season.

We survived the 120° heat of the drought by lying under wet towels. The moisture attracted hundreds of honeybees.

The wind was too strong. I would have to fly onto the ground with power on and the tail up. Once again I put the plane on the approach. I kept nursing the power, holding the aircraft in stable flight against the wind. Now we were just above the ground, but too far from the touchdown point. I eased on more power and we lifted slightly. The wind picked up—more bank. I thought we were on track, but the rain was pouring down so hard I couldn't see. I looked out my side window and found the faint line below. I bled off a little power, and the right wheel began to rumble. We were down! But the plane was trying to swing too soon, and neither the rudder nor the brake would stop it. We shot off the landing strip and through the grass toward the trees where the lions were lying. Leaning heavily on both brakes, sliding the main wheels, I prayed there were no holes. The outlines of trees loomed in the storm ahead. But the headwind helped to slow us. Finally we skidded to a stop.

Delia was out before me, dragging stakes and tie-down lines from the aft hatch. I cut the engine and jumped into the stinging rain. Together we secured Whisky Golf against the storm.

We pitched our billowing pup tent beneath the wing of the plane and heated steaming cups of tea and soup over a backpack burner. Wind and rain lashed the tent all night long, the plane rocked and creaked, but we were warm and content inside our sleeping bags.

During the following weeks we flew all over the Central Kalahari, collaring lions farther and farther from camp. Delia often drove for hours, on a compass bearing, to meet me on some distant pan or dry riverbed. When we couldn't find lions from the air, we landed Echo Whisky Golf near an antelope herd and spent the night camped under her wing, watching and listening for some pride on a hunt. And always, when we returned to our base camp area, we searched for the Blue Pride.

Time was running out. It was already nearing the end of March and the rainy season would only last another month or so. When the antelope began to leave the valley, we would be less and less likely to find these lions. They were our most important group because they were nearest camp and we knew them best, but we hadn't yet equipped them with radio transmitters. Maybe more of the pride had been killed by hunters or ranchers. Whenever we came back to camp we anxiously asked Mox if he had heard them. *"Wa utlwa de tau bosigo ya maabane?"*

"Nnya" was always his answer.

But one morning, after sleeping out with the plane on Passarge Valley, we taxied into camp and found Mox standing at the fire, his face beaming. He pointed to lion tracks imprinted everywhere on the ground, then led us through camp on a pantomime broken with bits of English and Setswana. He stalked to the bath and kitchen bomas, then pawed at the gap between two wooden posts where a line full of biltong had been drying. He pinched his ear, then pointed to the blue in my shirt and pulled at his chin with his fingertips, meaning it had been the lions with the blue ear tags and a couple of young males with scraggly manes. In his final act, he drew a J in the dust with his finger and patted his backside; one of the males had a J-shaped scar on his hip. It was the same two males who had serenaded Delia, Roy, and me from the airstrip some weeks before.

The three of us piled into the truck and drove the 400 yards to Bush Island, where Mox had last seen the pride. We drove slowly, giving the lions ample opportunity to see us and the Land Cruiser in advance. They sat up as we approached, and we watched their expressions and postures for signs of fear or aggression.

We need not have worried. As always, Sassy and Blue came straight for the truck, chewing its tires and peering at us over the half door, their whiskered muzzles and resin-colored eyes just an arm's length away. I was tempted to reach out and pick off a tick above Sassy's eye, but then thought better of it. The two scruffy young males, whom we named Muffin and Moffet, lay a few yards away from the lionesses. Despite their youth—they were only about four years old—they had apparently laid claim to the Blue Pride and its territory. Only time would tell if they could hold it against older and larger challengers.

We sat with the pride while they lay in the shade of the truck. It was just like old times. Except that Sassy, Gypsy, Liesa, Spooky, and Spicy were now full-grown adults. Together with Blue and old Chary, they made quite a pile of lions.

After we darted the Blue Pride, more than sixteen individuals, from five different prides holding territories along the Deception, Passarge, and Hidden Valley fossil rivers, were wearing radio transmitters that we could home in on from the plane or truck. Through the association of collared lions with their pride-mates, many of whom we had already ear-tagged, we had direct contact with more than thirty-six individuals. We also put transmitters on six brown hyenas, members of the Deception Pan clan, and another clan near Cheetah Pan.

Now that all the collars were in place, all we had to do was keep track of the lions daily with the airplane during the remainder of the rainy season, and then document their migration. Throughout the dry season, the brown hyenas would remain close to the valley, making it easy to find them from the air. We began to feel the tensions of the past few weeks drain away. We had done it! The stage was set for our aerial telemetry studies of lions and brown hyenas.

On the morning that we had finished collaring the last hyena, we proclaimed the rest of the day a holiday and headed for camp. When we parked between our two tents, we were met with a rush and flurry of wings as Chief, Ugly, Big Red, and forty other hornbills settled on the guy ropes, clucking for their daily ration of mealie-meal. "Horn billy-billy," Delia sang out to them as she started down the path into the shade of our tree-island home. Ugly settled on her shoulder, pulling at her earring with his curved yellow beak, and Chief struggled to hold a perch on her head. The whole flock followed her toward the kitchen, where Marique, the Marico flycatcher, swooped to the ground in front of her, shaking his wings and vibrating all over, begging for his share of the handout. I had to wait my turn after a hundred or so camp birds, William the shrew, and Laramie the lizard. Then I, too, was remembered and fed.

Late that afternoon we sat on the riverbed near camp, watching the orange sun turn silvery grass-heads into a sea of fire before it slipped below West Dune. Soon after, the valley echoed with the mournful cries of Captain, Sundance, Skinny Tail, Gimpy, and other jackals, their calls a lullaby to the lonely beauty of the Kalahari going quietly to sleep around us. The color drained from the sky, and the silhouette of West Dune quickly faded in the dusk. The click-click-click—like marbles striking together—of barking gekkos and the plaintive scoldings of plovers announced the coming of night. Finally, as the rush of cool air drained off the dunes, we headed back to camp.

Mox had built a fire and laid out antelope steaks for us before going to his camp. "*Go siami, Ra,*" he called softly, meaning that he was finished for the day and was wishing us goodnight. The soft scuffle of his boots on the clay path to his camp sounded friendly, comfortable, at home. I was glad he was with us.

We weren't ready to eat, and for a long while just sat silently staring into the leaping fire. When the flames had died down a bit we could see beyond, and lying on the chips next to the woodpile a few feet away, were Muffin and Moffet. Sometime earlier they had joined us, and now they watched and listened as we quietly talked. We had to remind ourselves that they were wild lions. What we felt at such times could not be expressed with any one of the usual emotional terms. It was an amalgam, really, of several emotions: excitement, gratitude, warmth, companionship.

Later they stood up and stretched, and then walked to one of the same trees that Bones had so often scent-marked. No more than ten feet from us, they turned, raised their tails, and jetted scent into its branches. Carrying flashlights, we followed them as they ambled into the kitchen. They seemed as big as horses, standing in the three-sided reed boma. Muff cocked his head and put his muzzle on the table, and I could have put my hand on his head as his fleshy tongue lapped up the meat Mox had put out for supper.

Meanwhile, Moff was smelling the shelves. When he reached the twenty-five-pound bag of flour, hanging high on a post, he clamped his jaws on it and pulled. The sack ripped open and white flour showered his muzzle and mane. He stepped back, sneezing and shaking and flinging the flour all over the kitchen. Then he grabbed the bag and strutted from camp, leaving a long white trail behind him. Muffin followed him, and after they had pulled the sack to pieces, they lay quietly next to the kitchen, like big mounds of sand in the moonlight. Miming their soft coos, we walked quietly to within six feet of them and sat there, listening to the squeaks and rumbles of their stomachs. Half an hour later they stood up, roared, and then walked north up the valley.

* * *

Each morning at sunrise we lifted off into the still, cool air to locate our collared lions and hyenas. I had turned Delia's seat around so that she sat facing the tail and could use the plane's food box as a work table. She would tune to the frequency of a lion or hyena, and by listening to its signal in her earphones and switching it back and forth between the two antennas, she would direct me toward the animal. When the signal peaked, we were directly over it. By taking compass bearings on two or three geographical features, we could plot our position on aerial photographs of the Deception research area and have a record of the subject's exact location for that day.

With position coordinates established, we would drop to just above the ground. Diving and turning steeply, heavy g forces pressing us into our seats, we tried to spot the lions. Delia took notes on habitat type, the number of lions in the group, what, if anything, they had killed, and the area's prey concentrations, all while facing backward toward the plane's rollercoaster tail. I don't know how she did it; I would have lost my breakfast in five minutes. Yet she continued to fly with me every day for more than two and a half months.

When I had gained a bit more experience and could safely fly at low levels while tape-recording my observations and operating the radio, I began making the flights alone. While I was in the air, Delia visited the hyenas or worked on data in camp. A couple of years later, after we had bought a base station radio for the plane, we often stayed in touch with each other between one lion position and the next, so that if I'd had to make a forced landing, she would have known where to look for me.

One of the most exciting pictures that began to emerge from the radio-tracking was the relationship between the Blue Pride and the Deception Pan clan of brown hyenas. No matter where the lions were in their wet-season territory, the hyenas found them. The Blue Pride rarely made a kill that Star, Patches, or one of their clan-mates did not find. It became obvious that the brown hyenas depended heavily on the lions for food and that the clan's territory almost completely overlapped the Blue Pride's wet-season territory; even their scent trails often coincided. From the air we could see the valley and riverbed as a big gameboard, with the hyenas uncannily monitoring the predatory movements of the lions so that they could get at a carcass as soon as the predators had moved off. They were all players in a contest of survival.

At the outset we had been concerned that the plane would frighten the lions, making close aerial observations of them impossible. We needn't have worried. Very soon we could fly at grass-top level twenty-five or thirty yards away, without disturbing them. At this range, if we were quick, we could easily see a lion's radio collar and often the color of its ear tag. The lions' reactions to the plane varied. Muffin often made funny faces, rolling his eyes up without lifting his head as we glided over. Satan would crouch and sometimes playfully chase the plane a little way. Or if we sailed over his head, he would rear on his hind legs, pawing the air. Occasionally, when they were resting on the riverbed, we would land and taxi over to them and then picnic in the shade of the wing while we watched them.

* * *

Though their manes were not yet fully developed, they each weighed more than 450 pounds, and Muffin and Moffet made it obvious that they intended to hang on to their claim to the Blue Pride territory in Deception Valley. They strutted up and down the riverbed every night and early in the morning, bellowing, scraping, and spraying their scent on trees, shrubs, and grass clumps along the way.

One morning, however, a rift developed in the male alliance when Blue came rambling into camp with the males in tow. She was in heat and doing her utmost to beguile her two brawny suitors. She slinked and swayed bewitchingly before them, dusting their noses with the tuft of her tail. When two male lions court a female, usually one gives way—or they share her favors. But it soon became evident that in this case the issue had not been settled.

After lying for several minutes near the plane, Blue began to move toward Mox's camp, and together Muff and Moff approached her hindquarters as if to mount her. Instead they bumped shoulders. With growls and snarls, the two males stood on their hind legs cuffing, biting, and clawing each other. Blue ran to the other side of the tree island and cowered behind a bush. Muffin reached her first and whirled to face Moffet. Again they fought, and this time Blue made for the thick bushes at the edge of the riverbed.

Muffin came away from the second round with his left eyebrow split and blood draining over his face. The two males snuffled through the grass, each trying to find the female first.

It might have ended at this point if Blue, the reward, had not chosen that moment to peer out from behind the bushes. Muffin saw her and began trotting toward her. But before he had gone halfway, Moffet charged in from the rear. They fought viciously, rolling over and over, uprooting grass and shrubs as they raked and battered each other with heavy forepaws.

When they broke up, Muffin took final claim to Blue—by now thoroughly intimidated by the fighting—and lay down facing her in the hot sun. Moffet had gone to a shade tree to rest. Blue grew more and more uncomfortable in the heat, and she began to look toward the place where Moff was resting. But when she rose to join the other male, Muff curled his lips, wrinkled his brows, and growled menacingly. She cowered and was held captive all morning, panting heavily in the sun. The situation was finally resolved when Moffet sought more luxurious shade farther away. After that Muff allowed his lioness to rise and they both moved to the spot Moffet had abandoned.

For several days, while Muffin courted Blue, and for another week after that, he and Moffet were separated, even though before this, it had been unusual for them to be apart at all. Ten days after their scrap, we were awakened early in the morning by Muffin's bellows as he approached camp. After spraying scent on the small acacia tree in the kitchen, he moved north along the riverbed. Another lion answered his calls from farther up the valley, and the two moved toward each other, bellowing continuously. When Moffet emerged from the bush near North Tree, the two males trotted toward each other. They rubbed their cheeks, bodies, and tails together again and again, as if trying to erase the conflict that had come between them. Then they lay down together in the morning sun, Muffin's paw over Moffet's shoulders. It would take more than a rift over a female to break the bond between them.

*   *   *

We had spent years crashing through the bush in our truck to gather single tidbits of information on lions and brown hyenas. Now that we were using the plane and radio-tracking equipment, a stream of data began flowing into our field books. We knew where Muffin, Moffet, Blue, and the rest of their pride were on any given day and how far they were from the Springbok Pan pride, as well as from four others. To learn whether or not one of the lions had made a large kill, we simply took to the air, tuned its frequency, and flew over its head. I could depend on finding each of our collared animals from the air virtually 100 percent of the time, and could usually tell who they were with, in what habitat, how far they had traveled in the night, and whether or not they had cubs. It took no more than an hour and a half to two hours to find all of our collared animals in the rainy season. It was a field researcher's dream.

17

Gypsy Cub

Delia

...the things which will not awaken are giving life to those
that do...and thereby shall live again this spring...and always...

—*Gwen Frostic*

ALOFT IN EWG one morning, Mark circled the two male lions again,
confirming what he saw below. Lying under the same tree, no more
than three feet apart, were Muffin and Satan, rivals from the Blue and
Springbok Pan prides. Each rested his chin on an outstretched paw,
not moving a muscle, glaring intently at his opponent across the boundary of their adjoining territories. Moffet was nowhere around.

After Mark got back to camp, we drove over to the lions and found
them still trying to stare each other down. It was now midday and the
shade had moved, leaving them in the hot sun. Slowly Muffin's eyelids
began to droop. His head nodded drowsily and then slipped to one
side. Immediately a deep growl rose from Satan's chest and Muffin's
head snapped up to meet the challenge.

The stare-down continued into the afternoon. Whenever one of them
grew uncomfortable and had to change position, a low growl would
grow in his throat. As it increased in volume he rearranged his hindquarters, barely moving his head, and never taking his eyes from the
other.

Just after sunset both males slowly got to their feet, growls tearing
from their throats, neither daring to look away. Step by deliberate step,
they backed cautiously away from each other, finally turning and
disappearing into their respective territories. Not a shot had been fired,

but they had tested each other's strength just the same; it had been a draw.

Male lions who form an alliance with siblings or peers, as Muffin and Moffet had done, are more successful at gaining and maintaining possession of a pride and its territory than are single males.[1] The odds would be against Satan if he ever confronted Muffin and Moffet together.

*   *   *

The boundaries of adjoining pride territories were not entirely discrete, and there was some overlap: Members of the Springbok Pan Pride and the Blue Pride occasionally hunted on Cheetah Pan, at the border of the two territories, as long as the other group was not around. Yet the male lions, in particular, spent a great deal of time and energy during the rainy season defending their territories. They roared, raked, scraped, sprayed, and fought, if necessary, to maintain the claim to their areas and, ultimately, to the prey and reproductive females each encompassed. Muffin and Moffet spent hours roaring and scent-marking their boundaries, and they were especially vociferous in the period right after they assumed control of the Blue Pride territory.

Mark played a dirty trick on Muffin and Moffet one morning when they lay sunning themselves on South Pan. Earlier, we had tape-recorded Moffet's voice roaring an answer to Satan. Now Mark parked the truck about ten yards from where Muffin and Moffet rested peacefully, their heads on their paws, eyes closed, soaking up the sweet warmth of the new sun. He held the recorder to the open window and switched it on.

When he heard his own voice, Moffet leaped to his feet and whirled around to face the truck. Mark turned off the recorder instantly, but there was no switching off poor Moffet. Thoroughly agitated by this strange voice, and squeezing out thunderous roars from deep within his belly, he took several steps toward the truck, then stopped, head erect, ears perked, eyes searching. When he got no answer, he roared again, and looked back at Muffin, as if to say, "Come on! Get with it! Some fool's trying to take over our territory!" But Muffin, his head still resting on his paws, looked indifferent. After Moff had finished his fourth chorus of bellows, he walked to where Muff lay and roared again. As though he had no choice in the matter, Muffin stood up and, somewhat half-heartedly, added his bellow to the performance. After that, both of them roaring, and pausing only to scrape-mark, they set off at a fast walk, right past the truck, toward their phantom intruder.

A few nights later Muffin and Moffet were on their southern beat when Satan's roar came drifting over the dunes. They stopped abruptly, listening, and then roared in return, raking their hind feet through the sand. For three hours, while slowly moving closer together, the males called back and forth across Cheetah Pan.

Several hours before dawn, Satan stopped answering the challenge, and a silence settled over the valley. Muffin and Moffet each killed a hartebeest from a small herd on the territorial boundary between the Blue and Springbok Pan prides. They were feeding on their kills when Satan stepped into the clearing behind them. He stood watching from twenty yards away until Muffin and Moffet turned. Their eyes burned with aggression.

With tremendous roars and a shower of grit, they jumped over the carcasses and slammed into him. The charge drove Satan back several yards, his hind feet ploughing furrows in the loose sand. Claws extended, he lashed out with his heavy paws, snapping Moffet's head to the side. Then, rearing to full height on his hind legs, his wide-gaping mouth exposing long canines, Satan turned and took Muffin head on. Looking like massive prizefighters, they bit, pummeled, and slashed each other's shoulders, manes, and faces. Great cords of muscle stood out like steel cables across their backs.

By now Moffet had recovered from Satan's blow, and he attacked him from the rear, biting and clawing his back while Muffin hammered his head with both forepaws. With his enormous strength, Satan whirled and sent Moffet rolling into the thornbushes, but Moffet struggled to his feet, and he and Muffin charged Satan once again, driving him into the base of a tough desert bush. Heavy branches two inches thick splintered like matchsticks.

Muffin pressed his frontal attack, but Satan was punishing him severely, stabbing deep into his shoulders and chest with his long canines. Meanwhile, Moffet was again mauling Satan's back and flanks, crisscrossing them with open slashes. Though the bush partially protected Satan's rump from Moffet, it would not allow him to escape.

Muffin's face was gushing blood from a gash that ran from his right eye to the end of his nose. He was weakening under Satan's penetrating bites and thunderous blows, and his sides were heaving with exhaustion.

With Muffin weakening, Satan moved away from the bush. But as soon as he exposed his hindquarters, Moffet caught his left hind foot between his teeth and bit down with tremendous force. Satan roared with pain, but confronted as he was by Muffin, he could not divert his attack. Moffet held on, and this seemed to give Muffin new strength. He pressed in on Satan, biting and beating his head with a series of blows that sent tufts of black mane and broken branches flying into the air and blood splattering all over the ground. Satan's deep snarls and roars were gradually losing their power and changing in pitch to a near whine. Moffet now clamped his jaws around Satan's lower spine, and biting hard, he crushed the nerves and vertebrae with a dull grating sound. Satan slumped to the ground.

The brothers stood over the fallen lion for a minute. Then, panting heavily, Moffet turned back toward their hartebeest kills, with Muffin staggering after him.

For a long while Satan lay unmoving, his stertorous breath gurgling in his throat. Flesh and mane dangled from his torn neck, and blood oozed from his broken spine. Then, very slowly, he raised himself on his forelegs and began struggling away to the south, dragging his useless hindquarters. But he managed only fifteen yards before he collapsed again, urinating blood and gulping air. Again and again he half raised himself and crawled toward his territory. But each effort cost him more of his waning strength. Finally, with a great shudder, he collapsed and took a last deep breath.

When the new dawn arose Satan was dead.

*   *   *

Sitting backward in the cockpit, I tried to keep my attention on the telemetry instruments in front of me, but from the corner of my eye I could see the white tail of the plane dipping and swerving just above the treetops.

"Hang on. I'll make one more turn. Try to find them," Mark called over the intercom.

I clutched the seat, and the plane banked slowly over the crest of West Dune. The ugly warning horn squawked on and off as Mark held the plane on the edge of a stall. Fighting the urge to close my eyes, I scanned the ground under the acacias for a sign of Sassy and Gypsy, who for some time now had been separated from the other females of the Blue Pride.

"There they are—at the edge of that clearing!" I shouted.

"Okay—that's where they've been for the last few days. We'd better get the truck and have a closer look."

This style of flying, for hours every day, eventually began to unnerve me, but even later, when Mark began doing aerial locations alone, I didn't feel much better. This was high-risk flying, and with his attention divided between the telemetry work and the plane, there was a greater chance that he would have an accident. But he insisted that for both of us to be in the plane locating the lions was a waste of man-hours.

Now we flew back to camp, loaded the radio gear into the truck, and followed Sassy's signal up the face of West Dune. The long bodies of the lionesses were sprawled in a patch of drying grass on the crest of the dune. Except for the constant flicking of their tails to ward off the flies, they did not budge as we approached.

When we were six yards away, we stopped. A tiny head with woolly ears and dark eyes peered over Sassy's belly. Another pair of soft round ears and sleepy eyes appeared, and then another, until a line of five wee faces stared at us. Sassy and Gypsy, whom we had known since they were youngsters, now had infants of their own.

The cubs toddled around their mothers on stumpy, unsteady legs, stumbling into one another and falling backward onto their plump, fuzzy bottoms. Their straw-colored fur was peppered with freckly brown spots. When eventually they settled down again, three of them suckled Sassy and the other two went to Gypsy.

The mothers were about four years old, and as far as we knew, these were their first litters. They kept their cubs in the "nursery," a grass thicket with an unusually tall, spreading Terminalia tree at its center, near the top of the dune.

We were particularly excited to find the cubs because, in order to develop recommendations on how to conserve Kalahari lions, we had to know more about aspects of their reproductive biology: how often these lions bred, how many cubs they had, how the mothers fed their litters through the long dry season, and the number of cubs that usually survived from each litter.

Studies in the Serengeti of East Africa have shown that lionesses are notoriously poor mothers. Only after they have had enough to eat themselves do they allow their cubs to feed on a kill. They often abandon their young, sometimes for no apparent reason, other than that they seem to prefer to socialize with their pride-mates rather than face the responsibilities of motherhood.

Although prey is relatively abundant for most of the year on the Serengeti Plain and life is generally easier for predators than in a desert like the Kalahari, only twenty percent of infant lion cubs survive to adulthood.[2] Of those that do not survive, one-quarter die of starvation, often because their mothers simply fail to lead them to kills. Another quarter die from predation or accidents, and one-half die of undetermined causes. According to George Schaller, adult lions live many years, have a fairly low death rate, and do not rear very many young.

We thought things might be different in the Kalahari. If mortality among adult lions was higher, if their lifespan were shorter in the harsh desert environment, maybe they would take better care of their offspring. We stayed with Gypsy and Sassy as much as possible, hoping to gather this and other information.

Lionesses in a pride often come into estrus, breed, and give birth synchronously, in any season of the year. Then females with cubs will frequently separate from the pride to form a small group of their own until the young are old enough—at about four months—to keep up with the movements of the adults. At birth, lion cubs weigh only about three pounds and are almost totally helpless; their eyes usually do not open until between their third and fifteenth day of life. Gypsy's and Sassy's cubs were probably between two and three weeks old.

For the rest of that day the mothers lay with their infants in the shade of the nursery tree. Mostly they slept, a ball of cubs cuddled under Sassy's neck or next to Gypsy's forelegs. All five, snuggled together, were about the size of Sassy's head. Now and then a cub would waddle over to suckle one of the mothers, and the others would follow. Neither Gypsy nor Sassy ever appeared to notice which cub she was nursing. In the Serengeti, pride-mates communally suckle one another's young; now we knew the same was true in the Kalahari. They would suckle for five to eight minutes before wandering a few feet away or falling asleep at the mother's side.

At sunset, Sassy rolled onto her stomach and alertly scanned the three-mile strip of riverbed visible between the dunes. Gypsy, sensing the mood, lifted her chin and watched. Then abruptly they both stood up, rubbed their faces together, and stretched their long backs like bows, pushing their forepaws through the sand. Then, without looking back, they walked northward. Three of the infants followed their mothers through the grass for a short way, but Sassy and Gypsy quickly disappeared in the bush. The cubs all scrambled deep into their thicket beneath the nursery tree, where they would hide until their mothers returned. There was nothing to suggest to any predator that the lion family was there, except for the pugmarks left between the grass clumps on the face of the dune.

By exchanging roars with the other females, the two mothers joined the rest of the pride in the north end of the valley for a springbok hunt. When they had finished feeding, just after midnight, they returned to the nursery. Their soft coos brought the meowing infants tumbling from their hiding place. While the cubs wobbled about between their mother's legs, Gypsy and Sassy lapped their faces and backs, their rough, heavy tongues pushing them to the ground. Rolling each infant over, the mother licked the cub's underbelly and beneath its tail while tiny paws pushed at her muzzle. Then the two females began to nurse their youngsters.

Toward morning Muffin and Moffet came by the nursery and lay down next to Sassy. One of the cubs tottered up to Muffin and stuck its tiny face up to his giant whiskered muzzle. He ignored the pesky infant until it walked between his two front legs and turned to snuggle beneath the shag of his full mane. Mildly irritated, Muffin slowly raised the right side of his upper lip, wrinkling that side of his face and showing his long canine in a crooked, half-hearted snarl, as if this puny cub couldn't possibly be worthy of more threat. The tot turned its ears back, scampered to Sassy and pushed under her chin, looking back with round eyes at the crotchety old Muffin.

The males took no part in rearing the cubs, and perhaps only visited the nursery so that they could follow the females when they left to hunt.

Though both females nursed the five young cubs, it soon became apparent that Sassy was a much better mother. When the least bit of bickering arose from the tangle of tiny bodies at Gypsy's teats, she would often swing her head around and snarl and then roll onto her belly or walk away, leaving the infants crying for more milk. Soon all five cubs would be fighting over Sassy's four teats. Meowing loudly, the odd one out would go to Gypsy, and sometimes she would nurse it, sometimes not.

As the days passed, Gypsy stayed away from the cubs for longer and longer periods. Sated and apparently content, she would lie around all day with her other pride-mates. Meanwhile, Sassy was doing more than her share to raise both litters.

One day when the cubs were about eight weeks old, we found that Sassy and her three young were gone. Gypsy was lying on her back nursing her two cubs, but when they fought briefly, she bared her teeth, wrinkled her nose, and hissed wildly at them. Then she walked away and they were left gazing after her, their ribs showing through their scruffy coats.

Gypsy joined the pride on Leopard Trail and lay around with them for the rest of the day. The next morning, instead of returning to her cubs, she relaxed in the shade with her head snuggled along Liesa's back. Except for her swollen teats, she showed no sign that she had two hungry infants waiting for her.

Both of Gypsy's cubs were in poor condition, but one was especially scrawny and weak; it would not survive much longer without milk. Flying in Echo Whisky Golf the next morning, we found Sassy and her cubs with the old lioness Chary several miles from the nursery. By circling low in the plane we could see that Chary had four cubs of her own, several weeks older than Sassy's. The two females were lying together, peacefully nursing their young.

We switched channels on the radio and found Gypsy with Liesa, almost ten miles from her cubs. Later, when we drove to the thicket, we found that the weakest cub had died. Alone and emaciated, the other one was hunkered down between two forks of the tree, watching us with frightened eyes. If we did not feed it, the cub would probably be dead within twenty-four hours, but there was still a remote chance that Gypsy would come back for it. After much agonizing, we finally rationalized that, although it would be a lot of trouble to have a lion cub in camp, we would also learn a great deal. If Gypsy did not come back to her infant by the next day, we would adopt it.

The next morning the roving mother had moved even farther from her cub, and we knew it probably would not survive another day without milk. We took a cardboard box and an old blanket to the nursery to rescue the abandoned infant, but to our surprise, we found Muffin lying under the tree with it. The cub staggered on trembling legs to the big male, feebly pushing its tiny muzzle into his belly, groping for milk-filled teats that were not there. For several minutes it just stared up at Muff, dizzy with hunger, and then it stumbled back to the tree and with its head hanging down, it swayed forward and back, bumping, bumping, bumping its forehead into the trunk, over and over again. Its wasted body made its head and paws seem over-large. Finally, in the last stages of starvation, it just stood there leaning the top of its head against the bark.

We weren't sure how Muffin would react if we tried to take the cub, so we decided to come back that night when he had moved off to hunt. In the meantime it might starve to death, but at least no other predator would harm the infant so long as Muff was there.

It was an unusual day for late in the rainy season. Instead of moderate temperatures with a light breeze, it was muggy and still. The hornbills sat quietly in the trees, their bills open, wings cocked out from their bodies, trying to cool themselves. Nothing moved but the flies, which buzzed our faces or sat on our towels, rubbing their grimy feet in apparent glee.

Late that afternoon, a long black tunnel of low clouds rolled in from the southeast. Skimming rapidly over the dunes, it filled the valley and rushed toward camp. The sun dipped lower in the western sky, and the clouds turned brilliant shades of pink and mauve streaked with gold. But when they swirled overhead, the air became a frenzy of wind and sand.

We hurried to ready camp for the storm, zipping the tents, tying down the plane, and putting equipment boxes up on blocks—Mox was on leave in Maun, so there was no one to help us. Suddenly, the wind slapped the trees, thunder cracked, and tongues of lightning split the wounded sky.

Lashed by rain and hail, we finished tightening the guy ropes on the tents. Mark yelled over the roar of the wind, "Get in the truck— we've got to get out from under these trees!" We jumped into the Toyota and drove twenty yards from camp. The ziziphus and acacia trees reeled about madly above the tents, and sheets of rain scudded parallel to the ground. We could barely see the camp or the plane through the gale.

"There goes Echo Whisky Golf!" Like a wild horse, the plane reared up against its tie-down ropes, the starboard wing high in the air. The line holding the tail wheel snapped, and as the aircraft weathercocked into the wind, the port wing tore its stake out of the ground. The plane slewed around and slammed into a fuel drum and the fence.

Mark bounded over the fence and grabbed the starboard wing tip to keep the aircraft from flipping over while I staggered against the wind toward Echo Whisky Golf.

Mark shouted, "Grab the other wing and hang on or we'll lose the plane!" I stood on my toes to reach the wing. We clung to the plane in the blinding rain. The powerful wind caught under the broad airfoils and lifted our feet from the ground for seconds at a time. My arm and back muscles throbbed with pain, and I worried that I would lose my grip.

The lightning was a buzzing blue hue in the sky, and dangling from the metal wing, I felt like a lightning rod. The sleeping tent gave one last mighty heave against the wind, and collapsed in a tattered heap, draping itself like a wet spider web over the poles.

A few minutes later the wind slackened slightly, and we were able to tie drums of fuel to the wings to help stablilize the plane. We quickly inspected for damage and discovered that the port-side stabilizer had been crumpled by a drum.

Again the wind slammed into us, this time from the north, and once more we hung onto the wings, pounded by wind and hail. By now I was extremely weak. My arms felt as if they were being pulled from their sockets, and the cold sent sharp spasms through my back and shoulders. Just when I knew I could hold on no longer, the wind eased a bit. My fingers slipped from the wing and I sat down in the mud, utterly exhausted.

Mark came sloshing through the water and helped me into the truck. "Well done, Boo," he said, putting his arm around me. "We'd have lost her for sure if you hadn't hung on." He wrapped his shirt around me and hurried away to restake the plane. I felt a warm stickiness on my leg; when I reached down I discovered it was blood. I switched on the flashlight and saw a deep gash in my calf, probably made by the fence. I tried to stop the bleeding with tissue.

Mark jerked open the door of the truck. "I'm going to try to fix the tent. Come and help me if I whistle."

I sat shivering uncontrollably, hoping that he would not need my help. Although the wind had slackened, the rain was still pelting down, and the plane stood in a growing lake of muddy water. Moments later I heard the faint shrill of Mark's signal. I jumped from the truck, losing my sandals in deep ooze.

He heaved up the tent's center pole and we both struggled through the muck to retie the guy ropes. A few poles were broken, one side sagged, and the floor was covered with inches of muddy water, but at least it was a roof over our heads.

When Mark lit the lantern, he saw my bloody leg. I tried to tell him what had happened, but my teeth only chattered. He wrapped me in a dry blanket, and then bound my wound. When he started for the door I asked, "Aren't you going to dry off?"

"First, I'm going to get some hot food." And he ran out into the storm again.

Minutes later he was back, carrying a tray with mugs of steaming soup and tea. The dripping tent was beginning to warm from the lantern, and we sat on the tin trunks drinking our hot soup, feeling quite cozy.

Five hours after the storm had hit, it finally ended as suddenly as it had begun. All that remained was an occasional grumble of thunder and great drops of water that plopped from the trees onto the tent. We sat listening to the quiet and warming ourselves. A jackal called from North Bay Hill. Then from south along the valley, a lion roared— and we remembered the cub.

We grabbed the driest blanket, filled a canteen with hot water, and drove across the valley toward the dune. The thirsty earth had already soaked up a lot of the rain, though much of the riverbed was still under water. There were split and scattered trees everywhere in the fractured woodland. Using the spotlight, we finally found the nursery. Muffin was gone. Next to the tree the crumpled cub lay like a soggy rag doll, his eyes staring sightlessly into the night.

*　　*　　*

As far as we knew, Gypsy never went back to the nursery. She went on hunting and sleeping with the Blue Pride females until her udders lost their milk. From her behavior, it appeared that Kalahari female lions were no better mothers than those in the Serengeti, but it was too early to draw any firm conclusions. Gypsy, young and inexperienced, had been just one example of how not to rear lion cubs in the Kalahari. New mothers are often poor mothers and improve with experience.

The dry season was beginning, and we continued our study of maternal care of desert lions by watching Sassy and Chary and their seven cubs. Now that Sassy had joined Chary, an old and experienced mother, perhaps their litters would have a better chance of surviving.

Sassy's cubs were now about two months old, Chary's about three. They all played and fought as brothers and sisters, rolling over and over, mouthing and pawing at one another's faces and forequarters like kittens. They often attacked their mothers, and sometimes Sassy joined in the game. Chary, although patient with all the cubs, never joined in the tomfoolery.

In carnivores, as in all animals, play is not just for fun. The types of behavior important for hunting—stalking, chasing, jumping on a moving object—require coordination and practice, and they are the very ones exhibited in play. Lion cubs do not have to learn all the motions for hunting; they are born with most of this information in their genes. But by play-fighting and play-hunting, the cubs polish the skills needed for bringing down a moving prey.

One afternoon as a cub rested next to Sassy, a fly lit on the tuft of her tail. Lying with his chin on his paws and his eyes crossed, the cub watched the fly roam around the tassel. Sassy flicked her tail and the youngster pounced on it, rolling over in a somersault. Another cub joined in, batting and pawing at the tail snaking around on the ground.

Sassy jumped to her feet, whirled around, and pawed playfully at her cubs' heads. They reared up and swatted at her, and she took off through the grass. Instantly, all the other cubs joined in pursuit.

Eventually old Chary fell into the long line of lions racing after Sassy. In and out of the thorn shrubs they twisted and turned, swatting at whomever they encountered. Sassy, stopping abruptly, took a long, thin stick in her jaws and pranced through the tumbling cubs, her head and tail held high. The youngsters pulled at the baton, rolling over and over in the sand, trying to yank it from her teeth. None of them stood a chance of out-tugging the adult. But then Chary, her sagging back swaying, grabbed the other end. The lionesses romped and chased each other, pulling and tearing at the skinny stick as they swung about. Eventually the cubs all lay down in a row, and watched their mothers rolling and fighting over the twisted twig. When only a plug of shredded wood was left, Chary and Sassy gave up the contest and, panting, sauntered to their shade tree. When she passed the truck, Chary turned her ears and avoided looking at us. I could have sworn that she was embarrassed by her brief loss of control.

Later on, Chary and Sassy killed a young hartebeest in the woodlands of West Dune. They had fed for twenty minutes when Muffin and Moffet trotted up, chasing the females from the carcass. The lionesses returned to the cubs' hiding place, but instead of nursing them, they started walking slowly in the direction they had just come from, and cooing softly, encouraging their young to follow them to the kill. Muffin and Moffet completely ignored the youngsters and made no protest when they fed with them, but Chary and Sassy never got another bite.

Now that the cubs were old enough to eat meat, the pattern of their daily lives changed. The females would leave their litters hidden in the grass to hunt and to feed themselves. They would then return and lead their young to the kill—sometimes several miles away. Whenever Muffin and Moffet found the females on a carcass, they chased them off, but they always shared the meat with the cubs. Chary and Sassy also continued to nurse their young, but less frequently and for shorter periods.

The mothers seemed to be making every effort to insure that their offspring were well cared for. But with the skies clearing, the grass dying, and the antelope dispersing as the dry season began, their kills were becoming fewer and smaller. And the cubs were getting larger and hungrier.

18

Lions with No Pride

Delia

and my tribe is scattered . . .
—*Stanley Kunitz*

OFTEN, In those wild open spaces of the Kalahari, even a tent seemed too confining. So some nights we pulled our canvas cots onto the ancient riverbed to sleep under the stars. The fresh smell of drying grasses and the soft, cool air were more effective than a sedative. By this time we knew almost every species of bird or insect that squawked or clicked around us, and these familiar sounds, along with the cry of a jackal, would send us off to sleep. Periodically during the night, I would check on the position of the Southern Cross, following its gentle sweep across the lower sky, before going back to sleep again.

Once, about four o'clock in the morning, a loud rustling in the bushes suddenly opened my eyes wide. The massive form of a lion loomed in the starlight not five yards away, and he was walking directly toward us.

"Mark! The lions are here!" I whispered urgently, feeling around on the dark ground for the flashlight.

Buried deep in his sleeping bag, his voice thick with sleep, Mark mumbled, "Don't worry, if they get too close we'll move inside the tent." At that moment the lion was actually standing at the foot of Mark's cot, looking down at us.

"Mark," I said, trying not to move my lips, "they're right *here*. Get up!" I found the flashlight, slowly raised it, and switched it on. Moffet's amber eyes blinked in the light. Now Muffin walked in from the shadows of camp and stood two or three yards behind him. We were lying at the base of one of their favorite marking bushes.

Mark poked his head from his sleeping bag and peered at the lions over his toes. Moffet squatted over his hindquarters and began scraping back and forth with his hind paws while dribbling urine loudly onto the ground between his legs. He was marking his territory.

Even if it was Muffin and Moffet, I didn't like the idea of being between them and their tree. I struggled out of my sleeping bag and, without taking my eyes from the lions, began moving toward our tent, sixty yards away at the other end of the tree island.

As I passed the head of Mark's cot, he was feeling around on the ground for his clothes. "I can't decide what to put on," he said, standing there nude and half asleep.

"What difference does it make, for heaven's sake!?" I hissed between my teeth. The two lions watched Mark stumbling around, gathering up his clothes and his sleeping bag. Finally, I grabbed Mark's arm and pulled him toward the tent. At the edge of the trees we looked back at Muffin and Moffet, rubbing their heads together, paying no attention whatsoever to our bungling getaway. I wondered, myself, what all the fuss had been about.

*   *   *

The rains of 1978 had been generous, but they ended prematurely. The blustery dry-season winds had begun much earlier than usual, and the grasses had turned to straw much sooner. The skies were dull grey from wind-blown dust and sand, and the savanna looked as dry in June as it had in August of previous years.

As always, the short grasses on the shallow, heavy soils of the ancient riverbed dried faster than the vegetation on the dunes. The plains antelope moved from the river channel to the bush and tree zones on the duneslopes, where the leaves stay green longer. Slowly the herds broke up, and the smaller groups dispersed into thousands of square miles of rolling bush savanna.

Each dawn we hurried to Echo Whisky Golf to look for the lions from the air. We were afraid that if we missed a single day's radio-location, they might travel to some distant part of the Kalahari where we could never find them again. But weeks passed and the prides still had not migrated. True, they had abandoned the dry river channel, now that the antelope had gone. Muffin and Moffet no longer marched up and down Mid Pan marking their scent trees, and Blue and the other females stopped visiting camp. Yet the Blue Pride lions were not very far from their wet-season territory. With most of the large antelope gone and no water to drink, how were the lions surviving? By following them with the truck and radio gear at night, we began to get some answers.

*　　*　　*

Moffet had been separated from Muffin and the females for several days and had not eaten. Moving through the thornbush east of the valley, he quickly broke into a trot, his head low as he zigzagged after a chicken-sized korhaan scurrying through the grass ahead. When he was ten feet from the bird, it took flight, but Moffet lunged forward and, standing on his hind legs, swatted it down with a wide forepaw. Lifting his lip, he chewed into the feathery breast of the bird, sneezing and shaking his head to clear the down from his nose. Minutes later, feathers still clinging lightly to his mane, he began hunting again.

At first we did not take Moffet's new sport of bird hunting seriously, for surely a 450-pound male lion did not intend to feed himself on such morsels. Later that same evening, however, he killed a four-pound springhare and chased a mongoose to its burrow. The diet of the lions was changing drastically.

The Blue Pride still preyed on the occasional giraffe, kudu, or gemsbok in the dune woodlands, but because these large ungulates were scarce and widely scattered over the savanna, the lions hunted smaller animals more often. Instead of killing 500-pound gemsbok, as they had done during the rains, now they fed on fifteen- to twenty-pound porcupines, steenbok, honey badgers, bat-eared foxes, or kori bustards. But these prey are hardly enough to make a meal for one or two lions, and they certainly wouldn't feed an entire pride. The seven females of the Blue Pride who lay around together in the wet season, always touching and reassuring each other, were forced to break up into smaller groups, so that when a kill was made there would be enough meat to go around.

From the air we found that Chary and Sassy and their cubs had again separated from the main pride and were roaming near Crocodile Pan, about five to six miles east of Deception Valley. Instead of hunting on the borders of fossil riverbeds, they prowled the bush savannas and woodlands, where a few bands of gemsbok and giraffe, and more small prey, were available. The lionesses had to travel from five to ten miles almost every night to find food.

Gypsy and Liesa hunted near Paradise Pan; the rest of the pride roamed interdunal valleys two to three miles west of Deception. The Blue Pride had splintered into small groups of females, with Muffin and Moffet trekking from one group to the other. Because of the time needed to locate the different fragments of their pride, the males had to spend more time apart from their females and do much more of their own hunting.

The pride's home range had more than doubled in area, to about 600 square miles. Still, the lions had not really migrated in the true sense of the word, but had simply expanded their range greatly to the east and west. We drew a map with colored dots showing their daily locations and their patterns of distribution over the range. It looked as if it had been sprayed with BB shot.

The other prides had responded to the dry season in the same way, by breaking up and traveling greater distances away from the valley in search of prey and by hunting smaller animals. The drier it became during the winter months, the more the diet, range movements, habitat utilization, and social system of the lions differed from the wet season. We began to fly on moonlit nights to better document these changes.

*   *   *

Midnight takeoff: Cast in soft moonlight, the pewter desert fell away below us. Except for our gas lantern, set out on the airstrip to guide us home, not another light on earth could be seen as we sailed over the quiet, forgotten world of the Kalahari. Our faces glowing eerie red from the cockpit instrument lights, we followed the night movements of the lions and hyenas below.

Straining to recognize the subtle landmarks beneath us, we found Happy of the Springbok Pan Pride one night, on the boundary of the Blue Pride's territory. Within two weeks after Muffin and Moffet had killed Satan, another male, Diablo, had taken over the Springbok Pan Pride. The females had adjusted to their new male, and in recent weeks we had even seen Happy and some of the others mating with him. But now, as we circled overhead in the moonlight, we could see that Happy was within a few yards of Muffin and Moffet, who were patrolling the border of their territory. We were curious to know whether the two males would chase this foreign female back into her own territory or mate with her—if she was in estrus. In the Serengeti, male lions will court females from other prides, but we had never had the opportunity to observe this in the Kalahari. We flew back to camp, and then drove south to look for the lions.

When we found Muffin and Moffet, they were walking fast through brambles near Cheetah Pan, their noses to the ground. They stopped abruptly and looked up; Happy's eyes met theirs at less than thirty yards. The two lions stared intently at her for a few seconds, their tails twitching. Happy stood above the males on a low, scrub-covered sand ridge.

The lioness walked slowly forward, her head tall above the grass, her ears perked. Chests rumbling and tails lashing, Muffin and Moffet sprang to their feet and chased her for over 100 yards. But Happy was too fast for them, and when they broke off the charge she stopped just out of reach. They stared aggressively at her, raking their hind paws through the grass and roaring.

Again Happy walked cautiously toward them, and again they chased her, roaring and swatting the air just behind the tuft of her long tail. After each chase she ventured closer to them, but they seemed less and less inclined to pursue her. When she managed to get within twenty yards of them, Muffin and Moffet lay down side by side and watched what amounted to a feline burlesque.

Her hindquarters swaying sinuously, eyes half closed and jaws parted, Happy slunk toward the mesmerized males. Muffin quickly stood and strutted toward her, but she galloped away. When he stopped, she turned and wound her way toward them again, this time passing within a few yards of their noses. Muffin stood as tall as he could and, with all the savoir-faire he could muster, swaggered toward Happy. She lowered her hindquarters suggestively, inviting him to mount her. But when he stepped to her rear she suddenly spat and cuffed him hard across the nose. Muff roared and drew back, his ears flat and his long canines exposed, as Happy minced away, her tail flicking flirtatiously. After a few more attempts by both males to gain her favor, Muffin and Moffet seemed to tire of the game, and they walked back north into their own territory. Happy followed about thirty yards behind, apparently unconcerned that she was on foreign soil.

We knew that Liesa, Gypsy, Spicy, and Spooky of the Blue Pride were finishing a warthog kill on the crest of West Dune. Muffin and Moffet, with Happy trailing by fifteen yards, were moving directly toward them.

Since it was not unusual for male lions in the Serengeti to associate occasionally with females of another pride, it had not been totally surprising to see Muffin and Moffet interact with Happy. However, we knew that Serengeti pride females form closed social groups that do not accept new female members or tolerate foreign females in their territory.[1] There the pride is sacrosanct: a stable social unit of closely related lionesses and their young, who associate with the male or males who help defend the territory. A lioness may be kicked out and become nomadic, but these nomads do not join other prides. In the Serengeti, a single pride lasts for generations, with the same kin line, and at any one time, it has in its membership great-grandmothers, grandmothers, mothers, daughters, aunts, and female cousins.

Now Muffin, Moffet, and Happy padded steadily toward the crest of West Dune. We followed in the truck, preparing our flashes, cameras, and the tape recorder for the coming fight between Happy and the Blue Pride females.

By the time we could see the four Blue Pride females in the spotlight ahead, they had finished the warthog and were casually licking one another's faces. The two males greeted the lionesses, smelled the skeleton, and then lay down a few yards away. Happy sauntered past Spicy and Spooky and lay down next to Muffin and Moffet. Incredibly, there was not the slightest sign of aggression on the part of any of the lionesses. We switched off the tape recorder and pulled the cameras back inside the truck. It was astonishing: A foreign female had ambled into the heart of the Blue Pride camp, and its members had hardly noticed!

For the next four days, Happy was courted, first by Muffin and then by Moffet, just as though she were a Blue Pride female. During the heat of the day Muffin lay as near to her as he could, watching her every move. If she sought out better shade, he strutted so close beside her that their bodies rubbed together. Sometimes he would initiate copulation by standing at her rear. More often, however, she would walk back and forth in front of him, her tail flicking and hindquarters swaying, or she would brush her body along his before crouching in front of him. When he stood over her to copulate, he nibbled at her neck and she growled and flattened her ears. As soon as Muffin had finished, he would step back quickly to avoid getting clouted by Happy's paw, for invariably she would whirl around, snarling fiercely, and swat at him. Then, lying on her back, her legs extended, she would roll over and over in the grass, her eyes closed in apparent ecstasy.

They mated in this stereotypic fashion every twenty to thirty minutes for part of two days and all of two nights. Small wonder that Muffin did not object when Moffet took over the courtship at sunset on the third day.

During the day Happy rested—Muffin or Moffet always at her side—under the same bush as Spicy, looking very much as if she belonged. Then, on the fifth night, she walked south alone and returned to Diablo, Dixie, and the others of the Springbok Pan Pride.

This mixing of females between prides had never been reported in lions. Was this wandering lioness an aberration, a passing "stranger in the night"? Was her behavior unique? We could hardly think so. Since Happy had been so readily accepted by the Blue Pride females, it appeared that such exchanges of lionesses between prides might occur quite regularly.

*   *   *

The desert winter ended overnight—there was no spring. In late August there was a gradual warming of the days, but the nights remained bitterly cold. Then one silent morning in early September, the temperature suddenly shot upward.

When the hot-dry season had settled over the Kalahari, the thermometer often reached one hundred and twenty in the shade during the day; at night it fell to as low as forty or fifty. Differences of more than sixty degrees, sometimes even seventy, were not uncommon in a twenty-four-hour period. The relative humidity was lower than five percent at midday, and the sun beat down unmercifully, burning the last traces of life from the vegetation. The blossoms of the acacia and catophractes bushes—the blanket of pink and white magic that usually spreads over the Kalahari in the driest time of the year, providing succulent food for the antelope—never appeared that year. Here and there a puny flower hung with its withered brown face to the ground, only to shrivel and fall to the sand. The wind blasted across the scorched valley and the dry, brittle grass disintegrated, leaving stubbles sticking up from the cracked earth like broken broom heads. We had survived four dry seasons, but this was the worst.

By October there were almost no large antelope left on the dunes and in the sandveld around Deception Valley. During the rainy season over fifty percent of these animals concentrated on the ancient riverbed; now less than one percent wandered across its barren surface.

Chary and Sassy were still nursing their five- and six-month-old cubs, yet they had not had a drink of water for five months. In order to get meat for their growing families they searched farther and farther east, toward the game reserve boundary, where pockets of antelopes browsed the woodlands. They often traveled more than fifteen miles a night for several nights before they managed to kill a lone gemsbok.

Then one morning Mark found the two mothers and their young, together with Muffin and Moffet, outside the game reserve. They had crossed over the boundary into cattle country, as Bones had done, and again it was hunting season. Old Chary, with her sagging back and somber ways, was wise. She had survived many dry seasons, and probably a drought or two, by hunting outside the game reserve; she seemed to know the dangers.

A cow must seem the ideal prey to a lion: fat, slow and clumsy. But even though Chary led Sassy and their cubs to within 300 yards of cattle posts, they never killed a single domestic animal, as far as we knew. Instead, they preyed on the antelope moving out of the reserve to find water. Of course, neither Chary nor the others would have been rewarded for their discretion, had they been seen by ranchers.

Muffin and Moffet did not always stay with the females, and they were neither as old nor as wise as Chary.

We were answering a lot of the questions about Kalahari lion conservation: They could survive at least eight months with no drinking water, and instead of migrating in a single direction, they dispersed into huge ranges to find enough dry-season prey. They left the protection of the reserve not to find water, as we once suspected, but to get enough to eat. In all, the nine females of the Blue Pride increased their range by 450 percent, from 270 square miles in the wet season to over 1500 square miles in the dry months; the Springbok Pan Pride increased their pride area by 650 percent. These tremendous increases in range inevitably took the prides into areas where they were in danger of being shot.

With this expansion of range, the territories that they had defended so vigorously—even to the death—only weeks before, appeared to break down. The overlap in pride areas, which had existed to a minor degree, increased tremendously, and now it was Diablo who padded down Leopard Trail through the Blue Pride's old territory. Meanwhile, Muffin and Moffet roamed far into the range of the East Side Pride, often more than twenty miles outside their wet-season territory on Deception Valley, for more than two months at a stretch. When they did come back near their wet-season home range, they only stayed for two or three days at most before heading off again.

A dry-season silence fell over Deception. The lion roars and jackal calls no longer drifted down the valley at dawn. It wasn't just that the lions were too far away to be heard; even when they were close to the valley they did not call. From our camp on the valley floor we neither saw nor heard any signs of the lions. No wonder that for years we had accepted the common belief that they migrated to some unknown place for the dry months. Without the airplane and radio-telemetry equipment, we would never have known that some of them still prowled within the reserve, at times less than a mile from camp.

In the hot-dry season, the prides broke up into even smaller groups than in the short winter of June, July, and August. Two lionesses, at most, shared the kills they were able to make together, and they were often on their own. Muffin and Moffet were with the females only twenty percent of the time, compared to fifty-seven percent in the wet season. By contrast, Serengeti pride males are with females seventy to ninety percent of the time throughout the year.[2] Muffin and Moffet were often as far as forty miles from their Blue Pride lionesses.

The social organization of Kalahari lions under these extreme environmental conditions was turning out to be very different from that of East African lions. The most significant difference was the behavior of the females. We soon learned that Kalahari lionesses switched prides and pride areas frequently during dry seasons, as Happy had done on a number of occasions.

As Chary, Sassy, and their cubs traveled in and out of the game reserve, they occasionally met and socialized with members of the East End Pride, the Blue Pride, and other prides, and with some females we did not recognize. It didn't seem to matter which pride the lions had belonged to before the drought, they appeared to develop these new social affinities easily. These associations between individuals of different prides were usually temporary, unless the local concentration of large antelope prey was dense enough to allow the group to stay together and still find enough for all to eat. This happened on occasions when antelope concentrated on a flush of new grass after a fire.

We could hardly wait for the results of each new day's aerial locations. Every observation was a new bit of insight into the flexible social behavior of the lions under the environmental pressures of the Kalahari dry season. The flow of social and asocial events—who associated with whom, how many were in the group, and the nature of these relationships—was very dynamic.

Some of the lionesses transferred between groups more often than others: Happy associated temporarily with lions of four different prides eighteen times in nineteen months, and she eventually ended up roaming with Spicy of the Blue Pride. We were even more amazed one morning when Kabe, an ear-tagged female from the Orange Pride, strolled out onto North Pan. We had not seen her for three years, and she was traveling with a young male and two young females from the Springbok Pan Pride. A few days later she abandoned her young companions and joined Dixie of the Springbok Pan Pride—but inside the Blue Pride's old wet-season territory. If all this seems somewhat confusing to read about, imagine how we felt observing it at first hand, after seeing these lions grow up, hunt, sleep, and play together in their own prides for years. The whole lion social system, which we had spent years figuring out, seemed to be coming apart at the seams.

Without exception, all the lionesses we monitored associated with members of different prides. The cohesion and pride structure that were so permanent and fundamental to the social organization of Serengeti lions had temporarily disintegrated in the Kalahari population. It was a startling example of how a species can adjust its social system to extreme environments.

We could no longer be certain that the females of a pride were related; it was impossible to know the family origins of the older ones, whom we had not observed from birth. We had always assumed that Chary, the oldest, had grown up in the Blue Pride, but she may have been born and raised in the East Side Pride. And we could not ascertain the paternity of cubs born under these conditions, for the females of the Blue Pride mated with males from four different prides.

Chary, Sassy, their cubs, Muffin, Moffet, and many of the other lions continued to prowl outside the game reserve boundary. Perhaps when the rains came again, some of them would return to their original pride areas. Perhaps—but for now there wasn't a cloud in the sky.

19

The Dust of My Friend

Delia

In a rising wind
the manic dust of my friends,
those who fell along the way,
bitterly stings my face.
—*Stanley Kunitz*

THE DRY SEASON of 1978, like all the others, had a few good points, in spite of the dust and flies: The grass had died down, so it was easier to follow research subjects; we didn't have to worry about preparing camp for rain; and the animals in our tree island, attracted to the water and mealie-meal, had become more numerous and tame.

One of the new arrivals in camp was a grey-backed bush warbler we called Pinkie, a tiny fellow who could fit in the palm of your hand. With pink legs like toothpicks, a plump posterior, and an up-turned tail, he looked homemade.

Nearly every day Pinkie hopped around inside our sleeping tent, pecking under the trunks and boxes and behind folds in the canvas, in search of insects. A clutter of books, journals, and papers stacked at the head of our bed was Pinkie's favorite hunting ground for flies and beetles.

One afternoon when we were resting, Pinkie hopped from a book onto Mark's bare shoulder, then skipped across his chest and down his belly to his navel. He stood on tiptoes for a moment, craning his head this way and that and peering inside. Mark's stomach began to shake with laughter, but staring benignly, Pinkie rode it out. Then suddenly he sent his sharp little beak down, true as an arrow, right into Mark's navel. I don't know what he was after, or if he was rewarded, but he seemed quite satisfied as he hopped across the floor and out the door of the tent.

By now, there were seven Marico flycatchers in camp, including Marique, and on cold nights they slept all in a row, snuggled together on an acacia branch. The ones in the middle stayed warm and cozy, but after a while, those on the end would get chilly. In what looked like a scene from Disney, the outsiders would jump up, their eyes still half closed, hop along the feathered backs of their buddies, and wiggle their way into the better-insulated center spot. Soon they would be fast asleep again. A little later, the birds at the end of the row would find themselves cold, and they would repeat the performance; and so this continued all night long.

By far the fastest-moving character in camp was William, the shrew. He had Mickey Mouse ears, frizzy whiskers, and a long, incredibly dexterous rubber-hose nose. William, never still, was always darting along his own private paths with quick, herky-jerky starts and stops, like someone driving with one foot on the accelerator, the other on the brake. His nose constantly twitching, he zoomed in and out of the bushes, competing with the hornbills and flycatchers for mealie-meal.

One of William's routes through camp took him under our chairs in the ziziphus tree "tearoom." Since shrews have a high metabolic rate, they have to eat a tremendous amount every day. For this reason William was always in a hurry; still, he paused now and then to tickle our toes with his elephantile nose as he passed by. He was one of camp's main attractions.

At times we had a number of mice in camp, but the population plummeted after Dr. Rolin Baker of Michigan State University requested that we make a collection of Kalahari rodents for the university's museum. We didn't have the time to spend on such a project, so we taught Mox how to live-trap mice, sacrifice them humanely, and make study mounts of them. We agreed to pay him for each specimen, plus an extra tip for every new species he collected.

Mox set about his new responsibility with a great deal of enthusiasm and pride; at last he was involved in the science of the project. He hid traps in the corners of the tent and behind the tea crate. Every morning, after he had finished his other duties, he took a pair of pliers from the toolbox in the truck, and stalked from one trap to the next, gathering his specimens. He would take all morning to stuff three or four mice, but when he had finished, they would be perfectly shaped and very natural in appearance.

At noon one day we were reading under the ziziphus tree, when I heard Mox clear his throat behind us. He was standing at attention, proudly displaying his latest collection of rats and mice. They were all precisely arranged along a board, their feet tucked under them, their tails hanging down. These were his best yet; except for their cotton-filled eyes, they might have been sleeping. I started to tell Mox what an excellent job he had done, when I noticed a nose jutting out from the row. In the middle, his long snout stretched across the board in front of him—forever for Michigan State—was William.

*　　*　　*

Since the Frankfurt Zoological Society had provided us with an airplane, we hoped they would continue to fund our project, but as 1978 drew to a close and the new year began, we were out of money once again. Richard Flattery, the Standard Bank manager, kindly arranged a temporary loan with no collateral, he knew full well we had nothing to offer and he never raised the issue. To save money, we grounded Echo Whisky Golf and waited until January to go to Maun, when surely there would be word of a grant waiting for us in our mail.

Long before we were packed, Mox stood waiting at the plane, ready for the flight to Maun. He was dressed in his best, which had improved considerably since he first came to work with us. A big black comb crowned the back of his head, and he wore dark sunglasses with wide blue-and-red rims. Mark's jeans, plastered with patches, covered his spindly legs, and he wore the tennis shoes that Mark had retired after the wild dog tried to bury them. He was going to the village for the first time in more than three months and he was as excited as that territorial springbok watching his females return after a long dry season alone.

After landing at the airstrip in Maun, we took Mox to the Standard Bank, where we paid him his usual wage, plus the money for his rodent collection. The total was something over 200 pula—about $250. This was more cash than he had ever seen before. Together with Richard Flattery, we did our best to encourage him to open a savings account, but he seemed to have an inherent distrust of banks. When we insisted that this was the safest place for his money, he turned and hurried outside into the yard, where some goats and donkeys were grazing.

We caught up with him. "Mox, what's wrong?" I asked him gently. He kept his eyes on the ground for some time. Then he slowly looked up at me.

"Cowboys."

"Cowboys?"

"Ee, *cowboys*." He held out his right hand, its index finger and thumb like a six-shooter, his face in a frown. In broken English he explained.

Some months before, he had apparently seen a cheap cowboy movie, shown at the village center by one of the Peace Corps volunteers. In the film, the bank had been robbed. And although Botswana had yet to have its first bank robbery, nothing we could say would change Mox's mind. He was convinced that at any moment, masked men could ride up to the Maun bank on thundering horses and, in a cloud of dust, make off with all the money inside. He was sure his savings would be safer hidden in his mother's *rondavel*.

We borrowed Richard's Land Rover and drove Mox to his clay-and-thatch home. All the young children rushed to greet him, admiring his sunglasses and dancing around him. He patted each on the head. We arranged to meet him at the same place in two days' time and drove away.

Anxious about news of a grant, we hurried to collect the mail from our box at Safari South. In a stack of two-month-old Christmas cards was a cable from Frankfurt Zoological Society. We walked to a quiet corner of the yard, where I tore open the envelope. The message had been badly scrambled by the wire operator at the post office, but the gist of it was that we had been saved again: The society intended to fund us fully for the next two years.

Mark swung me high above his head. "You know what?" he asked. "Get yourself all fixed up—I'm taking you out to dinner."

And so we celebrated Christmas (a month late), our sixth wedding anniversary, and our grant, by dining at Island Safari Lodge on the banks of the Thamalakane River. The innkeepers, Yoyi and Tony Graham, gave us a free bottle of champagne and a cottage for the night. It was hard to believe—a tablecloth, wine glasses, waiters, a real shower, a real bed. We were more in love with each other, and with our work, than we had been when we first stepped off of that train in Gaborone, so many years before.

After two days of letter writing and shopping, we stopped in front of Mox's mother's rondavel, ready to head for the airstrip and Deception Pan again. A young girl was stirring a pot of steaming mealie-meal next to a fire, while several other children played in the sand. They all stood quietly watching as we walked to the hut. No one spoke, and when we asked about Mox they looked at us blankly.

An older teen-age girl I recognized came out of the house. No, she knew no one named Mox, she told us flatly, as if bored with the question. Several neighbors gathered around our truck. They all shrugged their shoulders—no one had ever heard of Moxen Maraffe.

For two days we drove around looking for him. Twice more we stopped by his mother's hut, and although there was no sign of him, we had the feeling that he was hiding inside. Mox had simply decided to disappear, and his clan was helping him. We finally gave up and drove away for the last time.

At first we were hurt and angry. We could well understand if Mox wanted to quit his job. Living in the desert far from his family was not a lively life for a young bachelor. But he had meant a lot to us, and we had thought he felt the same way. At least he could have told us he was quitting, instead of just disappearing. One of the hunters in Maun, though, told us that the fact that he could not face us with bad news was, in a way, his parting sign of affection.

Mox had acquired considerable recognition among the villagers. Not only did he fly in an airplane, but he was the *kgosi*—the chief—who worked with the people who shot lions and then brought them to life again. He was no longer the village buffoon. Respect and a new identity—these were the most important things he had brought back from the Kalahari, but they were of no use to him as long as he remained isolated.

Although we always asked about him when we were in Maun, we never saw Mox again.

* * *

After receiving the grant from the Frankfurt Zoological Society, we flew to Johannesburg in January 1979 to buy new tents and supplies and to have the plane inspected. On our first night we decided to go to town, perhaps to see a show.

Lofty towers, spires, and slowly spinning restaurants soared high above the city's nightlife. There were so many glaring lights, the stars were lost. Horns, engines, shouts, and sirens. Fumes and crowds.

Mark took my arm and pulled me away from a dark alleyway. I stepped on the remains of a greasy bag of fish and chips. Until I had lived on the desert sands I had never noticed how filthy city sidewalks were.

We stayed close together, stopping, dodging, turning to avoid bumping into people on the sidewalk. Suddenly, as we neared the movie theater, we recognized a face. I grabbed Mark's arm and we ducked into a small bookstore and peered over a shelf; one of the few people we knew in Johannesburg passed by. Then we looked at each other.

"Why did we do that?" Mark asked.

"I don't know."

We kept much too much distance between us and the person ahead of us when we stood in line to buy our tickets. Once inside we found two seats alone in a corner of the theater. The space around soon filled up with talking, laughing moviegoers, and when the movie began, the laughing and talking didn't stop.

"Let's get out of here."

Back on the street, we found a small café with sidewalk tables nestled among potted trees—real trees. We ordered two glasses of white South African wine and sat in silence, watching the city's nightlife.

*   *   *

The next morning we went into a gift shop to buy some small presents for the people in Maun who had helped us so much over the years. Shelves sparkled with rows of fine china, lead crystal, and silver. An attractive green-eyed woman in her thirties offered to help us. We declined the various items she suggested as not appropriate for Maun or for our budget.

"Are you from Botswana?" she asked.

I explained that we had been living in the Kalahari for six years, studying lions and brown hyenas.

"Oh... my father once lived in the Kalahari," she replied.

"Really? What was his name?"

"You probably wouldn't have known him—he passed away some years ago. His name was Berghoffer—Bergie Berghoffer."

For a second Mark and I were unable to speak. "You—you're Bergie's daughter!" I stammered.

We had wanted to contact Bergie's family for more than five years, to show in some small way our love for him and our appreciation for all he had done to help us. But we hadn't known either of his daughters' married names.

She introduced herself as Heather Howard and called her husband, Mike, down from upstairs to hear the flood of stories we had to tell about Bergie. They remembered his talking about "crazy Yank friends who had pitched up in the Kalahari, with nothing but a Land Rover, to study the wildlife." They had always wondered what had become of us. Sadly, we had to decline their invitation to dinner that night; we were returning to Botswana that afternoon. We promised to call them on our next trip to the city.

But we didn't contact them on our next trip, or on the one after that. Whenever we had to go near their shop we worried that we might accidentally run into them and have to explain why we hadn't phoned. We couldn't understand our behavior. Though we longed to see people, we avoided doing so. Mark and I each felt the other was the only person on earth who understood this idiosyncratic social behavior, and our contentment with each other only exaggerated the problem of dealing with other people.

It was almost a year after we had first met Heather and Mike that we finally called on them again. On a sunny afternoon we drove through the rolling green fields of the South African highveld to their home, beyond the outskirts of the city. It was very good to see them again, and they never inquired about the long lapse between our visits. Perhaps they understood better than we; after all, Bergie had spent much of his life alone in the wild.

Heather was pleasant, but pensive. We chatted for a time, and then she explained that, in his will, her father had asked that he be cremated and his ashes scattered in some quiet, grassy glade someplace in the wilderness. In all those years since his death, she said, the family had never felt that the time was quite right. Now that we had met again, they believed that Bergie would be pleased if we would join them in granting his last wish.

We walked through the meadows to a creek that rushed and swirled over rocks. There was a gentle breeze and there were butterflies. As I tossed his ashes to the winds, I could see Bergie's face smiling at me; we were setting him free again.

Some of his ashes caught on a spider's web that stretched between tall, waving reeds. I turned and looked at the distant, smoky haze of the city sprawling beyond the hills of green. I doubted if Bergie—or any of us—would be in the wilderness for very long.

*   *   *

In February 1979 we flew back into the Kalahari, our plane loaded with equipment and supplies. After several days of unpacking, sawing, and hammering, we stood back and looked at our new camp, with its five tents. The little yellow mess tent, trimmed in brown, was nestled in the center of the tree island, near the ziziphus tree. Inside was a dining table complete with tablecloth and chairs, and on each side there were orange-crate buffets that held pottery dishes, baskets, and glasses. A path wound through the trees to the sleeping tent, which held a real bed that Mark had made from packing crates. The office/ lab tent had a large working table, bookshelves, a typewriter, file cabinet, and another table, to serve as a desk. There was a storage tent with a gas freezer and refrigerator and a new three-walled reed kitchen boma.

If only Bergie could see us now.

20

A School for Scavengers

Delia

... in short, we see beautiful adaptations everywhere...
—*Charles Darwin*

WE HAD CONTINUED our brown hyena research along with the observations of lions all during the 1978 wet season and the winter months that followed. Star was more than eleven years old now, and her once thick coat of long, dark hair had thinned, exposing bare patches of coarse grey skin. Most of her blonde cape was gone, battle scars stood out on her leathery neck, and her teeth were worn to nubs from years of crushing bone. She seemed a little slower getting up—a little stiff, perhaps—and a little more apt to rest during a night's foraging.

From the air, Mark found her radio signal in the same spot on West Dune for four consecutive days. This was unusual for a brown hyena; a desert scavenger cannot afford the luxury of such roots. We could think of just two reasons why Star had not moved: She had slipped her collar or she was dead.

With the radio receiver in the truck, we homed on her transmitter on the duneslope west of camp. As we eased through the scratchy thornbush, the signal grew stronger; but there was no sign of Star. I steeled myself for the moment. Any second now, we would find her body, torn and broken on the sand, her bones picked clean by vultures.

Mark stopped the truck, switched off the ignition, and pointed ahead. About fifteen yards away, Star's weathered old face peered at us above a small scrub. Shaking chalky sand from her coat and flicking her tail, she walked to an opening in a small sandy mound, lowered her head to the hole, and made a low purring sound. Out wobbled three tiny cockleburs of charcoal fur—not only was Star alive and well, but she had cubs in a den only 300 yards from camp! Their dark eyes looked up at their mother, and she nuzzled them with her big muzzle as they stumbled around her feet.

At last we had another opportunity to observe a brown hyena mother caring for her young. We feared that Patches and Shadow had abandoned their cubs because we had tried to study them, but Star was so totally accustomed to us that we felt sure our presence would not disturb her. We named the female cub Pepper, and the two males Cocoa and Toffee.

Although by this time we knew a great deal about brown hyena feeding ecology, we still did not understand their social system. It was a mystery to us why they lived in a clan. Since they were scavengers and did not need each other for hunting large prey—as do other social carnivores—why did they associate in a group? Why did Patches, the dominant female, share food with Star and Shadow, when she could take it all for herself? Why would the clan share a common territory, if they did not need each other in some way?

Star had enlarged an existing springhare hole for her den. Three deep trenches in the sand led to separate tunnels underground, each concealed by a thicket of acacia bush. During the day she slept in the patchy shade about fifteen yards away, and every three to four hours she summoned the cubs by purring at the entrance. They toddled from the den and greeted her with wild enthusiasm, crawling around and around her, all the while squeaking hoarsely. They tottered about "grinning," their ears flattened and their tails curled over their backs, and Star licked and nibbled at each of them. Then she lay in one of the cool sandy runs and nursed them for twenty to twenty-five minutes.

When they were only three weeks old, the cubs began playing outside. At first this consisted mostly of stumbling into each other and falling down. But when they could keep their balance well enough, they practiced muzzle-wrestling and neck biting. Star seldom joined in the play, but lay there patiently while they tried with all their might to bite off her ears, nose, and tail or pounced on her round, dusty belly. Unlike lion and human mothers, Star never lost her patience. When it appeared that she could take no more of their mischief, she rolled them onto their backs, and while they squirmed to get away, she groomed them. As soon as they could escape, they would scamper away and begin chewing on one another again.

Just after dark, she led them into the safety of the den, and there they remained while she walked for miles in search of food. But since she had to return to nurse her young every four or five hours, she was unable to spend as much time foraging as the other hyenas, or to range very far from the den. This limited the amount of food she was able to find during the months that she was raising her litter.

One night when the cubs were six weeks old, Star gingerly clamped her powerful jaws over Pepper's back and carried her down the airstrip, across the valley floor, and into the bush on North Bay Hill, where she installed her in a new den. She then returned for Cocoa and Toffee. We did not know why Star moved her cubs, but it is common for some carnivores, such as jackals and wolves, to move their infants to two or three different lairs during their development.

Whatever the reason, it provided us with an excellent opportunity to investigate the interior of a brown hyena den. Armed with flashlights, notebooks, and measuring tapes, we walked to the abandoned site. When we reached the area, Mark squatted to examine the sandy spots around the entrance.

"What are you looking for?" I asked.

"Tracks. We'd better make sure that a leopard or warthog hasn't moved in here since Star left."

We searched through the hundreds of tiny brown hyena tracks for any sign of a new, larger predator.

When Mark was satisfied, he said, "Looks okay. You go into that entrance, I'll take this larger one."

I crawled head first into the open trench and then into a tunnel about two and a half feet high. By lowering my head and shoulders I could just squeeze inside. I pointed the flashlight into the pitch-dark. Ahead, the tunnel ran straight for about twelve feet and then made a turn to the left. I kept thinking that if a warthog or leopard had taken up residence in these dark corridors, it must be feeling very threatened, with us grunting, coughing, and crawling toward it from opposite directions. I could imagine angry eyes lurking around the corner ahead.

Flat on my belly, pulling with my hands and pushing with my toes, I inched forward. Now and then my head hit the roof and sand rained down on my neck and back. Still leaning on my elbows, I crawled down a gentle slope, shoving the flashlight ahead of me.

When I neared the end of the passageway, I stopped and listened. I could hear Mark's muffled bumps and scrapes drifting over from another run. Slowly, I shone the flashlight around the corner, half expecting the hiss-growl of a trapped leopard. I snatched it back. When nothing happened, I pulled myself forward and peered around the bend.

In front of me was a central chamber about five feet in diameter and three feet high, with hairy grey roots hanging down from the ceiling. This was apparently where the cubs had spent most of their time; there were little depressions in the sandy floor where they had slept. Three small tunnels and two larger ones led from the chamber.

I still could not see Mark, but sounding as if we were talking in a barrel, we called descriptions of the den back and forth to each other. We determined which tunnels were connected underground, and we measured their dimensions.

I was impressed by how clean the den was; Star was an excellent housekeeper. There was no dung or litter lying around, only a few bones, and there was no odor except for the dank, musty smell of earth. The skull of a young giraffe and a gemsbok scapula were the only furnishings.

"Hey! There's something biting me!" Mark yelled from the other tunnel. I didn't know whether he meant a mouse or a leopard, but then I began to feel fiery stings all over my body. I was so startled that it never occurred to me to turn around in the chamber and exit head first. Instead, I began belly-crawling backward, upslope, as fast as I could go. Frantically pushing with my hands, pulling with my toes, and constantly bumping my rear on the ceiling, I finally reached the entrance. Standing in the sunlight and fresh air, we found that we were covered with fleas.

We stripped off all our clothes, doused ourselves with water from the canteen, and skulked back to camp. For once, I was glad that Mox was not there to greet us.

There may be several good reasons why a female brown hyena moves her infants to a new den—perhaps to provide the growing cubs with a larger home or to protect them from predators who have discovered the first one—but I remain convinced that, at least in part, it is an attempt to avoid the flourishing flea population.

At two months of age, Pepper, Cocoa, and Toffee played for longer periods at sunset, scampering up to ten yards away from Star and their new den. However, at the slightest rustle in the grass—or even at the sight of a crow overhead—they always ran back to their mother's side or disappeared into the den.

When Star was ready to forage, she would stand up and shake

herself, and then walk away without a glance at the cubs. Now that they were slightly older, she made no special effort to tuck them away safely in the den. Pepper and Cocoa would gallop after her for about fifteen yards, then they would run back to the den; Toffee, always more cautious, watched from the safety of the entrance. All three would stand silently until they could no longer hear Star's footsteps in the dry vegetation, and then they played or explored around the den area for ten or fifteen minutes before going inside. At this age, the cubs were only slightly larger than house cats and good prey for lions, leopards, cheetahs, or jackals.

By the time Pepper, Cocoa, and Toffee were two and a half months old they had plump, round bellies. One night Star took Cocoa by the neck and walked west through the bush. Following close behind in the truck, we saw her move from North Bay Hill down onto the valley floor, and then northward, the cub dangling from her mouth like a limp rag.

Mark had found Moffet under the Topless Trio that morning, and now Star was headed along the dark riverbed directly toward the lion's position. Through binoculars we could just make out Moffet's large body lying perfectly still under the tree. Lions often stalk and kill brown hyenas, and unless Star changed course, she would walk directly into him. She might be able to escape, but she would probably drop Cocoa in the process.

I raised the glasses and watched anxiously as Star carried her cub closer and closer to the lion. Brown hyenas do not seem to have very keen eyesight, and unless Moffet moved, she probably wouldn't see him until it was too late. The night air was dead calm; his scent might not reach her until she was just a few yards away from him. Star continued on her way, oblivious to the danger ahead.

Moffet rolled over and gathered his feet under his heavy body, his big head raised and his eyes locked on Star, plodding toward him over the riverbed. From having observed lions stalk brown hyenas previously, we guessed he would wait until she and her cub were within twenty or thirty yards, then he would charge. By the time she could react he would be practically on top of her.

But when Star was only eighty yards away from him, she stopped and peered ahead. Then she turned abruptly and made a wide detour around the lion. Moffet dropped his muzzle onto his paws and apparently went back to sleep.

Star trekked north for over two miles, and during the whole time

Cocoa never stirred. The moon had not yet risen, but the calcrete shoal on the riverbed reflected the bright starlight, and we could easily see the hyena's dark form moving through the dry grass. Turning northeast onto the dune, she wound her way through thick thornscrub. She continued for another half mile, stopping now and then to look and listen. We could not understand why she was taking Cocoa so far.

We broke through the next stand of tall brush into a large clearing, and quickly switched off the engine. We stared ahead, dumbfounded. Before us lay an enormous den complex comprised of several great mounds of grey sand over fifteen yards long. Standing on each mound were young brown hyenas of different ages, and obviously belonging to different mothers. Here were the missing cubs, the ones we thought Shadow and Patches had abandoned. All the clan's young were at one communal den—the first such den ever seen by humans!

This, at last, was the answer to all the questions we had been asking for years about the raison d'être of brown hyena society. These scavengers associate in a clan, sharing food and territory, because they raise their young communally in a supreme cooperative effort to contend with the harsh and fickle Kalahari environment.

It happens too rarely in science that, after years of effort, a new discovery practically falls into the researcher's lap. We sat speechless. Star lay Cocoa softly on the sand and stood back. All the other cubs came forward and smelled their new denmate. Cocoa did not seem afraid or timid, he lifted his small black nose and sniffed the assortment of cubs that greeted him. While Star went to bring Pepper and Toffee to the nursery, Cocoa explored his new surroundings.

*   *   *

The Kalahari environment, with its sparse and unpredictable food supplies, makes it difficult for a female brown hyena to find enough to eat for herself and her growing cubs. We were to learn later that usually only one female in the clan gives birth to cubs each year, and thus there is a limit to the number of young at the communal den. With all of the cubs safely inside, each female is free to roam alone for several nights until she locates food that can be carried back to the cubs. Since each mother does not have to return to a private den several times each night, the clan's collective foraging time is increased, insuring a more regular supply of food to the young. Every adult female, whether or not she has ever bred, brings food to the cubs at the den.[1] And some of the males provision as well. Because they must forage alone, yet rear their young communally, brown hyenas are a curious blend of the social and the solitary, reflecting the capricious nature of the land over which they roam.

*   *   *

With the discovery of the communal den, our lives took on a different routine: In the early morning Mark flew around locating the lions and hyenas, and then later we would drive to some of the lions nearest camp. In the early evening, while he was transcribing his notes from the tape recorder, I drove to the hyena den and watched them for part or all of the night.

I took along notebooks, a flashlight, cameras, a tape recorder, a sleeping bag, fresh bread, and thermos bottles of soup and hot tea. In the back there were extra cans of food and a jerrican of water, in case I had to stay longer than planned. When I arrived at the den there were usually no hyenas in sight, and I would watch the sunset and listen to the Kalahari night fall: A jackal would call on North Dune, a korhaan would give a territorial squawk, and hundreds of barking geckos would begin their nightly serenade. After dark I could see the flicker of Mark's campfire three and a half miles away.

One sunset, before any of the hyena cubs appeared from the den, the truck suddenly shook. Startled, I looked around to see what it could be. Just when I was beginning to think I must have imagined it, the truck moved again. I opened the door to see if a grass owl had landed on the roof. Nothing there. Again the truck shuddered, and now I was really getting spooked. Then I looked through the back window, and there I saw Moffet's big furry head slowly appearing over the tailgate. He stuck his nose into the bed of the pickup, sniffing at the toolbox and spare tire. He lowered his head again, taking the trailer hitch in his jaws, he rattled the truck as if it were a toy.

"Hey, Moff, stop that!" I called out the window. He gave one last tug and then walked over to within two feet of the open window. Raising his head, he looked deep into my eyes. I said, very softly, "Look, I was just kidding. If you want to shake the truck, go right ahead."

Moffet yawned enormously, shook himself, and walked over to the hyena den, where he jetted scent onto a small tree. He disappeared into the brush, and the J-shaped scar on his flank was the last I saw of him.

Those nights alone on that duneslope, with the stars hanging close overhead, were some of the most special of my life. And slowly I began to know the hyena cubs. Pippin, the oldest, was over three years and really a young adult. He foraged on his own, but still visited the den to play with the younger cubs. Chip was the next oldest, and he, too, foraged away from the den area. Sooty and Dusty, a younger brother and sister, remained at the den site all the time, along with Puff, a very young female. Finally, there were Pepper, Cocoa, and Toffee, the newly arrived infants.

The night after we discovered the communal den, I saw Patches come along one of the paths worn through the grass, carrying a fresh springbok leg in her jaws. All the cubs jumped up at the sound of her approach, their hair standing on end. The younger ones dove into the den opening; for all they knew, the sound was that of a lion or some other predator. When Patches was close enough to be recognized, the older cubs bounded toward her and circled round her in greeting for several minutes. She laid the bone on the sand and sniffed each cub that paraded under her nose, licking their ears and backs. After Sooty had paid his respects to his elder, he grabbed the springbok leg and rushed into the den, and all the others followed. Patches slept on the mound while the cubs fed inside.

Later the same night, Shadow strolled out of the bush and the cubs circled around her, kicking up fine dust until a white cloud hung over the area. She flopped onto one of the mounds, and Puff began to suckle, her paws kneading the soft udders. I had just concluded that Shadow was Puff's mother, when Dusty also began suckling. Because of the differences in their ages, Puff and Dusty could not both be Shadow's cubs; she was nursing at least one that was not her own. Later we saw Patches and Star nurse each other's young. Communal suckling had only been seen in a few other wild carnivores, including lions and wild dogs; it had never been recorded in hyenas before. It was further proof of the cooperative social system of the browns.

Since the lactating females in the clan nursed all the cubs, and since all the females brought food to the den, at first it was not obvious who was the mother of whom. Fortunately, we had detailed records of the previous pregnancy and lactation periods of the females in the clan. By comparing this information with the ages of the cubs, and by sitting long hours at the den, we were able to confirm family relationships. We knew that Pippin was Star's cub from a previous litter, making him the half brother (Pippin's father had been displaced by another dominant male) of her present cubs, Pepper, Cocoa, and Toffee. Chip was Patches's cub, and Puff belonged to Shadow; we did not know who was Dusty and Sooty's mother.

With Star, Patches, and Shadow bringing food, and Pippin visiting the den to clear loose sand out of the tunnels and play with the cubs, one might expect the communal den to be crowded with hyenas, but this was never the case. Adult females did not visit every night and they rarely appeared at the same time. If they did happen to meet at the den, they barely acknowledged one another. We never saw the clan's immigrant adult male, the cubs' father, visit them or bring them any food.

Because they had to spend so many hours foraging, most of the time there were no adults at the communal den to protect the young. Clan members slept under bushes or trees widely scattered throughout the territory, some up to five miles from the den. An adult might occasionally sleep near it during the day, but never closer than 200 to 300 yards.

The infants were protected by the den itself and by the presence of older, larger cubs. Pepper, Cocoa, and Toffee would wander up to twenty-five yards away in the tall grass, but at the sight or sound of any approaching animal, whether it be a porcupine or a lion, they would dive into one of the runs and dash underground. Minutes later, their ears, then their eyes, and finally their noses, would slowly appear like periscopes over the rim of the hole to find out if the danger had passed. If they saw that the older cubs had not taken cover, Pepper, Cocoa, and Toffee would bound out of the den to resume play.

Late one afternoon, a pack of eight wild dogs loped toward the den. The smaller cubs disappeared, but Chip, Dusty, and Sooty, who were three-quarters adult size, stayed on the outside and faced the predators. Standing on the largest mound, with every hair bristling, they looked quite formidable. The dogs circled the den area three times, occasionally trotting closer for a better view, but eventually moved away. However, when Moffet came by the den one afternoon, all the cubs, including Chip, Dusty, and Sooty, disappeared inside and did not come out again for more than an hour after the lion had left.

The cubs occasionally wandered too far from the protection of the den. When Puff was about the size of a small, stocky bulldog, one night she strayed farther away from the den than usual. Loud screams and sounds of a struggle came from the tall grass. By the time we arrived, a leopard was dragging her torn body toward an acacia tree. Even after Puff's death, Shadow, her mother, continued to suckle and provision the other cubs.

Every day at sunset, after the heat had broken for the day, the cubs would peer out of the four den openings. When they were satisfied that there was no danger, they would plod out of the runs and collapse on the mounds of excavated sand. Later, when cool air began to flow down the duneslopes into the valley, the cubs began smelling grass-stalks, twigs, old bones, and anything else they could get their noses into. These are important lessons for hyenas, who spend much of their time foraging in tall grass, where they cannot see more than a few feet ahead, and who must locate widely scattered pieces of carrion. As adults, they would also depend heavily on their sense of smell to warn them of dangers from lions and to maintain contact with one another by scent-marking.

Pepper, Cocoa, and Toffee tried to scent-mark long before their anal glands began to produce the viscous paste excreted by adults. Over and over again they raised their small fuzzy tails, turned, and squatted over a stalk, trying to leave their olfactory calling cards and always sniffing to see whether they had succeeded.

At about four months of age, soon after they were moved to the communal den, they discovered that they were able to make their own paste. They looked very proud of themselves, strutting around the denning area, raising their tails, and daubing white, gooey drops on everything—even on the tail of an unsuspecting adult and the legs of our camera tripod.

Games in which one cub tested itself against another were an important part of their development. From the first day they emerged from the maternal den, the infant cubs exhibited the same behavioral patterns in play-fighting as those used by adults in combat: muzzle-wrestling, neck-biting, nipping the hind legs, and chasing. Play was important in developing social ties among the young, and it probably sharpened the fighting skills that would later be necessary in the battle for status in the clan's social hierarchy.

Whenever Pippin arrived at the den, the cubs greeted him with great excitement, springing forward to circle him, pulling at his tail, and jumping up to nip at his ears. In typical big brother fashion, he led them on a merry chase in and out of the bushes, but he always let them catch him, as he tossed his head from side to side to avoid their snapping jaws.

If young cubs were not reared in communal dens, they wouldn't benefit from the protection and play experience provided by older cubs, and they wouldn't be as likely to learn from, and form social bonds with, all of the adult members of their clan.

* * *

By February 1979 we had been searching the dull skies for months for any sign of rain. February is usually the height of the rainy season, but there was seldom a cloud in the sky, and midday temperatures were breaking 110 degrees in the shade. Occasionally the massive head of a cumulus formation peeked over the eastern horizon, only to sink from sight, spilling its life-giving moisture on other lands. By the beginning of April, the clouds were gone and we had lost all hope. The rains of 1979 had failed completely. Except for one brief shower, it had been twelve months since the animals and plants in Deception Valley had tasted moisture, and there was no chance of relief for ten months to come. The Kalahari was locked in drought.

When the lions—now spread over immense areas beyond the dry river system—occasionally visited the valley, their kills were so small that nothing but blood-stained feathers, quills, or a horn or hoof were left for the brown hyenas. The cheetahs and wild dogs had disappeared when the last of the springbok had left the valley. Now ants, termites, birds, rodents, and an occasional steenbok were the only prey available for leopards and jackals, and most of these small prey were completely consumed. Star, Patches, McDuff, and Pippin walked over the grey desert twenty miles or more every night in search of food. They, too, lapped up termites and chased mice, hedgehogs, porcupines, and springhare. The food supply could only get worse during the coming dry season, and since there had been no rain, there were no wild melons to be eaten for moisture.

At eighteen months of age, Dusty and Sooty had begun to forage on their own. As Pogo and Hawkins had done years earlier, they tagged behind Star, Patches, or Shadow when one of them left the den. Pepper, Cocoa, and Toffee were often the only cubs left in their earthen home, which had become a quiet and lonely place. Sometimes several nights would pass while they waited patiently for something to eat. Star was the only female still lactating, and at six and a half months, the three cubs were still heavily dependent on her milk for nourishment and moisture.

One windy night, Star walked north along Leopard Trail to Bergie

Pan, crossed the riverbed, and moved northeast to the slopes of East Dune. By midnight she had traveled more than twelve miles and had not found anything to eat, for herself or to take back to her young. She poked her nose into the entrances to several rat and springhare colonies, but no one was home. She was tired and lay down to rest near a grove of broad-leafed lonchocarpus trees about a mile from the riverbed.

Muffin and Moffet had spent days roaming the sandveld areas east of Deception Valley, crossing the game reserve boundary into cattle country several times. This night, they moved back west into their former wet-season territory and started trudging up the face of East Dune. Having seen only one hartebeest, which had cantered away from them, they were lean and hungry.

Lying flat on her side, her scarred head and neck on the cooling sand, Star occasionally pawed some of the fine grains onto her wrinkled belly. Sometime later she heard a faint sound from downwind. Perhaps the wind had covered the lions' approach, or maybe she had slept too soundly. But when she jumped up it was too late. Muffin and Moffet rushed in, mauling her, and in seconds Star was dead.

21

Pepper

Delia

A little more than kin and less than kind.
—*William Shakespeare*

PEPPER, COCOA, AND TOFFEE had no way of knowing that their mother was dead. Hour after hour, night after night, they lay on the den mounds, their chins on their paws, watching the path on which Star usually approached. As the time passed, their lethargy deepened; they did not play. Every few hours they plodded slowly around the area, smelling once more the few splinters of dessicated bones. They spent the long hot days in the den's cool interior, conserving the moisture in their withering bodies.

Their bony shoulders stuck out at sharp angles, and their hair began to fall out. The days were incredibly hot and dry, as though there had never been water on the earth, but the nights, at least, were mercifully cool.

On the fourth night after her death, Pepper, Cocoa, and Toffee did not come out of the den. For three nights we sat watching the bare mounds in the moonlight, hoping for some sign of life. We had to find out, so we crept to one of the main entrances, knelt, and listened. No sound from inside, and no small hyena tracks in the sand. The cubs must have starved to death or died of thirst.

But when we stood up and turned to go back to the truck, a thud and a squeak drifted out of the hole. At least one was still alive—but for how long?

Around midnight, a loud rustling came from the grass west of the den, and Pippin, the cubs' half brother, stepped into the clearing, a freshly killed springhare dangling from his jaws. He laid the four-pound mammal in the sand, walked to the entrance, and purred loudly. Immediately all three of the weak and hungry cubs scrambled out to greet him with eager squeals. They circled Pip again and again and then ran to the springhare and dragged it toward the den. On their way they stopped to rush around him one more time, raising great clouds of dust in their excitement. Then, tearing and pulling at the food, they disappeared inside.

Pippin stood on the mound alone, his legs long and lanky and his body thin. Without moving his head, he rolled his eyes and gave us a long look, just as his mother, Star, had so often done. Then he shook his long hair, flicked his tail, and walked away into the bush.

"Mark! Maybe he'll adopt them," I whispered. Not only did Star's cubs now have a chance to survive, but this was a high point in our brown hyena study. Adoption in nature is extremely rare; group members in most species usually abandon orphans and devote their efforts to rearing a litter of their own.

Early the next evening, Dusty came to the den carrying a large piece of giraffe skin, which had quite a bit of fresh meat attached to it. Following a few steps behind was Chip, the older male cub. Squealing and grinning, their tails raised in greeting, Pepper, Cocoa, and Toffee rushed forward and crawled around their old den mates. Cocoa grabbed the skin and they vanished inside. If we had observed this scene without knowing the identity of the individual hyenas, we could easily have thought that this was a mother and father feeding their young. In fact, they were all cousins.

Over the following days, certain clan members began to care for Pepper, Cocoa, and Toffee. We now realized why we had been unable to establish who was the mother of Dusty and Sooty: She had died, and, like Star's cubs, they had been adopted by the clan.

Now that they were being fed by Patches, Shadow, Dusty, and Pippin, the cubs grew stronger. Though brown hyena young often suckle until they are ten to twelve months old, these three had been abruptly weaned from their mother's milk at the age of seven months. Despite this radical change in their diets, they seemed to thrive on the meat, skin, and bones brought to them, and their chances for survival looked better and better.

*   *   *

We observed the communal den for more than three years and made several interesting discoveries. For one thing, adoption occurs often in brown hyenas. In the period during which we watched the den, seventy percent of the cubs that survived were adopted orphans.

Most brown hyena females stay in the clan to which they are born; thus they are all related. Because we had been observing the clan for so long, we knew many of the relationships. Patches and Shadow were Star's cousins, making them Pepper, Cocoa, and Toffee's first cousins once removed; Dusty was an older second cousin; and, Pippin, we knew, was their half brother. The cubs had been adopted by their relatives.

Thus hyenas, long considered vermin and listed in a thesaurus as synonymous with "cur" and "viper," help other hyenas by feeding and adopting their cubs; not only are they very social creatures, they seem to be very selfless, as well.

But how selfless are they, really? In the midst of a drought, why did Patches, Shadow, Dusty, and Pippin provide Star's cubs with food that they could have eaten themselves? Why would they aid another's young at a cost to themselves?

Part of the answer can probably be found in the sociobiological theory of kin selection.[1] "Fitness" in the term *survival of the fittest* does not refer to how physically strong an individual is but rather to its survival and the number of genes that individual passes to future generations. Any animal, including humans, can increase its genetic fitness in two ways: directly, by producing offspring who carry one half of its genes; and indirectly, by enhancing the survival of its more distant relatives—cousins, nephews, nieces—who carry a smaller proportion of its genes.[2]

Since Pepper, Cocoa, and Toffee were Shadow's first cousins once removed, each cub carried some of the same genes she had. By keeping the three of them alive, Shadow would increase her genetic fitness. Shadow's only cub, Puff, had been killed. If Pepper, Cocoa, and Toffee had also died, Shadow would have lost one of her few chances to pass her genes to future generations. This is especially important since a brown hyena female does not have many opportunities in her lifetime to raise litters. Each of the cubs also carried one-quarter of the same genes as Pippin, their half brother; thus feeding them benefited him genetically.

So, according to the theory of kin selection, the clan members who were feeding the cubs were not being altruistic. In fact, the food they brought to the den to keep their cousins or half siblings alive was merely an investment in perpetuating their own lineage. Of course, the brown hyenas themselves did not realize why they were feeding the orphans. During the evolution of their social behavior, those who possessed the "helping" genes and participated in provisioning must have kept more of their relatives alive, than did those hyenas without the genes who did not provision; so the behavior evolved as part of the natural life style of brown hyena sociality.

Chip and Sooty, the cubs' male cousins, did not help feed them. And though they came to the den to play with Pepper, Cocoa, and Toffee, we suspected they were really looking for morsels of food. In fact, on many occasions Chip and Sooty actually stole the carrion that had been brought to the den for the younger resident cubs.

But why did Dusty, their female cousin, feed the cubs when their male cousins did not? They were all related to the same degree. The answer is probably that most females remain in their natal clan for their lifetime and benefit from an increased clan size, whereas most males emigrate and do not. Because cousins share only about one-eighth of the same genes, this alone may not be enough genetic incentive to induce a male to feed his cousins. He may benefit more by eating the food himself than by investing it in the survival of the few of his genes that are carried by them. Since he will eventually leave to become a nomad or to join another clan, he would not benefit directly from increasing the number of members in his natal clan. On average, half siblings share twice as many genes as cousins do, so that even though a male half sibling may also eventually leave the clan, by provisioning cubs he would increase his genetic fitness more than cousins would.

On the other hand, it benefits any female in the clan, no matter how distantly related to the cubs, to feed them, because she is probably going to remain in the clan. If she helps raise cubs, there will be more individuals in the clan to defend the territory and its food resources. Perhaps most important of all, there will be more females around to eventually care for her cubs.

Thus, brown hyena cubs are raised by all the females in the clan and by the males most closely related to them.[3] That the behavior is not purely altruistic should not make it any less special—that a birdsong has a function does not make it less beautiful. By understanding the selective pressures behind the evolution of behavior that at first appears to be altruistic in animals—and also in ourselves—we learn that there is a natural and necessary element of self-interest in these kinds of behavior.

I thought about these things as I watched Dusty, who did not yet have cubs of her own, bringing food to her younger cousins. Is there any true selflessness in nature—in man? Why had we come to Africa, worked so hard, for so long, under such adverse conditions? Was it only for the animals? Or was it partially for ourselves?

*   *   *

The cubs were being fed again, though in modest quantities, and once more they began to play. Late one afternoon Pepper roused herself groggily from her sleep. Yawning, she shuffled around the den area a few times and then bit the neck of Cocoa, who was still trying to rest, and chomped at his ears and tail. When he stood up to retaliate, she ran away at full speed, crashing through the dry grass and bush nearby. A few minutes later she dashed back to the mound, jumped high into the air above the den, and dove into one of its openings, a cloud of dust rising from the hole. A few seconds later, her black ears, eyes, and nostrils ringed with white calc dust peeped over the rim of the hole, apparently to see if everyone was watching. Everyone was. She sprang out of the den, ran over Toffee, and disappeared into the scrub again.

It was usually Pepper who came up with new games. One afternoon she minced over to the truck, slowly extended her muzzle to smell the bumper, and then galloped back to the den, hair standing on end. When she reached the mound, she looked back at the Land Cruiser, her eyes rolling. After that, walking very cautiously, she led her brothers single file to the truck. Standing side by side, they all sniffed the Toyota and dashed away. Each time, they gained more confidence, until finally they walked underneath the truck; I could hear them lapping, snuffling, and chewing every part of the undercarriage.

One night, after several hours of den observation, I started down the dune face, only to discover that no matter how hard I pumped the brakes, the truck would not stop. Gripping the wheel and swerving to avoid termite mounds, holes, and scrub bushes, I bounced down the sandy slope and shot out onto the flat riverbed. The rest of the drive to camp was uneventful, until it came time to stop. I slowed down, but misjudged my speed. Stamping the brake pedal against the metal flooring, I coasted through our usual parking spot and sailed past the sleeping tent, finally coming to a stop just three feet from the office tent. Mark, who had finished transcribing his notes and gone to bed, was up on his elbow peering out of the tent window with a flashlight. I parted the flaps and said as sweetly as I could, "Don't be upset, but I think the cubs ate the brakes." Indeed, they had chewed right through the brake hoses and drained away all the hydraulic fluid.

Since our study of brown hyena helping behavior would be the subject of my Ph.D. thesis, I observed the den for over 1000 hours, and the cubs became completely accustomed to me. One afternoon, instead of staying in the truck, I sat in the long grass on the edge of the clearing. When Pepper and Cocoa looked out of the den, they ambled up to me, stuck their snouts into my hair, and began sniffing my ears, neck, and face. I remained perfectly still, trying not to laugh when their cold, wet noses blew puffs of air down my neck and spine. Eventually they turned their attention back to the bones that were scattered around the mound at the den.

Sitting among the cubs, I could observe more details of their behavior and get a fresh photographic perspective. Pepper would even let me take measurements around her skull and neck with a tape while she sniffed around me. But I had to be very careful with all my gear; one time she grabbed my notebook from my lap and ran into the den with it. Fortunately, she dropped it just inside the entrance, so I was able to retrieve it without too much difficulty.

At times, Pepper would rake her oversized paw down my arm, just as she did to initiate play with Cocoa. And once, at eight months of age, she latched onto my little finger with her incisors and stared into my eyes, as if issuing a challenge. She could easily have bitten off my finger, so I dropped my gaze and quickly pulled my hand from her grasp. It would have been fun to play with her, but this would have interfered with the objectivity of our research. Besides, her powerful jaws could have been quite dangerous. Since I did not respond to their attempts to incite play, the cubs ultimately treated me as little more than an object of curiosity. Pepper would always shuffle to the truck when it arrived and smell me as I stepped out, but then she would usually ignore me for the rest of the observation period. I was always careful to return to the truck before an adult approached, for they were less relaxed when I was sitting in the open only ten yards from their communal den.

That dry season of 1979 was the toughest we had known because there had been no rainy season at all preceding it. By late September, the temperatures again soared above 120 degrees in the shade, and the daily relative humidity was less than five percent. The adult animals needed more food for their own moisture requirements, and they visited the den less often. Sometimes nothing was brought to the cubs for two or three nights, and for the first time since we had started our study of brown hyenas, both the adults and the cubs looked worn and haggard. Once again, Pepper, Cocoa, and Toffee remained in the cool den most of the time and seldom played.

Late one afternoon Mark and I were busy in the kitchen when, to our amazement, we saw Pepper walking toward us on the track. She was not quite a year old. Cubs usually don't leave the den until they are almost eighteen months old, and then they follow adults closely for three or four months before foraging on their own. Yet here was little Pepper out all alone, three and a half miles from the den. It must have been a frightening adventure for her; every hair on her thin body stuck straight out, making her look like a bottle brush. That she had begun foraging on her own so early in life was another sign that the cubs were not getting enough to eat from the adults.

Pepper came into the kitchen without a moment's hesitation. She walked to where I was stirring stew over the fire, smelled the wooden spoon, grabbed it, and tried to pull it from my grasp. I held on and won the tug of war, though I would have liked more than anything to have given her the whole pot of stew. For me, one of the most difficult aspects of observing the brown hyenas and the lions during the long drought was resisting the urge to help them in some way. But we were there to learn how they survived and whether the game reserve offered enough resources for their conservation, not to offer them comfort. We burned our garbage and buried it in deep pits in the sandveld, away from the riverbed. We kept food out of reach and tried to remember to empty the water basins—I felt guilty every time—so there would be nothing in our tree island for animals to drink. We did everything possible to discourage the hyenas from keying on camp as an oasis in the desert. When they learned that they couldn't get to the sources of the scrumptious odors, they usually ignored them.

All except Pepper, that is. After she had smelled all the shelves and boxes in the kitchen boma, she walked down the path toward the dining tent. She stepped inside the door, and before we could stop her, she grabbed the tablecloth and pulled all the dishes onto the floor with a loud clatter. Then, for the next hour, she snooped around every corner of camp, smelling the water drums, poking her head inside the tent doors, and standing on her hind legs beneath the hanging shelf. It was dark when she finally left, and we followed her in the truck to see how well developed her foraging skills were.

Her hair again on end, she walked north, sniffing here and there on the dry ground and lapping up a few insects. As she rounded Acacia Point, she stopped and stared north along the valley. In the spotlight we could see the large eyes of another predator walking toward us from about 100 yards away. Obviously frightened, Pepper backed slowly toward the truck, ducked under it, and hid behind a front wheel, peering out with widened eyes. Through binoculars we could see Chip, her cousin, approaching; she apparently had not recognized him from a distance. After Chip circled us and moved on, she came out and continued north along the valley floor.

Nearing the dry water hole on Mid Pan, Pepper was intercepted by two jackals. She stared at them briefly and then scurried toward the den, which was still two miles away. Emboldened by her fear, the jackals closed in, their noses to her tail. She turned back her ears and walked even faster. Apparently realizing that she was a rookie, one darted forward and nipped her on the rump. Pepper tucked her tail between her legs and drew in her hindquarters, but both jackals continued to snipe at her legs as she struggled along, now almost sitting on her backside.

This continued for several hundred yards, until Pepper abruptly came to a halt. As if she had just realized that she was twice their size, she drew herself to her full height, her neck hackles bristling, and chased the jackals all the way to the edge of the riverbed. We followed her until she returned safely to the den; except for a few ants and termites, she had found no food.

Several weeks later Mark was in Maun for the night and I was in camp alone. After dark I heard a racket in the kitchen and cautiously crept down the path to see who was there. Pepper rounded the corner and walked straight up to me, smelling my toes and fingers. I followed her as she sniffed around camp, and after she walked out of the island and went to stand on the flat riverbed. I sat down five feet away from her. She turned to look at me and then settled on her haunches, her head up. The moon was only a quarter full in the western sky, but I could see for several miles over the Kalahari. The dark line of the dunes was silhouetted against the lighted sky; the night was quiet. We sat together, two small forms on the desert floor. I had never felt so close to the Kalahari, to the natural world. After about ten minutes, Pepper flicked her tail and walked away without looking back. I wondered where *I* would go if I had to walk into the desert to find my supper.

Cocoa and Toffee also explored away from the den, searching for morsels. One night, just after leaving the den, Toffee, who had always been the most cautious, was grabbed and killed by a leopard. We found his remains stashed high in an acacia tree only 150 yards from the den. Sooty disappeared from the area, probably having emigrated from the clan. We found Shadow's wizened remains on the parched desert floor. She, like Star, had been killed and eaten by lions. McDuff, the dominant male, was dead of unknown causes. Only Patches, Dusty, Pippin, Chip, Pepper, and Cocoa were left in the clan. The drought was taking its toll.

Pepper and Cocoa began foraging away from the den with an adult, whenever they could find one to accompany them. Patches, the only adult female; Pippin, their half brother; and Dusty, their cousin, always allowed the cubs to tag along and shared any food they had with them. The young hyenas were entering subadulthood, and they must have learned a great deal about the scavenger's way of life on these excursions: the clan's territorial boundaries, the network of pathways through the home range, and how to find and appropriate carrion from other predators.

If they could not find an adult, the cubs wandered off on their own, usually returning to the den by midnight, when it was most likely that a relative would bring them food. Even through the worst of the drought, Patches, Pippin, and Dusty continued to feed the orphans whenever they could.

Mark was in our reed bath boma one night, taking a spongebath in the dark. He was bending over the basin, his head covered with soapsuds and his bare feet perched on a narrow board to keep them out of the sand. All of a sudden a tongue slithered across his toes. He let out a loud whoop and jumped back. At the same time Pepper's head jerked upward, hitting the table and upsetting the water basin. She turned and shot for the entrance, but she missed and rammed the log frame of the door with her head. This frightened her even more. Like a ricocheting seventy-pound cannon ball, she charged around inside the small hut, with Mark tripping and shouting in confusion as he tried to figure out what was in there with him. Finally, Pepper blasted right through the wall of the hut, leaving a gaping hole in the splintered reeds.

When they had both recovered their composure, Pepper shook off the bits of reeds clinging to her long hair and walked calmly down the path to the kitchen. There she picked up the empty water kettle—its handle still bore her mother's tooth marks—and walked off into the night.

22

Muffin

Mark

From his slim Palace in the Dust
He relegates the Realm,
More loyal for the exody
That has befallen him.
—*Emily Dickinson*

BRILLIANT SHAFTS of sunlight stabbed through the leafless trees in camp. The rays of white heat stalked us into every corner of our island. It seemed as if there were no longer any seasons—only heat. Each day was like the last, each week, each month the same. By September 1979, Deception Valley had received only four inches of rain in twenty months. The Kalahari was an expanse of grey earth stretching out to a grey sky, as if reaching for rain that would not come.

One of our thermometers, nailed to a tree in camp, rose past 122 degrees and stuck; surface temperatures on the open riverbed were near 150. Every day we poured small puddles of water onto our canvas cots and wallowed there, in a stupor, for hours. Or, like roaches seeking the cooler darkness, we pressed ourselves to the floor of the tent or sprawled on a corner of the ground sheet.

The heat clung to us like a leech, sucking away our strength. When we had to perform some task—anything—we dragged through it in slow motion, exhausted. Often, when we stood up from our cots, dark spots would spread behind our eyes; dizzy, we would lower our heads between our knees until the nausea passed. The hours of torpor scrambled our brains and confused us. Sometimes we had to stop and think, "Why are we still here, trying to survive the fifth—or is it the sixth?—dry season...the first—or second—year of drought?"

Because the relative humidity was so low and the evaporation so intense, we did not sweat; that is, the moisture from our bodies was vaporized before it could dampen our skin. We drank quarts of water, hot and smoky from being boiled on the fire, but it tasted good just the same. When the sun finally sagged behind West Dune, our skin felt chilled beneath a sticky film of dust mixed with dry perspiration salts. We could hardly wait to sponge-bathe. But when we stood naked and wet in the boma, with the wind rushing through the flimsy reed walls, we shivered—angry that we were always either too hot or too cold.

*   *   *

Often apart, sometimes together, Muffin and Moffet roamed for months over miles of wasted desert. At times they were as much as thirty or forty miles from any of their scattered lionesses, and we seldom found them with their females anymore. It was a time of social disintegration, of enforced isolation, for members of the Blue Pride. The most the lions could find to eat was a rat here, a springbok there, a porcupine or steenbok (if they were lucky), but never enough to allow them to stay together as a pride.

Like us, Muffin and Moffet spent the hot days in torpor, lying on bare sand and fire-blackened stubble beneath any sterile bush that offered a little shade. They constantly worked their mouths, pushing their tacky tongues out through dry lips, slowly rolling their vacant eyes toward us when we collected their scats for analysis. Their bellies were high and tight against their spines, their manes thinned, ragged, and lusterless, like the hair of bed-ridden invalids.

It had been more than a week since they had fed on anything substantial. With no large antelope left in their former Deception Valley territory, they had moved east, hunting every night through the open woodlands near the border of the game reserve. Now the two lions lay panting next to a bush, the only sign of life in sight.

When the heat subsided that evening they walked east to the boundary of the reserve and crawled through the wires of the fence. They could smell not only prey but also water. Ahead, cattle stirred uneasily inside a boma of felled thorn branches. The lions silently stalked forward.

Suddenly a white-hot pain shot through Muffin's leg. He roared and lunged against the trap, his foot twisting in its jaws. Biting at the steel, he rolled over and over, shredding the muscles of his leg as he pulled at the chain and the log to which he was bound. Moffet rushed to him and smelled the mangled foot and trap, but there was nothing he could do.

All night long Muffin fought to get away, panting and staggering as he hauled the heavy log in circles over the sand. Moffet watched nearby. In the morning, a native rancher on horseback rode up, lifted his rifle, and shot Muffin in the face and chest. Moffet turned and ran west toward the game reserve, the man and his pack of baying dogs tearing through the scrub after him. A rifle cracked repeatedly and bullets slapped into the sand around the fleeing lion.

Later that morning I flew over the spot where I had last found Muffin and Moffet, near the fence. They were no longer there and I could hear Muffin's signal coming from far away in the east. With a heavy feeling growing in my chest, I banked the plane and headed out of the game reserve toward his transmitter. I kept thinking that maybe he'd gone to the river for water or followed a herd of antelope, but I knew by the unusual clarity of the radio signal that he was no longer wearing his collar.

Sixty-five miles east, the signal peaked over the village of Mopipi near Lake Xau. By flying low, back and forth over the thatched roofs of the huts, I pinpointed the one with the collar and transmitter. When I looked down, the skin of a large male lion was pegged out in the sand near a low shed. People were scurrying through the village, all pointing up at the plane.

I landed in a clearing nearby and was surrounded by a throng of natives clapping their hands, waving, and laughing. Scowling and silent, hurting inside for yet another dead friend, I made my way through the crowd and into the village to the hut I had singled out from the air. A middle-aged black woman came warily to the door and peered around its edge at me. "Someone from this house has shot a lion," I said. "You must have the collar that was around its neck."

Unable to understand what had led me to her rondavel, for a moment the woman seemed confused and frightened. I managed a weak smile to reassure her. Then I saw Muffin's worn and blood-stained collar hanging behind her on a post, its transmitter still functioning. She gave it to me and I asked her how many lions had been shot. She said she didn't know, but that her husband had brought only one skin to the village from the ranch. I told her I would be back in a few days to get the details of Muffin's death from him.

Airborne again, I tuned in Moffet's frequency, but I could not hear his signal. It was not easy to tell Delia that Muffin was dead and that I could not find Moffet.

For weeks I continued to search for Moffet from the air, but never heard his signal. I figured he must have been wounded and had wandered off to die. One of the bullets had probably smashed his transmitter.

23

Uranium

Delia

He rips earth open for her ancient veins
Of molten splendor; toppling floods give room
Faced by his dams; the lightning knows his chains . . .
He loves his handicraft and . . . scorns what doom?
　　　　　　　　　　　　—*Gene Derwood*

THE SMALL, ROUND water hole on Springbok Pan had been dry for
months. Its cracked grey bottom was patterned with perfect footprints
of those animals, large and small, who had come in search of water.
There were old prints, made when the water was fresh: A brown hyena
had knelt to drink, a lion had slid in the mud, a porcupine had swished
its bristly tail. Then there were the deep spoor left by those who had
plunged through the mud to the last stagnant puddles in the center and
the desperate hoof marks of a gemsbok who had pawed deep into the
sludge for the last few drops of seepage. Finally, there were the tracks
of animals who had come, smelled around, and left, with only the
memory of how it feels to drink.

The water hole was surrounded by large acacia bushes and small
ziziphus trees, and kneeling beside it, we were well hidden. We had
driven to Springbok Pan hoping to collect lion and brown hyena fecal
samples. Analysis of the scats was important because it supplemented
our direct observations of what the predators were eating during the
drought.

Suddenly a loud whop, whop, whop drifted toward us. Startled,
we looked up to see a helicopter circling the trees. We backed deeper
into the bushes, hoping we wouldn't be seen. We were confused,

threatened, curious, annoyed. What was a helicopter doing here?

The chopper blew up a storm of dust as it landed. The rotor wound down, and three young men dressed in baggy jeans stepped onto the riverbed. Blue plastic bags full of soil samples were tied to metal trays mounted on the skids of the aircraft. We introduced ourselves, and they explained that they were field geologists on contract with an international mining company.

"What are you prospecting for?" Mark asked.

The chief geologist answered, his nervous glance dropping to Mark's shoulder and then to the ground. "Uh—well, we're really not supposed to say—but, uh—diamonds," he stammered.

A heavy pressure filled my chest, and my palms began to sweat. The vision of a massive diamond mine, with its great open pit, mounds of tailings, conveyers, trucks, and shantytowns looming over the gutted ancient river valley, flashed into my head. Perhaps there would be a parking lot where the brown hyena den had once been.

"Do you have a permit to prospect here?" I asked.

The geologist answered too quickly, "We're not operating in Deception Valley; we use it only for navigation. We're prospecting in the southern part of the reserve."

After a few stilted comments on how beautiful they thought the Kalahari was, the men walked to their helicopter and took off. Afterward we found sample holes and blue plastic bags littered at intervals all along Deception Valley.

A few weeks later a red-and-white Beaver—a single-engine bush plane of a type used in Alaska for years—circled our airstrip several times and landed. As it taxied to camp, I could feel that same tight feeling in my chest.

The pilot and his navigator introduced themselves as Hal and Caroline, mineral surveyors from Union Carbide. Caroline had sandy hair, a broad smile, and freckles. Hal, who was from Michigan, was tall, dark, and exceptionally polite. He explained that they were using a magnetometer to search for uranium in the Kalahari. We asked them into camp for tea, to discuss their operation. No one had notified us that they were coming or told us what they would be doing in the reserve.

The hornbills, flycatchers, and tit-babblers gathered in the trees above our heads, raising their usual cheerful ruckus. Our visitors were amazed at how tame the birds were and told us with great excitement that they had seen a lion from the plane that morning. How wonderful it was, they said, to be in a real, pristine wilderness among such wildlife. Pouring the tea, I quashed the urge to glare at them. How long did they think the Kalahari wilderness would last if they discovered minerals in Deception Valley?

They proudly explained that for the next several weeks they and others would be flying along the pans and dry river channels in the game reserve. The ancient riverbeds looked particularly promising for uranium deposits. If it were found in significant quantities, a drilling team would follow, to investigate the possibility of establishing an open-pit mine in Deception Valley—perhaps right where we were sitting.

We were horrified. After nearly six years of living alone in the Kalahari, suddenly we were being inundated by people in aircraft, who sat drinking their tea and cheerfully telling us how they hoped to contribute to the destruction of all that we had worked to protect.

"That's quite an airstrip you've made for yourself," Hal remarked. "We were wondering if we could use your camp area as a fuel station— the choppers and planes could easily land here to refuel."

"No," I answered abruptly. "I'm sorry, but we're working with sensitive animals here. That would cause too much disturbance."

"Oh, I see. Well, that's too bad. It would have been a big help, but we understand your position."

I thought to myself, The last thing we want to do is *help* you strip-mine the Kalahari, you stupid SOB. Out loud, I asked, "Would you care for more tea?" and smiled far too sweetly.

After a few more minutes of small talk they said goodbye and took off again in their Beaver.

*   *   *

One of the most important considerations for the conservation of Kalahari wildlife was the critical need to preserve pans and ancient river channels like Deception Valley. During years of adequate rainfall, the old river bottoms were covered with nutritious grasses, primary food for plains antelope during calving. The woodlands surrounding the valleys were essential browse for giraffe, kudu, steenbok, and eland, and for grazing antelope, who must switch to browsing in dry season and drought. These ungulates attracted predators, most of whose ranges were centered along the dry river systems.

Fossil river channels meandering through the dunes represent only a tiny fraction of the entire range area, but they are one of the most crucial habitats in the desert. An open mine, with its associated development, in Deception Valley or any of the other fossil channels, would be a disaster for Kalahari wildlife.

And now, seemingly overnight, Deception Valley held great interest for the mining industry. Surface uranium deposits had been discovered in dry riverbeds in Australia; the same could be true of the Kalahari.

We could hear the planes and helicopters flying over the desert every day for several weeks. Our reports to the Botswana government, urgently requesting that the game reserve be spared mineral exploration, received no response. All we could do was wait. The skies finally quieted down, but we had no idea what the results of the mineral survey had been.

Then one morning a deep rumble sounded from beyond East Dune, and we saw a column of dust that stretched for miles above the savanna. Standing on the riverbed near camp, we watched a convoy of trucks, ten-ton trailers, and a twenty-five-ton drilling rig roll single file into the valley. Union Carbide had come to the Kalahari to drill test holes for uranium, to determine if a mine would be profitable. We met the convoy at Mid Pan and talked to the drillers about their plans.

Doug, the geologist in charge, a young man with a plump face and a hang-dog expression, scuffed at the ground with the toe of his boot as he spoke. He promised not to allow his truck drivers to speed along the riverbed, not to chase animals or to frighten any brown hyenas that came to their camps at night, and not to drive at night, when the lions and hyenas were on the move along the valley.

"I know how important your research is—the Wildlife Department has told me—and I'll try not to interfere with your work."

We were greatly relieved with his apparent concern and shook hands warmly before he climbed into his truck. But we soon learned that his offer to cooperate had only been an attempt to placate us.

For years we had tiptoed around the old river at five to ten miles per hour. Now, ignoring our pleas and protests, heavy vehicles roared up and down the valley at fifty miles per hour, day and night, along the same paths Pepper and Cocoa used. They chewed deep ruts in the fragile surface of the riverbeds, scars that will last at least 100 years. Over and over again we cajoled, begged, and finally threatened, until we were given assurances that the trucks would slow down and not drive at night. The promises were never kept. The few springbok and gemsbok that had come back to the valley to re-establish their territories galloped away from the riverbed.

Discarded drums, beer cans, and other litter were left at each campsite the drillers set up along the valley. Long strands of blue plastic ribbons marked sites worthy of further investigation at some later date. Fluttering from the limbs and branches of acacia trees, they were a driller's trademark, a laying of claim to the valley.

We drove to the rig every afternoon, wherever it was operating, and asked anxiously about the results. Drawing lines in the sand with his boots, Doug assured us that they had not found uranium in significant quantities. He would not show us the official graphs.

Eleven days after they had come into the valley, the long convoy of heavy trucks pulled up next to camp. They had completed their tests, and they told us that they had not found significant amounts of uranium. We watched them disappear over East Dune, on their way to another fossil river for more drilling. We wondered if we could believe them.

*   *   *

Our research was beginning to show what it would take to conserve the Kalahari. But were we too late? Would it all be lost to man's greed for more minerals and for cattle? We were a lobby of two against powerful forces of exploitation. We had learned a lot about this ecosystem, but that was not enough. Other people had to care. The Botswana government had to view the Kalahari as a precious natural heritage rather than just a tract of exploitable resources.

We would do whatever we could. For starters, we tore down as many of the blue plastic survey ribbons as we could find.

24

Blue

Delia

Green fields are gone now, parched by the sun;
Gone from the valleys where rivers used to run . . .
—*Terry Gilkyson*

BLUE STOOD on North Dune, the wind blowing into her face. The once sleek and powerful lioness was very thin, her waist drawn in, wasplike. Her hair had fallen out in several spots along her back, leaving circular grey splotches, and her gums were pale.

She lifted her head and called with a soft "coo"—to the east, to the south, to the west, her ears perked to catch an answer. For most of her seven years of life Blue had slept, hunted, fed, and bred flank to flank and nose to nose with at least some of the Blue Pride females. In other years her pride-mates would have answered her coo and sauntered from the bush to rub heads in greeting. But for a year and a half, during the same time that we had observed the brown hyena cubs develop, the drought had driven them apart. We had not seen her with another lioness for months.

Blue could not know that Chary and Sassy were over fifty miles away. With their seven small cubs struggling behind them, the two lionesses wandered the bleak plains east of the reserve. They occasionally met with strangers, both females and males, who accepted the cubs as if they were their own; the males even sharing food with the cubs, as Muffin and Moffet had done. Except for their faded ear tags, there was no way to know that Chary and Sassy had once belonged to the Blue Pride.

Now that the drought was in its eighteenth month, the pride had disintegrated; Blue was the only lioness left in its original territory. The others ranged over a huge tract of mixed woodland, bush, and grass savanna more than 1500 square miles in area, eating anything they could catch.

After losing her cubs through neglect, Gypsy had joined Liesa to wander far to the southeast of Deception Valley, preying on duiker, kudu, porcupines, and the smaller rodents. Happy, of the Springbok Pan Pride, and Spicy from the Blue Pride transferred back and forth among various prides, then formed an alliance of their own. Prowling in open bush and grass savanna due south of the valley, they preyed on springhare, steenbok, and other smaller mammals. In May 1979, each had given birth to two cubs within a few days of each other, as pride-mates often do. We could not possibly know who the fathers were, since the females had mated with males from three different prides.

The first time we saw Happy and Spicy nursing each other's young, it was an exciting moment for us. Previously, communal suckling in lions had been seen only among closely related females of the same pride. Happy and Spicy could not be close relatives since they were from different prides. This and similar observations had important implications for the evolution and ecology of cooperative behavior among lions, which was the subject of Mark's Ph.D. thesis.

We had not seen Moffet again, although whenever Mark flew, he turned the radio receiver to the lion's frequency. Once in a while there was a faint beep mingled with the static, giving some hope that Moffet still roamed somewhere in the desert. Yet when Mark turned the plane toward these phantom signals, they always vanished.

* * *

When we sat with Blue on North Dune I had the feeling she was grateful for our company. She slept in the shade of the truck, and with its doors open for the breeze, I could have nudged her with my foot. The tires still held a fascination for her, and lying on her back, legs in the air, she rolled her head to one side and gnawed gently on the rubber.

Ordinarily she would not have stirred until after sunset. But she was very hungry, and at four o'clock she began foraging north toward the dune woodlands. From the duneslope, she could see more than a mile north and south along the valley; not one prey animal was in sight.

She zigzagged in and out of bushes for more than two hours—
stopping, listening, watching—but found nothing to eat. Panting heavily, she lay down to rest for a time, then continued toward West Dune
at nightfall. Near an open sandy area, she stopped in mid stride,
lowered her body slowly until her shoulder blades jutted above her
back, and stalked toward a springhare hopping near its burrow. When
she was fifteen yards away, she sprang forward. But her prey saw her
coming. With dazzling speed and a series of quick turns, its bushy
tail flicking deceptively behind, it dashed toward its hole. Blue gave
chase, the sand showering off her feet. Her nose was at its tail when
it shot into the opening of its burrow, but it did not make a clean entry,
and in the second that it paused, Blue pinned the still exposed hindquarters with her paw. She seized the springhare with her jaws, pulled
it from its hole and chewed slowly, rhythmically, her eyes half closed,
as though savoring every morsel of her four-pound prey. Within five
minutes, all that was left were a few drops of blood and tufts of hair
on the sand. It wasn't enough meat to sustain her for very long; she
continued to hunt through the woodlands.

Blue found no more food that night, or the next, though she walked
eighteen more miles. She lay down often to rest and to scratch her
irritated skin. Every day the scabby bald spots spread over more of
her body and we feared that she had sarcoptic mange, a debilitating
skin disease caused by a microscopic mite. These parasites can be
present in a healthy animal without causing harm, but can flare up to
provoke hair loss when the animal loses condition, as from malnutri-
tion.

Despite her poor diet and condition, Blue's slender stomach began
to swell slightly and her nipples enlarged. We worried that it would
be difficult for her to have young during the drought with no other
pride-mates to cooperate in the hunting and with almost no large prey
in the area. It was with a feeling of pity that we found her one morning,
nursing two kitten-sized male cubs in a thicket of tall grass. At a time
when she could barely manage to feed herself, and with no water for
thousands of square miles, she could ill afford to lose the moisture
and nutrients needed to nurse a demanding family.

We went to see Blue every day to learn how she was raising her
young in the drought. One of the first things her cubs, Bimbo and
Sandy, focused their eyes on was our battered Toyota. Often the mother
and cubs would lie in the shade of the truck, and we would have to

be careful not to run over a tail or a leg when we drove away.

In the evening Blue left Bimbo and Sandy alone, tucked away in a thicket of tall grass and bush while she walked for miles in search of food. She would return hours later and stand fifty or a hundred yards away, cooing softly. The grass would stir, and the cubs would scamper out, answering her with rasping, high-pitched meows. As Blue washed them with her rough, pink tongue, they squawled and squirmed. Bimbo managed to struggle to his feet and attempt a getaway, heading for his mother's teats while she was busy with Sandy. But her huge paw bowled him over, and he found himself wrapped in her great tongue again. When the licking and nibbling was over, they all settled beneath a tree where Blue nursed them.

Food was so scarce that sometimes Blue had to leave her cubs for twenty-four to thirty-six hours in order to find a meal. Both cubs were thin, but Sandy, the smaller one, began to show signs of weakness. More and more often he simply sat in the grass, his eyes listlessly watching his brother bounding through the bush or playing with sticks. Whenever Blue ended a nursing period, it was Sandy who cried loudly for more milk.

One night when the cubs were about two months old, Blue cooed softly to them and started walking away through the grass. Bimbo and Sandy scampered along behind as she walked west down the dune face. But by the time she had gone halfway across the riverbed, the cubs were falling far behind. Bimbo trailed by twenty yards, Sandy by thirty, both meowing loudly. Blue stopped and waited, cooing, but when they caught up with her, she continued west without giving them a chance to rest. She led them to the top of West Dune, a trip of nearly three miles, and there she left them in a patch of tall grass at the base of a tree. Then she walked away to hunt.

Blue's last drink of water had been during a brief shower ten months before; her diet of springhare, mice, honey badgers, and bat-eared foxes was her only source of moisture. She often had to walk at least ten miles at night to find food, and more and more frequently she encouraged Bimbo and Sandy to follow her for at least part of the way. It was increasingly difficult for Sandy to keep up. He was barely two-thirds the size of Bimbo, his fur was thin, the hard angles of his tiny bones showing through his skin.

One morning we found Blue and Bimbo alone. Sandy had been left behind, or perhaps he had been killed by a leopard, jackal, or hyena.

Mother and cub lay together beneath a thornbush. The hot wind swept small waves of sand, soot, and ash between the blackened bristles of grass clumps left from the dry season's fire. Blue's ribs and pelvis were outlined under her skin, her gums were white, and her hair had thinned all over her back and belly. She nuzzled Bimbo; he stood on his hind feet and put his padded forepaws on her face. With her broad tongue she turned him onto his back and, while he licked her forehead, she nibbled at his shrunken body. Though she could barely feed herself, Blue gave no indication that she intended to abandon her one remaining cub.

She had been wearing her radio collar for eighteen months; the edges were frayed and the antenna curled like a stretched bed spring. It was more and more difficult to hear her faint signal from the air and several times we had been unable to locate her from the truck. Although we did not want to dart her with Bimbo so near, her old transmitter would have to be replaced. And it would also be a good opportunity to examine her physical condition more closely.

We waited until dusk, when mother and cub had fallen asleep under a large acacia bush; then, with the velocity control knob on the darting rifle set at its minimum and the silencer in place, Mark darted Blue from ten yards away. The dart lobbed slowly, and quietly penetrated her flank. She jumped up, lifting her feet one at a time high in the air, looking around on the ground as if a snake had bitten her. Bimbo watched his mother curiously for a minute, also looking around in the grass, and then both of them went back to sleep.

Fifteen minutes later, Blue had apparently succumbed to the drug; she did not wake up when Mark shuffled his foot in the grass to test her. As we eased out of the truck, Bimbo's head shot up and he gazed at us with piercing eyes. He had seen us standing in full view many times, but never so close and never walking toward him. While we slowly moved to Blue, he looked back and forth from his mother to us, but she slept more deeply than ever. If we were acceptable to her, we were apparently okay with him; he put his chin down on his paws and, from ten feet away, watched us treat his mother for the next hour and a half.

When we rolled Blue over and examined her closely, we found her in even worse condition than we had expected. She had lost almost all of the hair from her underside, and large patches on her flank and neck were covered with heavy scabs. It was most certainly a case of sarcoptic mange.

Treatment for this disease in the wild is quite complicated, because the animal has to be immersed in a dip of chemical solution to kill the parasite. We had neither the equipment nor the essential drugs.

"I think I have an idea," Mark whispered. "We can drain some oil from the truck engine and smear it all over her. If we do it well enough, and she doesn't lick it off right away, there's a chance it will smother the mites."

It sounded crazy to me, but I couldn't think of anything better. Mark crawled under the truck and drained three quarts of black oil from the engine, leaving just enough to get us back to camp. We poured it all over Blue, rubbing it into every inch of her fur with our hands. Bimbo cocked his head to one side and then the other as we rolled his mother over and coated her chest. When we had finished, Blue was a mess. Sand, oil, and ash had caked to form a sludge that made her look like the victim of an oil spill.

We bolted a new radio collar around her neck, made notes about, and photographs of, her tooth wear, and gave her an injection of antibiotic. From ten feet away we could see no signs of the disease on Bimbo's skin, so we packed all the equipment and returned to the truck; Blue was just beginning to lift her head and look around.

In two days Blue had licked most of the sand and grit from her coat, but a layer of oil remained, and she didn't seem to have suffered any ill effects from the treatment. In fact, she began to scratch much less often, and within a week the edges of the large scabbed patches turned pink with healthy skin. Once her hair began to come back, the healing process was remarkable. Within three and a half weeks after her oil bath, her coat had almost completely recovered and all of the wounds had filled in with new hair.

At three months of age, Bimbo was still almost completely dependent on Blue's milk and had not filled out to the stocky shape normal for his age. When he was nearby, he showed an interest in the few kills his mother made, but he was rarely with her when she hunted. Since most of her prey were small, Blue usually consumed them completely while he was in hiding several miles away.

One night Blue killed a female honey badger and its cub, and after eating the adult, she carried the young badger back to Bimbo. When she laid it on the ground he seized the back of its neck in his jaws and strutted around, holding his head high. Then he lay down, his paws around its back, and quickly ate the three pounds of meat. Clearly the time had come: Blue had to find more meat for her cub.

The next night, after leaving Bimbo at the base of East Dune, she hunted through the dry scrub of the interdunal valley, and then up the dune face. As she neared the crest she saw something she had not seen for a long time. Lowering herself, she stalked forward. Along the dune top, silhouetted against the purple night sky, a bull wildebeest led a long line of black forms through the bush. Shrouded in a cloud of dust, blacker than the night, hundreds of antelope wound their way along the dune.

The tip of her tail twitching slowly, Blue flattened herself in front of the herd. As the third wildebeest passed, she sprang forward, leaped onto its back, and threw her paw over its shoulder, her claws hooking deeply into the tough skin. The antelope groaned and bolted, dragging her along the ground and through sharp thornbushes. But she held on and threw all of her weight under the wildebeest's neck. When her struggling prey hit the ground, Blue released her grip on its shoulder and grabbed its throat in her jaws, pinching off its windpipe. Lioness and antelope lay together, the wildebeest kicking out at first, trying desperately to breathe, then growing still. Panting heavily, Blue chewed into the flank, licking the blood and eating the tender viscera.

A few minutes later she walked the two miles back to where Bimbo was hidden. She cooed softly and he scampered out, and his nose near the tip of her tail, he followed her back to the kill. Two jackals had found the carcass and were tearing at the meat; Blue loped forward and scattered them. Bimbo joined her and they began feeding, not hurriedly, but with relish.

Eventually, Bimbo slumped down against his mother and slept. For the first time in his life his belly was round and full.

*   *   *

Late the next night Patches, the brown hyena, smelled the pungent odor from the torn belly of the wildebeest. By midnight Blue and Bimbo had finished all but the skin, bones, and attached meat. Patches circled cautiously around the area until she was satisfied that the lions had gone. With hackles bristling and tail raised, she scattered the jackals, already snipping morsels from the carcass, and she began feeding on the large bones, sinew, and skin left by the lions.

Pepper and Cocoa, who had been subsisting on old bones and the occasional springhare, were inside the den. Suddenly an entire wildebeest leg, covered with skin and bits of red meat, dropped through an entrance in a shower of dust. With thumps and squeals, the two cubs pulled the leg into the inner chamber. Patches lay down on one of the mounds and slept; she had carried the heavy leg for nearly three miles.

The following night when we drove to the den, Pepper was lying on her side atop one of the mounds. Usually she would get to her feet and plod to the truck to investigate its interesting odors. But not this night. She merely cocked open one eye, peered at us briefly, pawed some dust carelessly over her stomach, and dropped off to sleep again.

Blue's wildebeest kill had fed two lions and three hyenas. But it was only one wildebeest, and in this part of the Kalahari they were rare.

25

Black Pearls in the Desert

Mark

An ecologist must either harden his shell and make believe that the consequences of science are none of his business, or he must be the doctor who sees the marks of death in a community that believes itself well and does not want to be told otherwise.

—*Aldo Leopold*

THE MORNING after Blue had made her wildebeest kill, Echo Whisky Golf and I beat the sun into the sky, trying to find the lions and hyenas before the turbulent desert winds came to life. With Blue's signal sounding in my earphones, I dropped over the trees into a dune valley, where I was surprised to find her and Bimbo feeding on a wildebeest. Where had it come from and why was it in our area? In all the time we had lived in Deception, we had seldom seen wildebeest, and there had been almost none in the last three years. It must have been an errant old bull who had left his herds 100 miles or more to the south. But since wildebeest are quite gregarious, I couldn't understand why he would have come all the way to Deception on his own.

I tuned Moffet's frequency and climbed higher, listening closely to the crackle of static in my earphones. It was hopeless, and so I changed channels to look for Geronimo of the Ginger Pride.

When Deception Valley lay curling away to the south behind me, I noticed plumes of dust, or possibly smoke, rising all across the savanna ahead and below. I had never before seen anything quite like this from the air. Flying nearer, I could see hundreds, thousands of black dots moving along in single file through the bushveld. Stunned, I shouted over the radio to our base camp, "Wildebeest! Delia, I've found tens of thousands of wildebeest! They're moving north!"

I pulled back the throttle and began to descend. Below, the files of antelope were winding through the bush savanna, like long strings of black pearls against the tan monotony of the Kalahari in drought. Although we did not realize it at the time, we had just stumbled upon the second largest wildebeest migration on earth.

In Maun we had heard the hunters reminisce about times when they had waited for hours on the main road to Francistown while hundreds of thousands of wildebeest crossed. But no one knew where they had come from or where they were going. Many people just assumed that the populations exploded in the years of good rainfall, and then died off in years of drought. Only a few months before the present migration, a countrywide aerial survey conducted by a foreign consulting firm had counted 262,000 wildebeest in their southern Kalahari range, a population second only to the Serengeti herds. But the research team had concluded that in Botswana these antelope never migrate.

Delia and I took off at first light the morning after I had spotted the migration from the air. Banking and turning 100 feet above the savanna, we followed the wandering trails south, away from the herds and deeper into the Kalahari, making notes on where the migration had begun, what range conditions were like there, what routes the population was following, how fast it was moving, where it was headed, and other details that we would need to describe this event.

In the five previous years, the Kalahari rains had been generous. The wildebeest herds had led a vagabond's existence, chasing scattered rain clouds and patches of green grass deeper into the southern part of the desert, far from the only lakes and rivers, 300 miles to the north. Even though they had nothing to drink for several months each dry season, they found moisture as well as nutrients in the fodder they consumed. The grasses on and around the hundreds of calc pans never completely dried out from one rainy season to the next. Each rainy season the wildebeest population had swollen with the birth of many new calves, and these grew sturdy and strong from the protein and minerals in the pan vegetation.[1]

This year, 1979, when the rains had failed, the grasses had turned from green to tan, and now, in mid-May, they were little more than bleached straw, crisp and brittle under the sun.

*   *   *

The wildebeest stood on a low sand ridge, their manes and beards and stringy tails flowing in the dry wind. It may have been instinct, or a behavior passed down for generations, but something told them to trek northward, to the only place they could get water to survive the drought. Probably for centuries, Lake Xau and Lake Ngami, the Nghabe and Boteti rivers, and the southern fringes of the Okavango River delta had provided refuge for Kalahari antelope in times of drought. The dust blowing from their tracks, the bulls and cows and their calves lowered their heads and began plodding north.

This was unlike the Serengeti migration, where herds of wildebeest often mass together in great numbers. Because they live in a marginal, semidesert habitat, the Kalahari population is more mobile and less concentrated to begin with; now it was moving in herds of from 40 to 400 antelope, scattered along a vast front measuring more than 100 miles from east to west.

Not all the herds headed in the same direction. One portion, probably more than 90,000 strong, had taken the route to the north; tens of thousands of others began walking toward the Limpopo River 300 miles to the east. Whether migrating north or east, once under way they spent little time feeding, for without moisture they could not digest what they ate. Their aim was to get to water, and perhaps to better forage, as quickly as possible. Without water they would starve to death in a savanna filled with grasses. And even when there is an abundance of fodder, it may lack sufficient protein and essential nutrients that antelope need for survival in drought. Trekking in the early evening, at night, and in the early morning in order to avoid heat and dehydration, for days the great long lines of wildebeest plodded onward.

The herds covered about twenty-five or thirty miles each night. From the air the dusty migratory trails looked like gnarled fingers reaching for the lakes and rivers. Some of the wildebeest had already come more than 300 miles, from southern and southwestern Botswana— even from across the border with South Africa. The desert was taking its toll of the very young and the old; they were left behind for the scavengers. The physical condition of each animal was pitted against the great distances that had to be traveled, with little to eat and nothing to drink, but evolution had prepared them for the trek, and the strong should survive.

Suddenly, the wildebeest stopped short. Confronted by something many of them had never seen before, they bunched together and milled about nervously. Stretched across their path were strands of high-tensile steel wire—the Kuki foot-and-mouth-disease control fence, extending for more than 100 miles across the northern border of the Central Kalahari Reserve. At its east and west ends it joins other segments of fence that hem in the desert with more than 500 miles of wire.

The wildebeest were cut off from the emergency water and riverine habitat that for eons they had counted upon in times of drought. Nothing they had ever learned, none of their instincts could help them deal with this obstacle.

Frustrated in their urge to continue north, with the river little more than a day or two away, the herds turned east along the fence. There was nothing else to do. Having walked for days with little nourishment and no water, they were already weakened, and now the long fence beside them added more than 100 miles to their trek.

As they plodded along the fence, they encountered many other herds, part of the same migration headed for the lakeshores and riverbanks. Each day they were joined by giraffe, gemsbok, and hartebeest, all needing water but trapped by the wire and posts.

Thousands upon thousands of antelope that had been spread out over vast areas of the savanna were now forced by the fences to take the same migratory route toward water. The grasses of the fragile rangeland were soon stomped and broken, ground into dust by the hooves of the first herds to pass. For those that followed, there was nothing to eat. Animals began to drop from hunger, thirst, and fatigue. A giraffe who could easily have stepped over the wire became tangled in it. He struggled to get free, but the coils of high-tensile steel sliced deep into his flesh until he pitched forward, breaking his foreleg at the knee. His hind legs still ensnared, he pawed at the ground for days, building small mounds of sand around him as he tried to rise again. He never did.

Eventually they came to a north-south fence called the Makalama-bedi, which forms a corner with the east-west Kuki line before running south along a portion of the east boundary of the game reserve (see map I, opposite). Here there was chaos among the wildebeest. To follow this second fence they had to turn south, directly contrary to the direction they needed to go for water. They stood in confusion, their heads hanging, until many began to sway and stumble and finally collapse. But they were tenacious and the end did not come easily. In contrast to the victims of the swift tooth and claw of a predator, the animals who fell prey to the fence died more slowly. Even as they lay, still pawing the dust, their eyes were often plucked out by crows and vultures while other scavengers chewed off their ears, tails, or testicles. A few thousand died along the fences, but the carnage had just begun.

Finally the wildebeest herds struck off south along the five-foot wall of wire. A day later they came to the end—the fence just stopped in the middle of the savanna, as though someone had forgotten to finish it. The herds swung around it. Now there was the sweet, unmistakable smell of moisture on the prevailing easterly wind. They followed the promising odor. But as soon as they had moved around the end of the fence at the border of the game reserve, they entered a safari hunting area. To reach the water they would have to risk being shot.

Two more days of walking, and the wildebeest who had survived the trek, the fence, and the hunters, shuffled out of the woodlands and onto a great plain black with thousands more of their own kind. The smell of water was stronger; it was only twenty-five miles away. They hurried forward.

Resident cattle, tended by natives from scattered kraals along the lakeshore, had stripped the once beautiful lake plain of every edible leaf and blade of grass. Now its surface was like concrete covered with several inches of grey powder, and the choking dust rose from the hooves of the wildebeest in the still morning air. Here and there a spindly bush stood rooted in the wasteland.

Wizened carcasses littered the plain. Dying animals lay on their sides, their legs moving rythmically, as if, in their delirium, they were still moving toward the water. Prime wildebeest bulls and cows, the breeding stock of the population, began leaving the long lines filing toward Lake Xau. Unable to take another step, their knees buckled, and their muzzles sank lower and lower, until their nostrils blew small potholes in the dust.

At dawn of the second day after leaving the game reserve, the surviving wildebeest had nearly reached the lake. But water was not

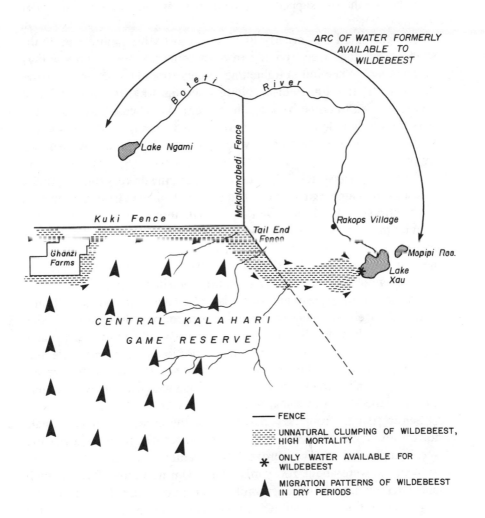

ARC OF WATER FORMERLY
AVAILABLE TO
WILDEBEEST

B o t e t i   R i v e r

Lake Ngami

Mckalamabedi Fence

Kuki Fence

Tail End
Fence

Rakops Village

Ghanzi
Farms

Mopipi Roa.

Lake
Xau

C E N T R A L   K A L A H A R I

G A M E   R E S E R V E

——— FENCE

::::: UNNATURAL CLUMPING OF WILDEBEEST,
       HIGH MORTALITY

 *   ONLY WATER AVAILABLE FOR
       WILDEBEEST

 ▲   MIGRATION PATTERNS OF WILDEBEEST
       IN DRY PERIODS

all they needed. There were no shade trees for miles, and there was nothing to eat. Time was short; they would have to drink and then get back to the shade and graze of the woodlands twenty-five miles behind them before the sun sapped them of their remaining strength (see map 2, opposite).

Suddenly the long, orderly lines began to splinter and circle; thousands of antelope were running from three trucks filled with waving, jeering men. Wheeling in tightening circles around the wildebeest, the driver of the five-ton Bedford bludgeoned his way through the herd, before turning to come back again. Several wildebeest hit on the first pass tried to hobble away. The driver turned sharply and, with clouds of dust boiling from the truck's wheels, he ran down the wounded one by one.

When they had struck down six wildebeest, the drivers slid the trucks to a stop and laughing tribesmen leaped out. Two men held each animal by its horns, and a third sawed through its throat with a knife.

With the sun rising higher and hotter, the survivors crossed a shimmering white salt pan and climbed a last barren ridge. Below them, not half a mile away, lay the blue waters of Lake Xau, pelicans and flamingos bobbing like flower petals on the surface.

To the north, more than 360 miles of river front and lake shores had once been available to the wildebeest during drought, as had a similar stretch of the Limpopo River, in the south. Now, fence lines and settlements had funneled a major portion of the entire Central Kalahari population into a tiny area, denying all but two or three miles of the riparian habitat to the 80,000 antelope who had come for water. The wildebeest had to drink here or die.

Sensing danger from the native huts on either side, the thirsty animals took a few tentative steps forward. The water was *there*. They could see it and smell it! They broke into a canter, heading for the lake. When they were about 200 yards away from it, bands of native men and boys with packs of dogs broke from cover near the kraals. The dogs were set upon the wildebeest and chased them in circles for long minutes. The dogs worked efficiently, holding exhausted animals by the hind legs, hamstringing and disemboweling them as they sank to the ground, the poachers rushing in with clubs and knives to finish them off.

Thousands of other wildebeest were prevented from going to the water by the disturbance. Some managed to reach the lake, where they

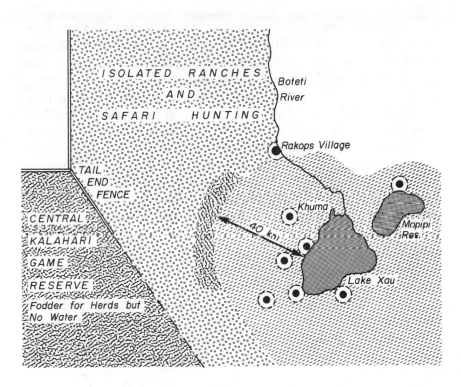

ISOLATED RANCHES

AND

SAFARI HUNTING

Boteti
River

Rakops Village

TAIL
END
FENCE

CENTRAL
KALAHARI
GAME
RESERVE
Fodder for Herds but
No Water

40 km

Khuma

Mopipi
Res.

Lake Xau

 SETTLEMENTS

 ONLY AREA WITH FODDER FOR
WILDEBEEST

 AREA OF INTENSE POACHING AND
OVER GRAZING BY CATTLE

 DAILY MOVEMENT OF WILDEBEEST
FROM FODDER TO WATER

collapsed into its coolness. But many were too weak to rise or even to drink, and their muzzles sank slowly into the shallow water and the mud.

Above in EWG, we watched through binoculars as the slaughter continued along the shore. Trembling with rage, I pushed the control wheel forward and we plunged toward the lakeshore. The poachers were preoccupied with their butchery and did not see the aircraft until it was at ground level, roaring across the plain toward them at 160 miles an hour. We flashed by inches above a pack of dogs attacking a young male wildebeest; at the last second they let go of their prey and dodged away. The dogs scattered in confusion and the antelope began to run. We chased the dogs until they fled to cover. Three of the men threw their clubs at the plane an instant before they dove into the dust, scuttling on their bellies through the thornbushes.

During the migration, I continued to make periodic low-level flights over the huts and the herds in the early mornings and on moonlit nights. It seemed to deter the poachers, and from then on we saw less harassment of the herds.

Ironically, even if the wildebeest were able to drink, the lake was a nemesis for them. Once they had tasted its water, their migration was frozen in that one area. Because of the native settlements they would go no further north, and thus were barred from the river beyond. Every day they needed to get back from the lake to the shade and grass in the woodlands before the sun grew too hot, and then to recross the plains and salt pan at night. It was an incredible fifty-mile round trip.

Day by day this distance between water and food increased, as the masses of antelope denuded the rangeland of grass farther and farther from the lake. For a while the wildebeest survived like this, but inevitably the time came when all the water they could drink at the lake and all the graze they could get from the woodlands did not supply them with sufficient energy to make the long journey in between. At this point the antelope began the inexorable slide toward starvation and mass death.

September, and the start of the Kalahari *hot*-dry season, could not have come at a worse time for the antelope. Temperatures shot upward, and phalanxes of dust devils swirled across the parched plains. With the sun rising earlier and setting later, the wildebeest could not avoid making part of their trek in the heat of the day. It was the proverbial last straw. Mortality soared and carcasses littered the plains, salt pan, and lakeshore. Without rain to draw the suffering animals out of the Lake Xau trap and back into the Central Kalahari, most of them would die.

*   *   *

Extensive fencing began in Botswana in the fifties, when beef exports from cattle ranching became an important industry and the need arose to control periodic outbreaks of foot-and-mouth disease (FMD) among domestic stock. During such epidemics European (EEC) countries refused to accept meat products from Botswana for fear of contamination. It is understandable that Botswana had to take action to protect this lucrative industry, and the Department of Animal Health was charged with the task of devising ways to control the disease. To date, that department has erected more than 800 miles of cordon fences through the country's wilderness areas; and construction of another 700 miles is currently under way.

Because cape buffalo and some species of wild antelope can carry the FMD virus, their populations were suspected of being a reservoir of the infection that periodically contaminated domestic animals.[1] The fences were erected to segregate the country's stock population from wild herds, to separate infected cattle from others, and to divide the range into sectors that could be quickly sealed off during outbreaks. By preventing the movement of infected animals from one area to another, the fences would, in theory, make the disease easier to control. But foot-and-mouth disease continues to develop and spread across fence lines in Botswana.

The use of fencing to control FMD has become a very controversial issue, and research veterinarians disagree on the effectiveness of this means of disease control. In spite of extensive experiments it has never been demonstrated that wild animals transmit the FMD virus to domestic stock.[2] The epidemiology of the disease is poorly understood, and no one really knows how it is spread.

The fences had been devastating Botswana's wildlife long before our study was under way. In 1961, and again in 1964, as many as 80,000 wildebeest died in the area of the Kuki-Makalamabedi fence corner and between there and Lake Xau. George Silberbauer, government officer of the Ghanzi District in 1964, estimated that a tenth of the population in the Central Kalahari Reserve was dying every five days while trapped behind the Kuki fence during drought.[3] It is not known how many survived the 1964 mortality. Dr. Graham Child, an ecologist with the Botswana Department of Wildlife, wrote that the 1970 die-off was, "the severest mortality in living memory."[4]

Bergie Berghoffer had tried in vain to save thousands of dying wildebeest. For weeks he had hauled water from the Boteti River to troughs he had made from dozens of steel drums and placed near the fence corner. "It was a bloody disgrace," he said. "You could walk up to the poor blighters, put your hand on them, and they would just fall over."

Zebra, which once used Deception Valley as part of their wet season range,[5] have been eliminated altogether. We didn't see one in seven years. Great mixed herds of gemsbok, eland, and hartebeest, which George Silberbauer described as covering an area three by five miles near the Piper Pans, have been reduced to a small fraction of their former wet-season concentrations.

After the reduction of the Central Kalahari antelope by drought and disease control fences, the survivors have become even more important to the predator community. If the large antelope all die off, lions, leopards, cheetahs, wild dogs, and scavengers, like brown hyenas, will suffer a similar fate. Though we cannot know how many carnivores there were before the fences decimated their food supply, their numbers must have dwindled considerably.

Since the Botswana Department of Animal Health began fencing the Kalahari, the Kalahari Bushmen have found it increasingly difficult to hunt and kill antelope for meat, one of the few sources of protein for them and other rural native Africans. Coincidental with fencing for disease control, the amount of protein in their diet has fallen precipitously, according to Dr. Bob Hitchock, a former rural sociologist for the government.

The wildebeest crisis must now be considered in a broader context than simply antelope versus the fence. It is part of a larger picture, the competition between man and wildlife for such limited resources as grassland and water. Alternatives for controlling FMD, such as a sophisticated vaccination program, need to be considered more seriously. (For more details, see Appendix A.)

* * *

In many ways, Botswana has had a positive attitude toward wildlife; indeed, about one fifth of the country is either national park or game reserve. Government officials had always been courteous to us and had granted us permission to conduct research in the Central Kalahari Reserve. But we were continually frustrated in our attempts to stimulate interest and action from the government in order to help the wildebeest.

We wrote letters and reports to the Wildlife Department describing the migration and mortality. We made recommendations, including the establishment of a game scout camp at the Lake Xau area to control poaching and harassment of wildebeest, and the maintenance of a corridor from the reserve to the lake so that the antelope could drink. But there was little response to our requests.

The drought persisted into October, and the wildebeest were dying at an ever-increasing rate.[6] We felt frustrated and alone in our efforts to save them. In all our years in the Kalahari, we had never seen such suffering among animals, such degradation of habitat. It all seemed so pointless. Had the antelope been allowed to distribute themselves throughout the miles of riverine habitat, as they had done for thousands of years in times of drought, many fewer would have died.

Almost everyone we knew told us to forget it. "Cattle is too big an industry; you'll never get them to take down the fences." One or two friends warned us we might get expelled from the country for raising the issue. But Botswana is a democratic republic and we did not believe that the government would expel us on account of a wildlife issue. We felt compelled to do something before it was too late. We knew that the reports written by Dr. Silberbauer and Dr. Child describing previous die-offs had simply been filed away. We were determined to ensure that a solution be found before the next drought. Since no one within the country would listen to our recommendations, we decided to try to publicize the issue worldwide, to enlist the support of prominent people outside the country who perhaps could encourage the Botswana government to review the problem.

*   *   *

One day, over our radio, we received an invitation to give a slide presentation for Prince Bernhard of the Netherlands, who would soon be touring some of Botswana's wildlife areas. Shortly after that, Dr. Richard Faust of the Frankfurt Zoological Society and a group from The Friends of Animals—our major sponsors—were due to arrive in Maun. It was unbelievable luck. Two major figures in global conservation would be practically on our doorstep—at least within a couple

of hundred miles of us—in a matter of weeks. We quickly wrote both
Dr. Faust and Prince Bernhard, describing the wildebeest crisis and
inviting them to camp, but we had almost no hope that the prince
would accept our invitation to visit us in the desert.

Even so, we began to wonder how we would deal with it if he did
accept. Where would he and his attendants sleep? We couldn't imagine
his sharing our one sleeping tent, with its packing-crate bed and piece
of foam rubber, with his staff. I also doubted that royalty would consent
to fly in Echo Whisky Golf, and our airstrip was certainly too short
for a much larger plane. What could we give him to eat and drink—
biltong and hot, smoky water? We worried most about the toilet. The
"thunderdrum," a bright red gasoline drum with a seat cut in the top,
stood in the middle of the riverbed.

Given these uncertainties, we decided that we had better do whatever
we could to prepare for the prince ahead of time, so that we would be
ready if he decided to come back with us after our meeting at Khwaii.
Delia washed the tent floors (sparing the spiders in their corners as
always), emptied old birds' nests from some of the kitchen pots, screened
the ants from the sugar bowl, and baked bread in the bucket oven. I
waxed the plane, hung burlap on some tent poles around the thunderdrum, and buried a bottle of wine, saved for a special occasion, beneath
the ziziphus tree.

On the day the prince was to arrive in Botswana, we flew to Khwaii
River Lodge, a posh resort on the east edge of the Okavango River
delta. We buzzed the lodge and landed on the long sod airstrip. A
Land Rover met us and the driver chauffeured us to a group of white-
washed rondavels set in a neatly manicured lawn. The dining house
was of dark timbers; beyond it lay the Khwaii River flood plain, with
scattered herds of lechwe antelope, and hippos lying like grey sub-
marines in the blue waters.

When we arrived at the lodge, we were relieved to learn that the
prince and his party were taking a game drive somewhere in the bush.
We were somewhat uncertain about the proper way to greet him and
hoped that we could find someone to ask. Should Delia curtsy, should
I bow? Should we address him as Your Majesty or Your Highness?
Odd considerations in the middle of the wilderness, but we didn't want
to appear gauche.

That evening as we entered the thatched dining house and self-
consciously made our way through the room full of people, we glanced
from face to face, hoping to somehow spot the prince. We were passing

the center table when a hand gripped my arm and a voice said, "My name is Bernhard—not Barnard, like the famous South African heart surgeon. You must be Mr. and Mrs. Owens." Suddenly we were facing the prince of the Netherlands, who had just let me know that I had misspelled his name in my letter about the wildebeest. A thin smile grew from the corners of his eyes and spread across his tanned and freckled face. His thinning hair was slicked straight back, and rather severe wire-rimmed spectacles were comfortable on his nose. I was reminded of a picture I had once seen of him as commander of the Dutch forces during the Nazi invasion.

"Anyway," he continued, "I'm glad to meet you. Please have your lovely wife sit here next to me."

During dinner he said casually, but with an unmistakable glimmer of anticipation, that he would be quite happy to fly to our camp in Echo Whisky Golf. Unfortunately, he would be able to stay only one day.

The next morning at six-thirty we met him and his secretary on the airstrip. At the dinner table the night before, the prince had given animated accounts of his old flying days: Gypsy Moths, short landing fields with hedges, and all. Now, as we took off, he looked at me and asked with a smile, "Do you mind?" I told him the heading for camp and gave him the controls. If he'd forgotten anything about flying he didn't show it, for he took us straight to camp.

After a quick cup of tea, we took to the air again and followed the route of the migration north to the Kuki fence. Dodging columns of vultures arriving for the day's feed on the carcasses, we followed the fence to its corner, and turned south. Miles away to the east, clouds of dust rose from the Lake Xau plains; thousands of wildebeest were returning to the trees after their long trek for water. The prince just shook his head, his jaw set grimly while we flew low over the black masses that speckled the plain as far as the eye could see. It was much hotter than it had been even ten days before, and the animals were dying at a much faster rate. We flew over a pall of destruction, death, and suffering. We hardly spoke on the way back to camp.

As we walked into camp after our depressing flight, Chief, one of our friendly hornbills, sailed from his perch and glided straight for a landing on the prince's pate. Over another cup of tea, Bernhard promised to help raise additional money for our research, and to inform the right people in Europe about the wildebeest crisis.

I dug the wine from beneath the ziziphus tree while Delia served up fresh bread from the bucket oven. Together with the hornbills, tit-babblers, and marico flycatchers, who hopped around the table, we had our lunch.

That afternoon Prince Bernhard met some of the lions, the brown hyena cubs, and the Pink Panther—who crawled into a hole and refused to come out. Just before dark, we touched down back at Khwaii Lodge. We gave our slide show that night, and went home the next morning to get ready to receive Dr. Faust.

Dr. Richard Faust is a man of phenomenal energy. He works all day seven days a week: from five to eight o'clock each morning as director of the Frankfurt Zoological Society; from eight to five directing the affairs of the Frankfurt Zoo; and from five to ten at night as director of the zoological society again. His trip to Africa was his first time off in seven years, and even then he was leading a party of the society's sponsors.

Now he stood on the truck's running board, his hair tossed by the wind, his face powdered with dust. We were driving from one wildebeest carcass to the next, counting, sexing, and determining the age of dead animals along the Lake Xau shoreline and salt pan. In the evening, we spread our bedrolls in the open just a few yards up from the water, and sat around our campfire. At sunset we had watched a young wildebeest standing in the dying heat waves on the bank above us, afraid to descend to the water to drink. Now the darkness was heavy with the stench of carcasses and the keening of night birds. No one spoke for a long time.

By ten-thirty our fire had burned down to embers. A subtle vibration began to fill the air. "Listen...do you hear something? There...like water rushing over rocks." The minutes passed and the sound became a rumble; a low moaning rose from the plains. "The wildebeest are coming!"

Black forms marched over the ridge above and a cloud of dust swept over us. I moved quietly to the truck and switched on the spotlight. A sea of antelope, their brilliant emerald eyes like spots of phosphor, was pouring over the bank, flowing around our campsite.

The legions passed us and entered the water, slurping and splashing as they drank. But each animal stayed in the lake for only two or three minutes. Then it was drawn into the black current of bodies eddying shoreward and then westward toward the plain. The return trek had begun almost immediately. All that way for a few gulps of water!

They had but a few hours to find shade before the sun began sapping the life from them again.

As I watched the antelope surge desperately into the lake and then out of it, I thought about the significance of the migration for the conservation of all wildlife in the Kalahari. The wildebeest, lions, and hyenas had taught us much: For all its great size, the Central Reserve does not provide adequate habitat for most of its highly mobile populations, both predator and prey. Without any permanent water holes in the reserve, and restricted by fences and human settlements from the country's only lakes and rivers, the antelope populations have almost no place to get water in severe drought. Though lions, leopards, brown hyenas, and other Kalahari predators can live indefinitely without drinking, they can do so only as long as there is prey available to provide them with enough food and moisture. We were convinced that if something wasn't done soon to allow these animals greater access to the Boteti River and Lake Xau, to stop the harassment by poachers, or to devise some other solution, much of the Kalahari's wildlife would disappear.

In the short term, the only hope for the antelope was rain—rain to grow green grass in the desert and to draw them away from Lake Xau. The only long-term solution was to designate the area between the reserve and Lake Xau as part of an even larger game reserve or, at least, as a protected corridor of access to the lake for the herds. The catch, of course, was that this would mean freezing the development of settlements and ranching in that area.

We were convinced that if tourism and other wildlife industries could be developed as a substitute for raising cattle, the overall living standard of local tribal people would be raised. At the same time a great natural resource would be conserved. But even while we were suggesting this to government departments in Gaborone, we knew there was little chance that they would accept the idea—too many important people have cattle interests in the Lake Xau area.

Dr. Faust was greatly moved by the wildebeest situation and pledged his continued support of our research. In the months following his and Prince Bernhard's visits, we spent many hours hammering away at the typewriter in our tent or working under the ziziphus tree, with the hornbills pecking our pencils. We drafted articles for magazines all over the world. We sent reports and circulars to other influential people, hoping that someone could convince Botswana to do something for the conservation of the wildebeest; or to convince the EEC, who insist upon the fences, to review their impact on wildlife.

We hardly dared hope that any of this would help change the attitudes of import-export businessmen or government officials. But we had done our best. Meanwhile we watched the bleak and smoky skies for clouds. There were none.

26

Kalahari High

Mark

You could feel the rain
before it came,
    the signals were that good.
        —*Rod McKuen*

ONE AFTERNOON in mid-October 1980, over two and a half years after
the drought had begun, there was a cloud. After months and months
of searing desert sky, a single pillow of water vapor stood alone above
the Kalahari, teasing us. Several hours later, others appeared. They
were scattered, but each grew darker and heavier in the sky to the
east, between the valley and Lake Xau.

When rain began streaking the sky beneath one of the clouds, Delia
and I ran to the plane. At 1500 feet, we flew beneath the soft, grey
belly of vapor, the rain splattering against the windshield and streaming
along the skin of the plane. We opened the windows and let our arms
trail outside. The cool wetness coursed down and off our fingertips
and the fresh smell rushed into the cabin. It was a Kalahari high.

We flew toward Lake Xau as other storm cells feathered the sky
with moisture. Over the lake plain there were no clouds, but below,
the black masses of wildebeest were bunching, looking west toward
the desert sky. Then the whole surface of the plain seemed to move,
as thousands upon thousands of antelope began galloping westward.
Some natural sense of order prevailed, and they began forming several
lines, some more than a mile long, each headed straight for a cloud
and its veil of rain.

It must have been the mist and spray around the airplane, or our excitement, that kept us from noticing at first, but, with a pang of despair, I realized that the ground was still grey with drought. We had been flying in "virga"—rain that evaporates in the hot desert air before it reaches the earth. Some of the wildebeest nearest the clouds, where the rain should have been, had slowed to a plodding walk with their heads lowered. Others had come to a standstill. Could a wildebeest feel utter dejection?

I throttled back, put down some flap, and for half an hour we dawdled from one cloud to another, watching the herds below. Finally, with the falling temperature in late afternoon, dense white columns of water began to reach the ground. Circles of savanna turned dark and wet. Puddles formed. The wildebeest gathered to drink and to eat what soppy grasses they could find.

Three days later it rained again, then again a week after that. Green grass shot up everywhere; the wildebeest quickly began eating their way back into the Kalahari and the game reserve. They were, of course, only the survivors of a much larger number that had migrated, but at least the rain bought us time—another year to convince Botswana's government and the rest of the world that Kalahari antelope are worth saving.

The sky was still stacked with cumulonimbus the morning after the first rain, and we took off in the plane to find the lions. Just as Sassy's signal sounded loud in our headphones, big drops began spiking against the fuselage. We passed the signal's peak, our foreheads pressed against the side windows, squinting through the mist at the trees flashing underneath. I banked steeply and came around again. When I saw Sassy she was standing near a big acacia with Chary and all seven of their adolescent offspring. Preying on wildebeest far outside the game reserve, often within a few hundred yards of cattleposts, the two females had brought all of their young through the drought. By any standard they had been good mothers. Now they all stood near Hartebeest Pan, licking the water from one another's backs and faces under the pelting rain. When we banked away the youngsters were stalking and chasing one another, and we could see their paw prints in the wet sand.

Blue and Bimbo were drinking at a water hole west of Crocodile Pan when we found them. We made notes on their position. I had just turned to fly back to camp when I spotted a large male lion resting near Blue and Bimbo at the edge of a grove of acacia bushes. He was nearly under the plane and just a blur, so I hauled EWG around for another look. But the aircraft was drifting heavily in the wind and there were dozens of thickets. We could not find him again.

It was just after sunset when the rains finally arrived in Deception Valley. Pepper was smelling a scent mark near the communal den while Cocoa rested under a bush nearby. The two brown hyena cubs stood, their ears perked, as the first rain they had ever seen began kicking up puffs of dust on the mound all around them. They began licking moisture from twigs, old bones, and the ends of their noses, and finally, lapping it from puddles on the ground. At two years of age, at last they had their first drink of water.

The next morning Blue and Bimbo were near Dog's Leg, in the upper end of Deception Valley. As I flew past I could see them lying with a big blond male, probably the same lion we had seen them with the day before. I was glad to see that, after almost two years on their own, they had found another companion. Perhaps when the rains were well under way and the antelope came back to the valley, the lions would walk the riverbed again. The nights and early mornings seemed empty without their roars rolling from dune to dune. I flew to camp and plotted their position on our aerial photographs. Then we set out in the truck to have a better look at the three of them.

We found them in an opening between two dense thorn thickets. The male was lying on his side and didn't even turn as I stopped the truck.

"I can't believe how relaxed he seems," Delia remarked. But then he swung his head around.

Delia raised the binoculars to her eyes. I heard her catch her breath.

"Mark! It's Moffet! He's alive! I can see the mark on his hip!" Though his collar and transmitter were gone, in his right ear were the remains of a red tag. He had seen Muffin trapped and shot and had himself been chased by a man on horseback with a pack of dogs; he may even have been wounded. But he had survived all that, and the drought, too.

After a time, Delia and I eased out of the truck. Cooing softly, mimicking the social call we had used for so long to reassure lions, we crept forward. Lying apart from Bimbo and Blue, Moffet was feeding on a porcupine clamped between his broad forepaws. He watched us intently, and then sighed and continued feeding. We settled beneath a bush about five yards away. It was like old times with him again.

Bimbo, now nearly two years old, still had a youngster's curiosity, even though he was a 200-pound subadult with the ragged beginnings of a mane. He slowly stood up and walked toward us. When he was five feet from us, he stopped and looked away. He licked his forepaw and smelled the ground and then took another tentative step forward, placing his paw carefully on the ground as if he were walking on eggshells. More than anything I wanted him to accept us totally, to show that his curiosity outweighed his uncertainty. If he touched us it would be a sign.

Another step. He leaned toward me, his nose and whiskers only three feet away from my face. He came still nearer, and I could see the reflection of the desert in his eyes, the flecks of golden brown in his irises as they adjusted to the changing light. Again and again, he pushed his muzzle forward, and then stood back, turning his ears slightly. After a last, clumsy attempt, he hastily put his nose into some leafy branches near my head and sniffed loudly, as if that had been his intention all along. Then he walked away. He had almost touched me, but something held us apart. The last barrier still remained.

We sat for a long time and watched Moffet slowly finish his meal. Then he rubbed the quills from his face and shoulders and licked his paws. When he had finished cleaning himself, he got to his feet and walked toward us, his mane swinging, his pink tongue sliding along his muzzle. He stopped near our feet, his soft eyes on us, then walked on to lie down with Blue beneath their shade tree.

In the midst of thousands of square miles of wilderness Moffet, Blue, and Bimbo were at least somewhat buffered from man's careless exploitation of the wilderness. Perhaps they, Pepper, Cocoa, and the others would be allowed to retain a part of the earth, to survive and endure.

But then my attention was drawn to something in a nearby tree that neither of us had noticed before. Tied in the branches were blue plastic survey ribbons, fluttering in the breeze.

Epilogue

Delia and Mark

The ecologist cannot remain a voice crying in the wilderness—
if he is to be heard and understood.

—*M. W. Holdgate*

BLUE, BIMBO, AND MOFFET are all that is left of the Blue Pride dynasty that once ruled a long stretch of Deception Valley. When we last saw him, Bimbo was a hefty young male with a scruffy new mane and a wanderlust. He would soon become a nomad, roaming far from Deception Valley in search of a territory and a pride of his own. In the meantime, he and his mother stay together in the pride's old home range, stalking through bushveld and woodlands to the east of the valley near Crocodile Pan. Now and then they meet Moffet, and the three of them hunt or rest together under a shade tree.

Moffet usually stays alone, preying on small animals and birds. He rarely roars, for he has no territory, but now and then he coos softly into the wind—perhaps listening for an answer from his old friend Muffin.

At the end of 1980 old Chary gave birth to three more young, probably sired by a male of the East Side Pride near the game reserve boundary. She and Sassy and their cubs range from twenty to fifty miles east of Deception, near Hartebeest Pans and beyond. In this picturesque area, where dense terminalia and combretum woodlands open out to parkland and rolling grass savanna, the lionesses prey on occasional kudu, duiker, hartebeest, and migrating wildebeest. Now that they have formed new associations with other males and females in these areas, they will probably never return to their old Blue Pride territory near Deception Camp.

Liesa and Gypsy are together near Paradise Pan, where, in 1980 Gypsy gave birth to three cubs. She was a much better mother the second time around, and by the time we left the Kalahari, her new cubs were healthy and growing quickly.

Spicy and Spooky, of the original Blue Pride, joined the Springbok Pan Pride, where Spicy reared her cubs with Happy's family.

Both Rascal and Hombre, who were young male cubs with the Blue Pride when we first met them, were later shot by ranchers near cattle posts just outside the reserve boundary. More than a third of all the lions we tagged or collared were shot by professional hunters, poachers, or ranchers before we left. We believe this mortality, directed mostly at males, is detrimental to the long-term welfare of the population. (See Appendix B for our recommendations concerning the conservation of Kalahari lions.)

Diablo, the dominant male of the Springbok Pan Pride, was ousted by three prime males that we called the All-Stars. He moved about twenty miles west of Deception Valley, where he associates with two young females. Happy, Dixie, Sunny, Muzzy, and Taco, the females of the pride, together with Spicy, Spooky, and two lionesses from other prides, briefly came back to their old stomping grounds during the short rains. However, they are now once again scattered over more than 1200 square miles of sandveld north and south of Deception Valley.

The entire Tau Pride, which had frightened Delia while she was checking for holes on the crude airstrip in Hidden Valley, were shot by ranchers when they left the reserve in the dry season.

*　　*　　*

Pepper has grown into a young adult brown hyena who still visits camp to steal the water kettle, just like her mother, Star.

Patches gave birth to four cubs late in 1980 and moved them into the clan's communal den, where Dusty and Pepper helped feed them. Dusty lost a litter of her own and began nursing Patches' cubs soon after they were brought to the communal den. Chip, the cubs' half brother, assists in rearing them by provisioning and playing with them, but Pippin, their more distantly related male cousin, does not.

Recommendations for the conservation of brown hyenas in the Central Kalahari are in Appendix C.

*   *   *

The rains that fell on Lake Xau in late 1980 temporarily drew the wildebeest back into the reserve after thousands more had perished. But the respite was brief. Apart from a few scattered showers, the drought continued through 1984. The herds still migrate to Lake Xau, which is now completely dry.

The boundaries of the Central Kalahari Game Reserve were drawn at a time when nothing was known about the shifting antelope populations. To conserve these animals, even in their present diminished numbers, a solution must be found for the migrating species.

We wanted the research on the wildebeest to continue, so we requested funds for the Deception camp to operate as a small research station. The Frankfurt Zoological Society agreed to finance the facility and a team of researchers to further investigate the wildebeest problem. Doug and Jane Williamson are following up on our preliminary research on the antelope with a more detailed study of their range ecology. They report that, in 1983 alone, more than 60,000 wildebeest died, just in the area around Lake Xau.

Publicity of the wildebeest issue has stirred considerable interest, and the Botswana government has received communications from all over the world expressing concern for the antelope. We have been told by an official in the Wildlife Department that the Botswana Ministry of Agriculture has agreed to allocate over one million pula to the Department of Wildlife and National Parks for research into the development of alternate water supplies for Kalahari antelope. The government has temporarily accepted the recommendation to freeze the development of settlements on the west shore of Lake Xau, thereby maintaining this migratory corridor for the wildebeest. The Kalahari Conservation Society has been founded in Gaborone, and discussions are continuing on the feasibility of developing water holes for wildlife inside the Central Kalahari Game Reserve.

Unfortunately, a game scout camp has yet to be established at Lake Xau, and poaching and harassment of the migrating antelope continues to be severe, the local people chasing the wildebeest with vehicles, setting dogs on them, and shooting, spearing, and clubbing them to death.

Recommended solutions to the wildebeest problem are described in Appendix A.

* * *

Although we never saw Mox again, we eventually learned that he was working on an ostrich farm at Motopi, a village thirty miles east of Maun on the Boteti River. According to the farmer, Mox often sits around the evening campfire telling stories about Bones, about being treed by the Blue Pride, and about our bungled attempts at radio-tracking Star for the first time. He still drinks beer and occasionally terrorizes the native women of Motopi; he also enjoys the well-earned title, *Ra de Tau*, man of the lions.

We are presently writing and publishing our research results and completing our Ph.D. degree studies at the University of California, Davis. Soon we will return to the Kalahari to continue our studies of Pepper, Dusty, Blue, Sassy, Moffet, and the other animals we knew for seven years.

* * *

We could have stayed in Deception Valley for the rest of our lives, filling up one field journal after another; its mysteries are for us an endless fascination. But such an indulgence would have accomplished little for the Kalahari. We had to process seven years of data, write, and publish our results for science and conservation. Just as important, we had to make the people of Botswana and the rest of the world aware of the wilderness treasures that lie in the Kalahari. None of this could be done from our tent camp.

We had lived through some difficult times in the desert, but the most difficult task of all was leaving Deception Valley.

Early one morning in December 1980, we bounced down the airstrip in Echo Whisky Golf and lifted into the desert sky. Boeing the spring-bok trotted out of the way and the hornbills in camp flitted about the trees. Neither of us could speak as Mark turned the plane north for a short run over the valley. We flew low over the trees where we had done surgery on Bones's broken leg, and over the hyena den, where we could see Pepper, resting beneath an acacia bush. We lingered over the shoulder of East Dune, where Muffin and Moffet had killed Star, and over the small clearing on Cheetah Hill, where Captain and Mate had raised Hansel and Gretel. Then we turned south onto a heading of 163 degrees and flew away from Deception toward another world.

APPENDIXES
NOTES
REFERENCES
ACKNOWLEDGMENTS
INDEX

APPENDIX A

Conservation of Migratory Kalahari Ungulates

IN DROUGHT, Kalahari ungulates, including wildebeest and hartebeest, migrate across the Central Kalahari Reserve and beyond its protection. The herds move toward the waters of Lake Ngami, the Okavango River delta, Lake Xau, and the rivers that connect these natural reservoirs. Besides water to drink, the antelope may also need more nutritious forage than can be found in the reserve during these dry times. Their migrations take them into areas where human settlement has increased over the last twenty years, and now there is direct competition between man and wildlife for such limited resources as water and grassland. In addition, fences erected to control foot-and-mouth disease (FMD) block the migrations and channel antelope populations into only a very small portion of previously available riverine habitat (see maps on pages 297 and 299).

Finding a solution to the conflict between people and the migrating desert antelope is not going to be easy. However, without immediate action, these ungulate populations may not survive the Kalahari's periodic droughts. The following recommendations deserve consideration:

1. Detailed research must be undertaken to qualify and quantify the role of cordon fences in the control of FMD. Little is known about how the disease pathogen is transmitted from animal to animal and, despite extensive experimentation, it has never been conclusively proven that wild ungulates can transmit FMD to healthy domestic stock.[1] Thus, no one can be certain that wild ungulates are a reservoir of infection that is held in check by fences.

Since the erection of the fences in the early 1950s, Botswana has experienced more than nine major outbreaks of FMD, which have spread over large sections of the country irrespective of the disease-control fences. In part, this has been due to the fact that FMD is caused by three different strains of virus, each of which may cause an outbreak in a different area almost simultaneously when environmental conditions are conducive. There is also good evidence that the virus may be spread through the air over considerable distances[2] or in damp soil clinging to vehicles,[3] which cannot easily be constrained by fences. So, there appears to be as much circumstantial evidence to suggest that the fences do not help control the development and spread of the disease as there is evidence to suggest that they do. What we do know, is that they have devastated migrating antelope populations all over Botswana.

However, taking down fences may not be a viable long-term solution to the wildebeest problem. There is great pressure for human settlement along the shores of Botswana's few rivers and lakes, so that even if the fences were removed, Kalahari antelope would probably be blocked from water in the near future. The government has never seriously considered proposals to develop tourism and other wildlife-based industries in these areas, which would probably raise the living standards of local people and at the same time preserve a great natural heritage. To prevent yet another wildlife disaster, research is needed that would determine how new fences that are being planned will affect wildlife.

2. Botswana should consider alternatives to fencing for disease control, the most obvious being a modern and efficient vaccination program. In the past, often fewer than fifty percent of all cattle in infected areas have been vaccinated during an outbreak. And, at least on one occasion, outdated vaccine was used. A new vaccine is currently under development in the United States that will provide cattle a lifelong immunity to foot-and-mouth disease.[4]

There are other alternatives that would reduce the need for fencing and make the beef-export industry less subject to the vagaries of FMD: the development of more meat-canning plants in areas where FMD is endemic; stricter penalties, such as confiscation of cattle, for movement of domestic stock from one area to another during FMD outbreaks; and patrols using light aircraft instead of fencing along borders of quarantined areas.

3. One partial solution to the wildebeest problem would be to extend a portion of the eastern boundary of the Central Kalahari Game Reserve to include the area around Lake Xau. If the government will not consider this, the area between the lake and the western shore of the reserve at least should be set aside as a permanent corridor for migrations, and further development of villages and ranches in this area should be frozen. The corridor would have to be maintained free from settlement even in years when migrations did not occur, but it could be used to support tourism, safari hunting, or other wildlife industry.

4. A permanent game scout camp should be established in the Lake Xau area to prevent the poaching of migratory antelope.

5. We do not highly recommend the use of boreholes in the reserve as a means of providing water for the ungulates. However, if such action is taken certain considerations should be made.

a. The abundance of a grass species gives no assurance of its quality as a forage for antelope. Research is needed to determine if graze and browse plant species in the Central Kalahari Reserve contain enough protein and mineral nutrients to sustain antelope during drought. If they do not, merely providing migratory antelope with water will not ensure their survival.

b. Artificial water supplies will tend to focus mobile antelope populations and, unless they are carefully managed, will result in regional overgrazing and desertification.

c. Artificial water supplies will also attract people, who will settle nearby with livestock and crops. This would prevent access for wildebeest and aggravate the already sensitive problem of what to do about such growing settlements as Xade, which has developed around a borehole inside the game reserve.

6. Pan and fossil riverbeds constitute only about eleven percent of the Central Kalahari Game Reserve, but their soils contain essential minerals and the grasses have a more favorable protein/fiber ratio than do sandveld grasses. Even though these habitats make up a small portion of the total range, fifty-seven percent of all ungulates counted inhabited them during the wet season in years when rainfall was ten inches or more. Our results show that these fossil riverbeds are important habitat for wildlife and should be protected.

7. The European Economic Community, and in particular Britain, which imports most of Botswana's beef, heavily subsidizes Botswana's cattle industry. EEC officials in Brussels rigidly insist on the veterinary cordon fences, despite the fact that no research has ever been done to demonstrate their efficacy in controlling foot-and-mouth disease; and regardless of the fact that in the thirty-odd years since the fencing program was begun, it has been directly responsible for the deaths of at least a quarter of a million wildebeest and untold thousands of other antelope. "Easy" money from these subsidies encourages overstocking that leads to overgrazing and desertification—and to further displacement of wild antelope populations. Many native people are becoming less able to meet their own needs for food and clothing, traditionally obtained from the products of wild animals, and they are growing very dependent on foreign-aid programs for subsistence. Meanwhile, the few wealthy cattle owners, those who own most of the country's domestic stock, are growing richer.

The European Parliament should revise this policy immediately. Foreign subsidies should encourage Botswana to develop wildlife industries, such as tourism, photographic and hunting safaris, game ranching, and others. The necessary funds could come from the annual beef "rebate" to Botswana, currently running at more than 14,500,000 British pounds. Wildlife industries are more sustainable over the long term because they are less likely to contribute to desertification, less expensive to maintain, and more likely to raise the standard of living of the general populace.

APPENDIX B

Conservation of Kalahari Lions

1. The Predator Control Act should be amended to require that before ranchers destroy predators, they must provide proof that the predators have actually killed their stock. An immediate investigation should be made by the appropriate authority, and if sufficient proof of damage exists, the destruction of the predator concerned should be the responsibility of the government authority and not the rancher. The official should confiscate the remains of the predator, the sale of which should be used to maintain a fund for the remuneration of ranchers for damage to stock by predators. Skulls of all predators should be aged by the Department of Wildlife for much-needed data on population structure and dynamics.

2. Based on our density figures, we recommend that Kalahari lion-hunting quotas be reduced by one-half and the price of a lion-hunting license for safari hunters be doubled. The skulls of all lions taken by safari hunters should be aged by trained personnel of the Department of Wildlife. (We remind the reader that this book is not a scientific treatment of our results. Complete and detailed accounts of density figures, range sizes, and habitat utilization will be published in appropriate journals.)

3. Safari hunters should be required to buy a lion-hunting license only if they have already shot a lion. Many hunters are shooting very young lions simply because they have been required to purchase a license in advance and cannot find older males. Because of this, hunting pressure on male lions does not diminish even when there are no older lions.

4. Laws against baiting and trapping of predators should be more strictly enforced. This practice results in the indiscriminate killing and maiming of animals that have not damaged domestic stock. Baiting is also an expedient method for drawing predators out of reserves and parks by poachers.

5. The Department of Wildlife should have two to three times more men in the field to improve enforcement of existing laws.

6. As discussed in detail in Appendix A, one of the most important measures for the conservation of lions and other wildlife in the Kalahari, is to readjust the

Central Kalahari Game Reserve boundaries to include at least the western shore of Lake Xau. Though, as we have seen, Kalahari lions can live for months without drinking water, many of them have to leave the protection of the game reserve to locate adequate prey during prolonged dry seasons and drought.

7. Until the above measures are taken, permission to shoot lions outside game reserves should be suspended in times of drought, when many of them must leave the protection of the reserve in order to survive.

8. A tourist facility should be developed at the southern tip of Lake Xau. Such a facility could include accommodation at a lodge overlooking the lake, canoe trips through the papyrus to bird blinds on the lake, trips by canoe up the river for fishing and game viewing, photographic safaris into the Makgadikgadi Pans Game Reserve and the fossil rivers of the proposed Central Kalahari National Park, and game-viewing flights from a local airstrip. The Lake Xau facility could be utilized by tour operators as part of a packaged deal for tourists to visit the Kalahari, the Makgadikgadi, and the Okavango. It would greatly stimulate the local and regional economy, while at the same time conserving and advertising a unique part of Botswana that is still relatively unknown.

APPENDIX C

Conservation of Brown Hyenas

1. Because they are scavengers in a semiarid ecosystem where carrion is often limited, brown hyenas naturally occur in small numbers. The species is seriously threatened by decreasing habitat, as more and more Kalahari wilderness is used for grazing cattle. The hyenas are commonly shot and trapped in Ghanzi, Tuli, and Nojani, and in most ranching areas. However, the threat they pose to domestic stock is probably exaggerated. With one or two exceptions, the few times we observed them hunting in Deception Valley, they never pursued anything larger than a rabbit. Even at the beginning of the rainy seasons, when springbok dams synchronously birthed their lambs, we never saw a brown hyena kill a springbok fawn. When one is found feeding on the remains of a domestic animal, it cannot be assumed that it is responsible for the kill. In many cases—perhaps in most— a brown hyena discovers or appropriates a carcass after a prey has been killed by other predators. Some ranchers in South Africa have begun to allow brown hyenas on their rangeland; the public must be educated to the fact that this species of hyena is predominantly a scavenger and does not usually endanger domestic herds.

2. Predators on which brown hyenas depend heavily for food, especially lions and leopards, are being shot at increasing rates by ranchers, poachers, and hunters. Conservation of major predator communities is essential in maintaining the present brown hyena population density in the Central Kalahari.

3. Veterinary cordon fences are killing off thousands of antelope that otherwise would represent a lasting food resource for brown hyenas and other Kalahari predators. Addressing this problem is critical to the conservation of the brown hyena.

4. Since the brown hyena is of little real threat to livestock, and since it is an endangered species, it should not be included in Botswana's Predator Control Act. Proof of damage to livestock should be required of ranchers before they are issued special depredation permits to kill brown hyenas.

APPENDIX D

Latin Names of the Mammals, Birds, and Snakes Mentioned in the Text

MAMMALS

aardvark	*Orycteropus afer*
aardwolf	*Proteles cristatus*
cape fox	*Vulpes chama*
caracal	*Felis caracal*
cheetah	*Acinonyx jubatus*
bat-eared fox	*Otocyon megalotis*
brown hyena	*Hyaena brunnea*
duiker, Grimm's	*Sylvicapra grimmia*
eland, cape	*Taurotragus oryx*
gemsbok	*Oryx gazella*
hartebeest, red	*Alcelaphus busalephus*
honey badger (ratel)	*Mellivora capensis*
jackal, black-backed	*Canis mesomelas*
kudu, greater	*Tragelaphus strepsiceros*
leopard	*Panthera pardus*
lion	*Panthera leo*
meerkat	*Suricata suricatta*
mongoose, slender	*Herpestes sanguineus*
porcupine	*Hystrix sp.*
rabbit, Crawshay's	*Lepus crawshayi*
serval cat	*Felis serval*
springbok	*Antidorcas marsupialis*
spring hare	*Pedetes capensis*
spotted hyena	*Crocuta crocuta*
squirrel, striped ground	*Xerus erythropus*
steenbok	*Raphicerus campestris*

warthog	*Phacochoerus aethiopicus*
wild cat	*Felis libyca*
wild dog	*Lycaon pictus*
wildebeest	*Connochaetes taurinus*

BIRDS

bulbul, red-eyed	*Pycnonotus nigricans*
bustard, kori	*Otis kori kori*
cormorant, reed	*Phalacrocorax africanus*
eagle, tawny	*Aquila rapax*
finch, scaly feathered	*Sporopipes squamifrons*
fish eagle	*Haliaeetus vocifer*
flycatcher, Marico	*Bradornis mariquensis*
goose, spur-winged	*Plectropterus gambensis*
hoopoe, scimitar-billed	*Rhinopomastus cyanomelas*
hornbill, grey (in Maun)	*Tockus nasutus epirhinus*
hornbill, yellow-billed (in camp)	*Tockus flavirostris leucomelas*
kestrel, rock	*Falco tinnunculus*
kite, black-shouldered	*Elanus caeruleus*
kite, yellow-billed	*Milvus aegyptius*
korhaan, black	*Eupodotis afra*
lily-trotter (African jacana)	*Actophilornis africanus*
night jar	*Caprimulgus rufigena*
ostrich	*Struthio camelus*
owl, pearl-spotted	*Glaucidium perlatum*
plover, crowned	*Stephanibyx coronatus*
shrike, crimson-breasted	*Laniarius atro-coccineus*
stork, white	*Ciconia ciconia*
stork, white-bellied	*Ciconia abdimii*
teal, hottentot	*Anas hottentota*
tit-babbler	*Parisoma subcaeruleum*
vulture, lappet-faced	*Torgos trachelioutus*
waxbill, violet-eared	*Granatina granatina*
weaver, masked	*Ploceus velatus*

SNAKES

boomslang	*Dispholidus typus*
cobra, Anchieta's	*Naja haje anchieta*
mamba, black	*Dendroaspis polylepis polylepis*
puff adder	*Bitis arietans*
spitting cobra (Black-necked)	*Naja mossambica*

Notes

4 Cry of the Kalahari
    1. Moehlman, pp. 382–83.
    2. Trivers, pp. 249–64.

5 Star
    1. Kruuk, p. 126.
    2. Owens and Owens, 1979a, pp. 405–8.

10 Lions in the Rain
    1. Schaller, p. 33.

12 Return to Deception
    1. Mills, 1978, pp. 113–41.
    2. Skinner, 1976, pp. 262–69; Mills, 1976, pp. 36–42.
    3. Macdonald, pp. 69–71.

14 The Trophy Shed
    1. Bertram, p. 59.

17 Gypsy Cub
    1. Bygott, Bertram, and Hanby, pp. 839–41.
    2. Bertram, p. 59.

18 Lions with No Pride
    1. Schaller, pp. 34–42.
    2. Schaller, p. 38.

20 A School for Scavengers
    1. Owens and Owens, 1979b, pp. 35–44.

21 Pepper
    1. Hamilton, pp. 1–52.
    2. Dawkins, pp. 95–131.
    3. Owens and Owens, 1984, pp. 843–45.

25 Black Pearls in the Desert
    1. Young, Hedger, and Powell, pp. 181–84.
    2. Hedger, p. 91.
    3. Silberbauer, pp. 20–21.
    4. Child, pp. 1–13.
    5. Silberbauer, p. 22.

6. Owens and Owens, 1980, pp. 25–27.
Epilogue
1. Williamson, in press.
Appendix A Conservation of Migratory Kalahari Ungulates
1. Condy and Hedger, pp. 181–84.
2. Hedger, p. 91.
3. Siegmund, p. 255.
4. Abelson, p. 1181.

References

Abelson, P. H. 1982. Foot-and-mouth vaccines. *Science* 218: 1181.

Bertram, B. C. R. 1975. The social system of lions.
Scientific American 232: 54–65.

Bygott, J. D., B. C. R. Bertram, and J. P. Hanby. 1979. Male lions in large
coalitions gain reproductive advantages. *Nature* 282: 839–41.

Child, G. 1972. Observations on a wildebeest die-off in Botswana *Arnoldia*
(Rhodesia) 5: 1–13.

Condy, J. B., and R. S. Hedger. 1974. The survival of foot and mouth disease
virus in African buffalo with nontransference of infection to domestic cattle.
Res. Vet. Sci. 39(3): 181–84.

Dawkins, R. 1976. *The Selfish Gene*. New York: Oxford University Press.

Hamilton, W. D. 1964. The genetic evolution of social behavior, I, II. *J. Theor.
Biol.* 7: 1–52.

Hedger, R. S. 1981. Foot-and-Mouth Disease. In *Infectious Diseases of Wild
Mammals*, ed. John Davis et al. Ames: Iowa State University Press.

Kruuk, H. 1972. *The Spotted Hyena*. Chicago: University of Chicago Press.

Macdonald, D. W. 1979. Helpers in fox society. *Nature* 282: 69–71.

Mills, M. G. L. 1976. Ecology and behaviour of the brown hyena in the Kalahari
with some suggestions for management. *Proc. Symp. Endangered Wildl.
Trust* (Pretoria) pp. 36–42.

Mills, M. G. L. 1978. Foraging behavior of the brown hyena (*Hyaena brunnea*
Thunberg, 1820) in the southern Kalahari. *A. Tierpschol* 48: 113–41.

Moehlman, P. 1979. Jackal helpers and pup survival. *Nature* 277: 382–83.

Owens, D., and M. Owens. 1979a. Notes on social organization and behavior
in brown hyenas (*Hyaena brunnea*). *J. of Mammalogy* 60: 405–08

Owens, D., and M. Owens. 1979b. Communal denning and clan associations in
brown hyenas of the Central Kalahari Desert. *Afr. J. of Ecol.* 17: 35–44.

Owens, D., and M. Owens. 1984. Helping behaviour in brown hyenas. *Nature*
308: 843–45.

Owens, M., and D. Owens. 1980. The fences of death. *African Wildlife* 34: 25–
27.

Schaller, G. B. 1972. *The Serengeti Lion*. Chicago: University of Chicago Press.

Siegmund, O. H., ed. 1979. *The Merck Veterinary Manual*. Rahway, N.J.: Merck & Co.

Silberbauer, G. 1965. *Bushmen survey report*. Gaborone: Botswana Government Printers.

Skinner, J. 1976. Ecology of the brown hyena in the Transvaal with a distribution map for southern Africa. *S. Afr. J. of Sci.* 72: 262–69.

Trivers, R. L. 1974. Parent-offspring conflict. *Am. Nat.* 14: 249–64.

Williamson, D. T. 1984. More about the fences. *Botswana Notes and Records*. In press.

Young, E., R. S. Hedger, and P. G. Howell. 1972. Clinical foot and mouth disease in the African buffalo (*Syncerus caffer*). *Ondersterpoort J. vet res.* 39(3): 181–84.

Acknowledgments

WITHOUT THE ASSISTANCE of many people, our research and the writing of this book would not have been possible. We have not been able to mention in the text all of those who believed in us and helped us throughout the years. We deeply regret this and wish them to know that we will always remember their contributions.

Our very special thanks to the Friends of the Animals and the Frankfurt Zoological Society, under the direction of Dr. Richard Faust, who gave us an airplane and other sophisticated equipment essential for working in such a remote area. The society financed the project from 1977 to 1983 and is continuing its generous support. The personal interest and encouragement of Dr. Faust and Ingrid Koberstein, his assistant, kept us going when times were tough.

We are also deeply grateful to the National Geographic Society for our first grant, and to the Netherlands branch of the World Wildlife Fund and the International Union for the Conservation of Nature for their generous financial assistance. H.R.H. Prince Bernhard of the Netherlands helped us secure funding and was influential in our efforts to publicize the Kalahari antelope issue.

Our sincere appreciation also to the Okavango Wildlife Society for a grant that allowed us to purchase our first radio telemetry equipment and to continue our research at a critical time. We are especially grateful to Chairman Hans Veit, and to Kevin Gill, Barbara Jeppe, and Heinz and Danny Guissman for their support.

We owe much to Al and Marjo Price and their family, who, through the California Academy of Science, contributed generously toward the operation of our project's airplane.

The late Dr. Beatrice Flad, a warm and sensitive person who was at the same time tenacious in her defense of wildlife, gave her life for conservation. We appreciate her financial support during the writing of our results.

Thanks also to Dr. and Mrs. Max Dinkelspiel for their personal contribution toward a trip home when we badly needed to see our families.

We are very grateful to the office of the president of Botswana and to the Department of Wildlife and Tourism for permission to conduct our research in the Central Kalahari Game Reserve, for accepting our criticisms, and for considering our recommendations on the conservation of the Kalahari. We realize that it is not always easy to resolve the competition between man and wildlife, and we thank those officials who are sincerely attempting to do this for the betterment of all.

Besides our sponsors, there are people who contributed greatly to the running of the project in crucial ways: our lasting appreciation and warm thanks to Kevin Gill who graciously gave us the full use of his home whenever we were in Johannesburg and for treating us to mellow evenings filled with good wines, fine music, and stimulating conversation. Our thanks to Captain Roy Liebenberg for teaching Mark to fly, and for his assistance with radio equipment; to Roy, his wife, Marianne, and their children, and to Bruno and Joy Bruno for allowing us to be a part of their families on many occasions. Dave Erskine and Rolf Olschewski hauled thousands of gallons of aviation fuel through the desert to our camp. Dave also made windsocks for the airstrip, helped us with photography, made logistical arrangements with mine personnel, and made critical observations on the brown hyenas on occasions when we had to be away. Bobby and Mary Dykes (Delia's twin brother and his wife) gave us unending support in printing, sorting, and cataloguing our photographs, in shipping spare parts for our airplane from the United States, and in helping with project correspondence. They even brought our lion and brown hyena radio collars to Africa in their suitcase.

There are still many frontiers in southern Africa where people must depend on one another—sometimes even for survival. In Bulawayo, Zimbabwe, the Archers—Geoffry, Ruth, Margaret, and Jean—kindly gave us the use of their home, wonderful meals, and innumerable cups of tea, and a "bush shower" that helped equip us for life in the *bundu*. We thank the Tom Lukes and Graham Clarks, also of Bulawayo, for their warm friendship; Mr. and Mrs. White of Salsbury (Harare) for their hospitality; Ted Matchel and Ian Salt of the Zimbabwean Department of Wildlife and National Parks for advice on prospective study areas.

In Gaborone, Tom Butynski and Carol Fisher Wong put us up for several weeks while we were outfitting for our reconnaissance through Botswana. And over the years, whenever we arrived from the bush, Pietman and Marlene Henning of Gaborone always took us in, fed us, and gave us rest and friendship.

For seven years, whenever we came into the village of Maun for supplies, our friends provided us with everything from truck spares to parties and advice. We will never be able to thank them enough, for theirs was the truest expression of the pioneer spirit that still lives in small villages on the edge of the bushveld. Our thanks to Richard and Nellie Flattery, Pete Smith, Eustice and Daisy Wright, Mark Muller, Dave Sandenberg, Hazel Wilmot, Toni and Yoyi Graham, Diane Wright, Dolene Paul, Dad Riggs, Cecil and Dawn Riggs, John and Caroline Kendrick, Larry and Jenny Patterson, P. J. and Joyce Bestelink and Kate and Norbert Drager. Special thanks to Phyllis Palmer and Daphne Truthe for kindly reading messages and telegrams to us on shortwave radio.

There is a group of people in Maun who deserve special recognition: the professional hunters, especially of Safari South. When we first arrived with our back packs and dilapidated old Land Rover to study wildlife, the hunters managed to conceal the doubts they surely had about us and from the very beginning made us feel welcome. It would have been very difficult for us to begin our project without their advice and never-ending support. They gave us our only radio and talked to us often during the hunting season—our only communication with the outside world; they towed our truck when it broke down near the village; they loaned us the use of their airplane for game censusing; they gave us tents, chairs, tables; they flew Delia into Maun when she was very sick with malaria and did hundreds of other favors. Our sincere thanks to Lionel Palmer, Dougie Wright, Willie Engelbrecht, Bert Miln, John Kingsley Heath, Simon Paul, Wally Johnson, Junior and Senior; Tommy Friedkin, owner of Safari South; manager Charles Williams, and David Sandenberg. Though we did not condone all of their hunting practices, most of the hunters remained our friends.

We had many illuminating discussions with Steve Smith, Curt Busse, and Carol and Derrick Melton, who gave us a standing invitation to visit their baboon research camp in the Okavango delta. The evenings with the baboons scampering around us, the Christmas turkey that was cooked in a washbasin, and our swims in the hippo lagoon will never be forgotten.

We thank Dr. W. J. Hamilton III for accepting us as graduate students, for his support and encouragement while we were in the field, for his great patience with us as we write up our results, and for all the laughs both he and Marion, his wife, have given us when we needed them.

De Beers Consolidated Botswana allowed us to purchase aviation fuel from their stocks and to buy supplies at the mine store.

We would like to thank Lake Price and Warren Powell, who assisted us in the field for three months in 1979. They never complained about the long hours over the plotting board, the lack of water, or the rats and snakes in their shredded tent; or about the "turd patrol" (collecting, crushing, and sifting lion and hyena scats). Their contributions and companionship were invaluable.

We are grateful to Gordon Bennett for the generous use of his airplane and company facilities on several occasions; to Cliff and Eva Thompson, Hans Pearson, and Phil Parkin for donations of equipment. For their gracious hospitality in South Africa, we also thank Frank Bashall, Schalk Theron, Allistar and Maureen Stewart, Willy and Linda Vandeverre, and Liz and Jane Cuthbert.

To our families we owe special appreciation for their encouragement through the years. Delia's mother and late father sent endless "care" packages. Mark's father and Delia's mother brought much needed cheer to camp on their visits. Again, we are sorry that the many interesting details of their visits could not be included in the book.

Our good friends Bob Ivey and Jill Bowman have been involved with our project from the very beginning. Their encouragement and enthusiasm greatly inspired us, and they read and commented on the entire manuscript. We owe them much more than they will ever realize.

We are also grateful to Dr. Joel Berger, Carol Cunningham, Dr. W. J. Hamilton III, Dr. Murray Fowler (Chapter 25), Helen Cooper, and Dr. Bob Hitchcock for their constructive comments on the manuscript. Helen Cooper (Delia's sister) also helped with the epigraphs. We appreciate the encouragement, assistance, and lasting patience of our editors Harry Foster (Houghton Mifflin) and Adrian House (Collins) and the support and encouragement of Peter Matson and Michael Sissons, our agents.

Doug and Jane Williamson took over the wildebeest research and the operation of the camp during a four-year drought. In our absence they have made lasting contributions to science and the conservation of Kalahari wildlife under extremely difficult conditions.

We owe special thanks to Mox Moraffe, who assisted us for three and a half years in the desert. In his own quiet and humorous way, he gave us support and knowledge.

Mr. and Mrs. Langdon Flowers of Thomasville, Georgia, kindly invited us to stay in "Breezinook," their log-and-stone house on Greybeard Trail in Montreat, North Carolina, while we wrote much of this book. We are so very grateful for the peace and inspiration we found in this retreat and for the opportunity to get to know them better.

And last, we would like to thank Dr. Joel Berger and Carol Cunningham for teaching us what true friendship can be again, after seven years alone.

Index

Abbott, Winston O., 164
Acacia, 20, 24, 83
Antelope, 23, 24–25, 26, 60, 137,
   149
  conservation of, 315, 319–21
  destruction of, 300–301, 302–303
  migration during drought, 293–
    301, 307–308
  See also names of specific antelope

Baker, Dr. Rolin, 246
"Bandit" (wild dog), 23, 136, 137–
  38
Berghoffer, Bergie, 32–35, 42–43,
  46–47, 251–52, 302
Bernhard, Prince of the Netherlands,
  304–306
"Big Red" (hornbill), 217
Bilharzia (disease), 99
Biltong (jerky), 85–86
"Bimbo" (lion), 286–90, 310, 311–
  12, 313
Birds, 17, 21–22, 64–65, 90, 93,
  94–95, 217, 245–46, 280, 305
  See also names of specific birds
Black mambas, 62–63, 93, 107
"Blue" (lioness), 2, 115, 118, 128,
  129, 146, 156, 157, 161, 180,
  216, 220–21, 310, 311, 312,
  313
  in drought, 284–91
  as mother, 286–91

  with sarcoptic mange, 286, 288–89
Blue Pride (lions), 147, 157–63,
  216–44
  in camp, 1–2, 143–46, 155, 189–
    90, 218, 235–36
  communication in, 157–60
  in drought, 276–78, 284–91
  after drought, 309–12
  in dry season, 237–38, 242–44
  and females of other prides, 238–
    41, 244, 285, 313–14
  hunting by, 128–29, 161–62,
    233–34, 237–38
  individual lions in, 115–16, 216
  migrations of, 130, 180, 236–37
  play in, 128
  rearing of offspring in, 226–34
  return in rainy season, 179–80
  rivalry with Springbok Pan Pride,
    223–25
  territory of, 156, 198–99, 223–25,
    237–38, 242–43
  See also "Bones"; Lions
"Bones" (lion), 2, 143, 146, 158
  in camp, 189–90
  death of, 194–97
  marking by, 159
  Owenses' rescue of, 121–27
  position in Blue Pride, 128–29,
    156, 162, 179–80
Boomslangs, 93
Boteti River, 14, 15–16, 17, 25, 99

Botswana:
  fencing in, 295–96, 298, 301–303
  fire in, 36–47
  hunting laws in, 197–99
  Owenses' arrival in, 6–8
  poachers in, 298–300
  Rhodesian conflict and, 171–73
"Brother" (lion), 120–21
Brown hyenas. *See* Hyenas, brown
Bushmen, 2, 17, 26, 42, 131–32,
  302
Bush warbler, 246–47
Byron, George Gordon, 83

"Captain" (jackal), 37
  competition with hyenas, 68
  as hunter, 29–30, 51–52, 53–55,
  62–63
  killing of snake by, 62–63
  and "Mate," 53–55, 59, 60, 61–
  63, 65–67
  in rearing of young, 61–62, 65–67
  scavenging by, 28–29, 50, 66
Cats. *See names of specific cats*
Central Kalahari Game Reserve, 17,
  315
  disease control fence, 295, 301–
  303
  drought on, 293–301, 307
  See also Deception Valley;
  Kalahari
"Chary" (lioness), 2, 115, 116, 117,
  118, 119, 120, 128, 129, 146,
  156, 161, 180, 216, 237, 310,
  313–14
  as mother, 232–34, 242, 284
Cheetahs, 5–6, 26, 135, 136
"Chief" (hornbill), 90, 217, 305
Child, Dr. Graham, 302, 303
"Chip" (hyena), 260, 261, 268, 272,
  273, 314
Cobras, 92
"Cocoa" (hyena), 254, 255, 256,
  257–58, 260, 261, 262, 263,
  265–67, 269–70, 271, 272, 273,
  290, 312

  adoption by community, 266–70
Combretum, 24
Cormorants, 17
Crocodiles, 99

Darwin, Charles, 253
Deception Valley, 3, 17
  drought in, 177–78
  fire in, 43–46
  history of, 24
  mining expedition in, 279–83
  Owenses' arrival in, 19–26
  vegetation in, 24–25
  See also Kalahari
Derwood, Gene, 279
"Diablo" (lion), 239, 314
Dickinson, Emily, 275
"Dixie" (lioness), 153, 244, 314
Dogs. *See* Wild dogs
Drager, Kate, 56
Drager, Loni, 56
Drager, Norbert, 37, 56
Drought, 263–64, 271, 275–76, 298,
  315
  brown hyenas in, 290–91
  end of, 309–10
  lions in, 276–78, 284–91
  wildebeest in, 292–301, 307–308
Ducks, 100
Duiker, 25
"Dusty" (hyena), 260, 261, 263, 266,
  267, 268, 269, 273, 314

Echo Whisky Golf (airplane), 200–205
  arrival of, 201
  initial flight of, 202–204
  problems with, 202–205, 210,
  214–16, 230–32
  in radio-tracking of lions, 208–22,
  225
  weather and, 214–16, 231–33
Egrets, 17, 100
Eland, 25

Faust, Dr. Richard, 200, 201, 303,
  306–307

Fencing, 295, 298, 301–303
Finches, 91
Fire, 36–47
  animals in, 49–50
  Owenses in, 43–46
Flattery, Nellie, 165
Flattery, Richard, 165, 247
Flycatchers, 90, 92, 217, 246, 280
Foot-and-mouth disease (FMD),
  fencing to control, 295, 302–303
Foxes:
  bat-eared, 28, 48, 49, 50, 60, 71
  cape, 71
Francistown, 9, 101, 172
Frankfurt Zoological Society, 201,
  248, 249, 250, 304, 307
Friends of Animals, The, 304
Frostic, Gwen, 222

Gaborone, 6–7
Gazelles, 22
Geese, 17, 100
Gekkos, 217
Gemsbok, 25, 49, 85, 137, 141
Ghanzi, 100
Gilkyson, Terry, 284
Giraffe, 25, 26, 83, 141, 161–62,
  295
Graham, Tony, 248
Graham, Yoyi, 248
"Great Thirst," 2
"Gretel" (jackal), 61–62, 66–67
Guinea fowl, 64–65
"Gypsy" (lioness), 2, 116, 118, 129,
  145, 146, 157, 161, 180, 216,
  239, 285, 314
  as mother, 226–29, 232

"Hansel" (jackal), 61–62, 66–67
"Happy" (lioness), 153, 238–41,
  243, 285, 314
Hartebeest, 25, 49, 71, 137
  conservation of, 319–21
"Hawkins" (hyena), 76, 77, 78, 80,
  81–82
Hawks, 94

Hitchock, Dr. Bob, 302
Holdgate, M. W., 313
"Hombre" (lion), 116, 128, 129, 146,
  162, 314
Hornbills, yellow-billed, 90–91, 94–
  95, 217, 280, 305
Howard, Heather, 251
Howard, Mike, 251
Hunters, 10–11, 191–99, 300–301,
  314
Hyenas, brown, 27, 69–82, 93, 253–
  74
  adoption among, 265–69
  cheetahs and, 135, 136
  communication of, 80–81
  competition with jackals, 51, 55,
  67–68, 72
  competition among males, 81–82,
  186–87
  conservation of, 321
  dens of, 188, 254, 255–56, 258
  in drought, 185–86, 290–91
  after drought, 311, 314
  eartagging of, 76–77
  greeting, 75
  helping behavior, 258–59, 260–
  61, 267–70
  humans and, 70–71
  identifying sex of, 76, 77–78
  leopards and, 133–35, 262, 273
  lions and, 135, 136, 219, 257,
  264, 273, 290–91
  muzzle-wrestling of, 79, 80
  neck-biting of, 78–80, 186–87
  "pasting" by, 74–75, 81, 262
  play of, 262, 269
  rearing of offspring, 253–63, 265–
  69
  reproduction of, 187
  as scavengers, 69, 71–73, 74,
  127–28, 185
  social behaviors of, 74–76, 78–82,
  253–64
  social order of, 75, 80, 81–82, 188
  teeth of, 73
  tracking of, 71, 126–27, 128, 132,

Hyenas (*cont.*)
  180, 182–88, 217, 253
  wild dogs and, 135–36, 261

Insects, 9, 43, 50, 60, 92
"Ivey" (hyena), 79, 80, 81–82, 135,
  184, 186–87

Jackals, black-backed, 22, 34, 218
  brown hyena and, 51, 55, 67–68,
    72
  in camp, 28–30, 94
  care of offspring, 61–62, 66–67
  hunting by, 51–55
  Kalahari vs. Serengeti, 51, 55
  killing snake, 62–63
  leopards and, 66
  lions and, 38–39
  mating behavior of, 53, 67
  at night, 28, 29–30
  scavenging by, 28–29, 50, 51, 66
  territory of, 58–59
  tracking by Owenses, 51–55
Johannesburg, 249–51

"Kabe" (lioness), 244
Kalahari:
  climate of, 24, 30, 36, 37, 57–60,
    87, 89, 141–42, 177–78, 179,
    213–15, 230–32, 236, 241,
    263–64, 271, 275–76, 293–94,
    309–10, 315
  drought in, 263–64, 271, 275–76,
    293–301, 309–10, 315
  dry season in, 36, 57–58, 141–42,
    177–78, 236, 241
  fire in, 36–47
  lack of exploration in, 2
  night in, 26–28, 29–30
  rainy season in, 59–60, 142–43,
    179, 230–32, 236
  vegetation in, 24–25, 48–49, 64,
    83–84, 293
Kalahari Conservation Society, 315
Kalahari lions. *See* Lions
Kays, Kenny, 103

Kays, Ronnie, 103
Kestrel, 22
Kin selection, 267
Kipling, Rudyard, 36, 99
Korhaans, 28, 49
Kori bustard, 54–55, 132
Kruuk, Hans, 79
Kudu, greater, 25
Kuki foot-and-mouth disease, 295,
  301–303
Kunitz, Stanley, 195, 235, 245

"Laramie" (lizard), 91, 217
Lear, Edward, 5, 171
Leopards, 49, 93–94
  hyenas and, 133–35, 262, 273
  jackals and, 66
Leopold, Aldo, 208, 292
Liebenberg, Marianne, 200
Liebenberg, Captain Roy, 200–201,
  202, 205–06
"Liesa" (lioness), 116, 118–20, 129,
  146, 161, 216, 238, 285, 314
Limpopo River, 295, 298
Lions, 72, 73, 144–63, 208–44
  bonding of males, 120–21, 221
  in camp, 1–2, 38–40, 93, 96–97,
    111–13, 143–46, 155, 189–90,
    218, 235–36
  communal suckling, 227, 285
  communication among, 157–60
  competition among males, 156,
    162, 220–21
  conservation of, 322–23
  coordinated hunting, 128–29, 160–
    62
  darting of, 114–21, 122, 154, 212
  after drought, 310–14
  in drought, 276–78, 284–91
  in dry season, 237–38, 242–44
  females mixing with other prides,
    238–41, 244
  in fire, 49
  humans and, 1–2, 38–40, 96–98,
    111–13, 143–46, 155, 189–90,
    211–12, 235–36, 259

hunting by, 160–62, 233–34, 237–38, 285–86, 289–91
hunting of, 194–99, 314
hyenas and, 135, 136, 219, 257, 264, 273, 290–91
identification at night, 27
Kalahari vs. Serengeti, 155–56, 160, 198, 226–27, 239
makeup of pride, 155–56
males not with females, 276
migrations of, 130, 180, 198–99, 236–37
play of, 129, 156, 233
range size increase, 242, 286
rearing of offspring, 226–34, 285–91
return in rainy season, 60, 143, 179
rivalry between prides, 223–25
roaring of, 137–38
sarcoptic mange in, 288–89
scent-marking by, 159–60
territory of, 155, 198–99, 237–38, 242–43
tracking of, 147–55, 198–99, 208–44
Lizards, 91, 217
"Lucky" (hyena), 76, 80, 81

Macdonald, David, 189
"McDuff" (hyena), 186–87, 273
McKuen, Rod, 200, 309
Makalambedi fence, 295, 296, 301–303
Makgadikgadi Pans, 11–16
Malaria, 167–69
Malopos (waterways), 16
Mamba, black, 62–63, 93, 107
Mange, 288–89
Maraffe, Moxen. See Mox
Marico flycatchers, 90, 92, 217, 246, 281
"Marique" (flycatcher), 92, 217, 246
Maun, 9, 10
  social life in, 100–109
Meerkats, 49, 71

Mice, 37, 52, 90, 91, 246–47
Mills, Gus, 185–86
Mining exploration, 279–83
Moehlman, Dr. Patricia, 61
"Moffet" (lion), 216, 218, 228, 237, 239–44, 257, 259, 311, 312, 313
  bond with Muffin, 221
  in camp, 235–36
  defending pride's territory, 223–25
  disappearance of, 278, 285
  in drought, 276–78
  females and, 220–21, 238–42
  kills Star, 264
Mongooses, 49, 71, 91, 93
"Moose" (mongoose), 91
"Morena" (lion), 153
Mosquitoes, 9
Mox (camp assistant), 104, 316
  in airplane, 205
  communication with, 108–109, 110
  disappearance of, 247–49
  hiring of, 108–109
  lions and, 112–13
  and William the shrew, 247–48
  work as assistant, 115, 118, 119–20, 122–23, 125–27, 136, 137, 139, 163, 178–79, 182, 189, 206, 217
"Muffin" (lions), 216, 218, 239–44
  and abandoned cub, 229–30, 232
  bond with Moffet, 221
  in camp, 235–36
  death of, 277–78
  defending pride's territory, 222–25
  females and, 220–21, 238–42
  kills Star, 264
Muller, Mark, 166
"Muzzy" (lioness), 153, 314

Nata road, 11
National Geographic, 57
Ngamiland, 100
Nxai Pan, 16

Okavango River, 9, 16, 304
Okavango Wildlife Society, 170
Omar Khayyam, 131
Ostrich eggs, 86–87, 184
Owenses, Mark and Delia:
  arrival of in Africa, 5–8
  avoidance of people by, 250
  Christmas celebration of, 63–69
  decision to go to Africa, 3–4
  Delia's encounters with lions, 38–
    40, 96–97, 211–12
  Delia's malaria, 167–69
  finances of, 4, 8, 47, 49, 57, 164–
    67, 169–70, 247–49
  in fire, 43–46
  illnesses of, 50–51, 57, 87–88,
    167–69
  leaving Deception Valley, 316
  Mark's first airplane flight, 201–
    204
  social life of, 100–101
  theft of supplies, 106–107
  travel problems of, 8–10, 12–15,
    16–17, 31–34, 150–52, 171–
    77, 202–204, 206–207
  truck accident, 206–207
Owls, 90, 93

Palmer, Lionel, 10–11, 12, 17, 88,
    100, 103, 152
  leading hunting party, 191–94
Palmer, Phyllis, 100, 170
"Pappy" (lion), 120–21
"Patches" (hyena), 76, 78, 80, 132,
    184, 273, 290–91
  as mother, 255, 260, 261, 263,
    266, 267, 314
Paul, Dolene, 101, 105, 109
Paul, Simon, 101, 105, 109
"Pepper" (hyena), 254, 255, 256–58,
    260, 261, 262, 263, 268–69,
    290–91, 311, 314
  adoption by community, 265–69
  and Owenses, 269–74
"Pinkie" (warbler), 245–46

"Pink Panther" (leopard), 133–35,
    306
"Pippin" (hyena), 261, 262, 263,
    265–66, 267, 274, 314
Plovers, 28, 49, 217
Poachers, 298–301
"Pogo" (hyena), 76, 77, 78, 79, 80
"Puff" (hyena), 260, 261, 262
Puff adders, 93, 126

Radio-tracking equipment:
  with brown hyenas, 181–88
  with lions, 208–44
  problems with, 178–79, 181–82
"Rascal" (lion), 116, 128, 129, 146,
    162, 196, 314
Rats, 37, 52, 53–54, 247
Reptiles, 50
  See also Snakes
Riggs, Cecil, 102, 104
Riggs, "Dad," 102, 104
Rodents, 37, 50, 52, 53–54, 90, 91,
    246–47
Rondavels (huts), 7

Samadupe Drift, 17
"Sandy" (lion), 286–87
Sarcoptic mange, 288–89
"Sassy" (lioness), 2, 115–16, 128,
    129, 145, 146, 156, 157, 161,
    180, 216, 237, 310, 313–14
  as mother, 226–28, 232–34, 242,
    284
"Satan" (lion), 153, 154, 214, 219,
    222–23, 224–25
Savuti Marsh, 16
Schaller, Dr. George, 155, 160, 227
Selebi Phikwe, 171
Setswana language, 100, 102
"Shadow" (hyena), 75, 79, 80, 81,
    132, 134–35, 184, 187–88
  death of, 274
  as mother, 254, 260, 261, 262,
    266, 267
Shakespeare, William, 111, 265
Shrews, 90, 246–47

Shrikes, 90
Silberbauer, George, 301, 302, 303
Snakes, 49, 126
  in camp, 92–93
  killed by jackals, 62–63
  theft prevention with, 107
"Sooty" (hyena), 260, 261, 263, 266,
  268, 273
"Spicy" (lioness), 116, 128, 145,
  146, 156, 161, 180, 216, 285,
  314
"Spooky" (lioness), 116, 146, 161,
  180, 216, 314
Springbok, 22, 23–24, 25, 49, 84,
  85, 95, 137
Springbok Pan Pride (lions), 153–55,
  212–13, 222–24, 238, 313, 314
Squirrels, 49, 71, 91
"Star" (brown hyena), 67–82
  death of, 263–64
  description of, 71
  eartagging of, 76–78
  and humans, 71
  and jackals, 68, 72
  leopard vs., 133–34
  and lions, 73
  as mother, 253–58, 263
  as scavenger, 69, 72–74, 76
  social behavior of, 74–76, 78–83
  teeth of, 73
  tracking of, 132, 181–84, 253
  wild dog vs., 135–36
Steenbok, 25, 50–51
"Stonewall" (lion), 153
"Sunny" (lioness), 153, 314

"Taco" (lioness), 153, 314
Tau Pride (lions), 212–13, 314
Taylor, Ann, 70
Terminalia, 24
Termites, 60, 92
Thamalakane River, 10, 101
Thoreau, Henry David, 21
Tit-babblers, 91, 280
"Toffee" (hyena), 254, 255, 256,

  257–58, 260, 261, 262, 263,
  271
  adoption by community, 266–69
  death of, 273

"Ugly" (hornbill), 217
Union Carbide mining expedition,
  280–84
Uranium exploration, 279–83

van der Westhuizen story, 164–67
Vegetation, 24–25, 48–49, 64, 84–
  85, 293
Veit, Hans, 170

Warbler, 245–46
Water:
  conservation of, 57–58, 88
  shortage of, 30–34
  transporting of, 89–90
  See also Drought
Waxbills, violet-eared, 90
Whitman, Walt, 48
Wild cats, 28
Wild dogs, 23, 133, 135–36, 137–
  40, 261
Wildebeest, 13, 25, 309
  conservation of, 315, 319–21
  destruction of herd, 298–301,
  301–302
  migration during drought, 293–
  301, 307–308
"William" (shrew), 217, 246, 247
Williamson, Doug, 315
Williamson, Jane, 315
Wright, Daisy, 101–104
Wright, Dougie, 103, 193–94
Wright, Eustice, 103–104, 108–109
Wright, James, 189

Xau, Lake, 298–301, 303–304, 307,
  315

Zambesi River, 100
Zebra, 13, 302
Ziziphus trees, 83

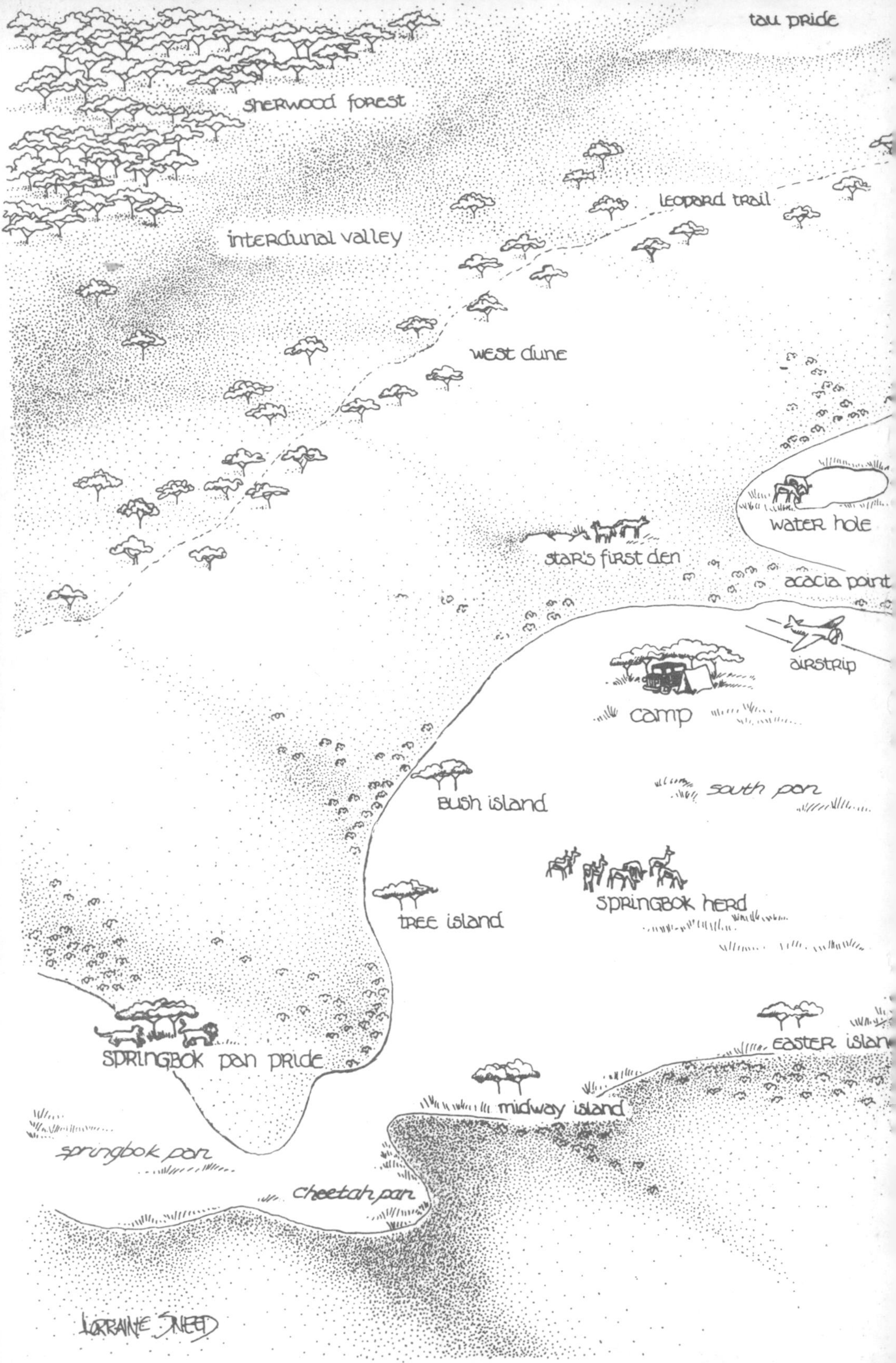
tau pride
sherwood forest
leopard trail
interdunal valley
west dune
water hole
star's first den
acacia point
airstrip
camp
south pan
bush island
springbok herd
tree island
easter island
springbok pan pride
midway island
springbok pan
cheetah pan
LORRAINE SNEED